Biomedicalization

BIOMEDICALIZATION

Technoscience, Health, and Illness in the U.S.

ADELE E. CLARKE,

LAURA MAMO,

JENNIFER RUTH FOSKET,

JENNIFER R. FISHMAN,

AND JANET K. SHIM, EDS.

DUKE UNIVERSITY PRESS Durham & London 2010

© 2010 Duke University Press
All rights reserved.
Printed in the United States of America
on acid-free paper ∞
Designed by Jennifer Hill
Typeset in Arno Pro by Tseng Information Systems, Inc.

Library of Congress Cataloging-in-Publication Data
appear on the last printed page of this book.

CONTENTS

In the late 1990s, a group that Professor Emerita Virginia Olesen called "the gang of five" began to gather weekly in a windowless conference room at the Laurel Heights campus of the University of California, San Francisco, one of twenty-five UCSF sites in the city. A massive former insurance company building where "desktop research" social science units and parts of the administration dwell, the Laurel Heights campus sits atop a San Francisco hill. It is linked by the Internet and extended shuttle bus service to the rest of UCSF, a transnational epicenter of biotechnology, healthcare, primary and tertiary hospitals, and a formidable array of industry-academia-state research collaborations.

The origins of the gang of five lay in a sociology dissertation writing group that began around 1997 and included Laura Mamo, Jennifer Ruth Fosket, Jennifer R. Fishman, and Janet K. Shim. Adele Clarke was chairing the dissertations of Mamo, Fosket, and Fishman and was a member of Shim's committee. Each of our empirical research projects was situated in late-twentieth-century biomedicine and what we saw as significant emergent phenomena: pharmaceutical drugs for prevention and enhancement, epidemiological and biomedical models of risk, commodification of biomaterials, and a new ethos of patient engagement with biomedicine. We were individually and collectively grappling with the inadequacies of medicalization theory and the concomitant disconnects among medical sociology and interdisciplinary fields such as science and technology studies, history of medicine, feminist theory, body studies, cultural studies, and others.

This was also the historic moment of the biotech bubble of the late 1990s, and we were amazed by the array of new biotechnological products and promises, start-up companies, venture capital firms, academic-industry alliances, programs in bioinformatics, stem cell research, and new medical specialty centers springing forth around us. As department chair, Clarke was routinely networked into the latest developments through the information flows of administration. This constituted her fieldwork site. Our politics of technoscience thus took seriously the assertion that we are all part of what Donna Haraway (1997) called "the New World Order, Inc."

The gang of five began collaboratively analyzing the shifting processes of biomedicine and the development of concepts and contours of what Clarke (1998, 275) framed as "the biomedicalization of life itself (human, plant, and animal) . . . often imaged as a juggernaut of technological imperatives." We generated a dense historical chart as our means of chronicling recent changes in biomedicine. This chart became the major empirical database for our project and is included in this volume (chapter 2). Next was a series of presentations followed by an invitation to publish. In 2000 our "Technoscience and the New Biomedicalization: Western Roots, Global Rhizomes" appeared (in French) in *Sciences Sociales et Santé*. The reception of our work was positive in the United States and abroad, and we continued meeting and writing. We wanted our audience for this work to include but not be limited to medical sociology. We submitted our key paper, "Biomedicalization: Theorizing Technoscientific Transformations of Health, Illness, and U.S. Biomedicine," to the *American Sociological Review*. It appeared in 2003 and is reprinted here as chapter 1.

The ASR article articulates biomedicalization theory as a synthesis and overview of changes in biomedicine and medicalization. There we discuss biomedicalization theory and its five broader processes in depth: (1) privatization and commodification, (2) risk and surveillance, (3) expanding technoscientific practices, (4) the production and distribution of knowledges, and (5) transformations of bodies and subjectivities. The missing links in the foundations of biomedicalization theory, if you will, have been the empirical cases that generated this extension of medicalization theory and its incorporation of transdisciplinary ways of knowing. As such, we proposed this volume as a way not only to showcase our grounded theorizing (Strauss 1987) of biomedicalization through our empirical research projects but also to provide a set of cases that materially demonstrate and elaborate the pathways and conditions through which biomedicalization and its processes operate.

Early in our endeavors, Virginia Olesen (warmly) criticized a draft paper of ours by asking whether there were any resistances, any moves to counter what she heard as "a tsunami of biomedicalized power"? To address her comments, in the next iteration of our paper, we spoke of old and new contingencies, stratifications, and the importance of how things play out in actual practices of knowledge production in everyday life. We pointed to how biomedicalization processes produce increasing inequalities by race, class, gender, and sexualities. And we expanded our lens to draw attention toward what we termed the workings of technoscience from the "insides" of biomedicine outwards—from its very organization toward its practices. We return to Ginnie Olesen's crucial question about resistances and countertrends here in our introduction, further urged to do so by our excellent Duke reviewers.

Yet the question remained: how and under what circumstances do the five processes constitutive of biomedicalization theory operate? Both practices and local processes, we argued, must be examined. We have done so in our empirical research, as have the others we invited to join us in this volume. These case studies are sites where we found particular bodies, identities, and corporealities becoming biomedical objects for (re)configuration in old and new ways. Context matters, not as a unidirectional cause but as situated, constitutive conditions of the techno-social shaping of meanings and practices.

Our goal for this book is to ground our published theory through research projects that allow the reader to reflect on and challenge our conclusions made in the broader theoretical framework. The book brings together scholars who in varied ways articulate how and under what conditions biomedicalization theory is useful and provocative. It is our ultimate hope that readers will take up biomedicalization theory in their own research and provide it with the empirical robustness to shift and change as needed to flexibly engage the ongoing transformations of biomedicine. We also seek to provoke studies of biomedicalization in its transnational travels.

ACKNOWLEDGMENTS

We are especially grateful to the many people who offered early feedback on our papers and presentations in the late 1990s: Vincanne Adams, Warwick Anderson, Gay Becker (since deceased), Steve Epstein, Carroll Estes, Eliot Freidson (since deceased), Donna Haraway, Sharon Kaufman, Donald Light, Ginnie Olesen, Howard Pinderhughes, Susan Leigh Star (since deceased), Charis Thompson, and Stefan Timmermans. We also thank those who helped circulate our collective ideas by inviting one or all of us to discuss biomedicalization at various conferences. These are acknowledged in our individual papers. Our gratitude to Brandee Woleslagle Blank for her organizational and electronic skills in preparing the book manuscript across several submissions is immense. We also thank Amber Nelson and Aleia Clark for compiling the index with skill and alacrity. Last we deeply appreciate Ken Wissoker, Courtney Berger, and the truly terrific anonymous reviewers for Duke University Press for seeing the merits of this project and offering insightful guidance and abiding patience.

Adele E. Clarke, Janet K. Shim, Laura Mamo,
Jennifer Ruth Fosket, and Jennifer R. Fishman

Biomedicalization

A THEORETICAL AND SUBSTANTIVE INTRODUCTION

The rise of Western scientific medicine fully established the medical sector of the U.S. political economy by the end of World War II, the first "social transformation of American medicine" (Starr 1982). Then, in an ongoing process called *medicalization*, the jurisdiction of medicine began expanding, redefining certain areas once deemed moral, social, or legal problems (such as alcoholism, drug addiction, and obesity) as medical problems (e.g., Zola 1972). In this book, we argue that since around 1985, dramatic and especially technoscientific[1] changes in the constitution, organization, and practices of contemporary biomedicine have coalesced into *biomedicalization*, the second major transformation of American medicine (Clarke et al. 2000, 2003).

Biomedicalization continues today and is organized around five key interactive processes:

1 a new biopolitical economy of medicine, health, illness, living, and dying which forms an increasingly dense and elaborate arena in which biomedical knowledges, technologies, services, and capital are ever more co-constituted;[2]

2 a new and intensifying focus on health (in addition to illness, disease, injury), on optimization and en-

hancement by technoscientific means, and on the elaboration of risk and surveillance at individual, niche group,[3] and population levels;

3 the technoscientization of biomedical practices where interventions for treatment and enhancement are progressively more reliant on sciences and technologies, are conceived in those very terms, and are ever more promptly applied;

4 transformations of biomedical knowledge production, information management, distribution, and consumption; and

5 transformations of bodies and the production of new individual, collective, and population (or niche group) level technoscientific identities.

These elements may be used as analytics and together constitute biomedicalization theory. The crux of this theory is that biomedicine broadly conceived is today being transformed from the inside out through old and new social arrangements that implement biomedical, computer, and information sciences and technologies to intervene in health, illness, healing, the organization of medical care, and how we think about and live "life itself." Medicalization practices typically emphasize exercising *control over* medical phenomena—diseases, illnesses, injuries, bodily malfunctions. In contrast, biomedicalization practices emphasize *transformations of* such medical phenomena and of bodies, largely through sooner-rather-than-later technoscientific interventions not only for treatment but also increasingly for enhancement. The panoply of biomedical institutions is itself being organizationally transformed through technoscience, along with biomedical practices (diagnoses, treatments, interventions) and the life sciences and technologies which inform them.

This book offers in-depth analyses and case studies of the development of and conditions through which biomedicalization processes move. We examine these along theoretical, historical, and substantive lines. Theoretically, we offer first our original article on biomedicalization, where the five key processes were initially elaborated (Clarke et al. 2003). Readers not familiar with our original arguments might begin there and then return to this introduction. A second theoretical essay (Riska) analyzes gender vis-à-vis medicalization theory and biomedicalization theory over the past several decades.

Historically, we begin with an in-depth analysis of the roots of biomedicalization in the United States from the rise of medicine qua institutional sector in the late nineteenth century to the present. Here we include first the extremely dense, previously unpublished historical chart that we developed, essentially our database for the generation of biomedicalization theory (Clarke et al., this volume). Second is an analysis of the visual cultures of (bio)medicine since the 1890s framed as three "healthscapes" (Clarke, chapter 2 of this volume). These healthscapes anchor our argument in American popular culture, where shifts from the rise of medicine to medicalization and to biomedicalization have been manifest—and palpably experienced by us all.

Substantively, the volume then offers an array of case studies of biomedicalization in the United States today emphasizing technologies of difference and of enhancement. The case studies focused on difference include an analysis of the business of (in)fertility medicine (Mamo), a study of the construction of technoscientific identities through MRIs (Joyce), an examination of stratified biomedicalization by race, gender, and class in relation to heart disease (Shim), new molecular epidemiological constructions of "persons at risk" (Shostak), and an account of BiDil as the first race-based drug approved by the FDA (Kahn). The case studies focused on enhancement include the use of Viagra as the biomedicalization of sexual dysfunction (Fishman), an account of the biomedicalization of obesity and its moral consequences for some patients of failed bariatric surgeries (Boero), the use of chemoprevention pills as the biomedicalization of risk of breast cancer (Fosket), and a study of the informatics of diagnosis in biopsychiatry (Orr). The range of sites of biomedicalization demonstrated by these studies underscores the breadth of its impact.

This introduction next situates biomedicalization theory within three major recent theoretical developments now coalescing across the social sciences and humanities and beyond—theories of biopower and biopolitics; theories of bioeconomy and biocapital; and theorizing of "vital politics" and "life itself." All of these both frame and complement biomedicalization theory and are requisite to fully understanding it.

From theory we turn to situating biomedicalization in practice, taking up the ways people are actually engaging "life itself" as new and potentially biomedicalizing alternatives are faced and negotiated in daily life. Because we agree with Rabinow (2003, 14) that the changes biomedicalization brings do not indicate "an epochal shift with a totalizing coherence but rather . . . fragmented and sectorial changes that pose problems," we

also discuss a wide array of countertrends. These trends not only complicate biomedicalization processes but at times directly challenge them.

We next turn to recent developments in medicalization and biomedicalization theory, including how biomedicalization theory is being taken up
and our hopes for its future use. Then we discuss the focus of this volume
on biomedicalization processes in the United States and the uneven and
changing relationships of gender, race, and class to (bio)medicalization
theory. Last we provide an overview of the book.

Bio(s), Life Itself, and Biomedicalization Theory

> There is an uneasy sense that something new is happening in the life sciences.
> (Sunder Rajan 2005, 28)

> We are inhabiting an emergent form of life.
> (Rose 2007, 7)

To grasp the processes and implications of biomedicalization theory, it is
necessary to traverse recent developments—from Foucauldian roots in
biopower and biopolitics to new theories of bioeconomy and vital politics.
The shift from medicalization to biomedicalization manifests the epistemic
shift from the clinical gaze initiated in the eighteenth century (Foucault
1973) to the emergent molecular gaze of today (Rose 2007). Biomedicalization theory is imbricated in the emergent capacities of the life and related
sciences to change the very meanings of life itself.

BIOPOWER AND BIOPOLITICS

The synthetic concept of biomedicalization asserts a number of changes
in biomedical domains that we signaled by adding the *bio* to medicalization (Clarke et al. 2000, 2003). The *bio* is intended to do several kinds of
work. First and foremost, it signals the increasing importance of biological
sciences to biomedicine—the life sciences and technologies broadly conceived as institutions, sets of practices, sites of knowledge production, implementation, and application.

Second, the *bio* signals that questions of *biopower* and *biopolitics* are integral to our project. In developing biomedicalization theory, we were particularly inspired by the work of Michel Foucault. Foucault (1975, 1984,
2008) argued that a new formation of power—biopower—emerged out

of the modernization and industrialization of Western societies during the eighteenth and nineteenth centuries in a shift from old sovereignty-based formations through the "birth of the clinic" to the development of "the medical gaze." For Foucault, the body per se is the site where multiple scales of the social are manifest—from macro to nano—making Foucault particularly important to medical sociology, anthropology, history, and beyond.[4]

Biopower is a "microphysics of power," taking particular forms of knowledge coupled with technologies to exert diffuse yet constant forces of surveillance and control over living bodies and their behaviors, sensations, physiological processes, and pleasures—individually and in terms of groupings and populations through governmentality. For Foucault, power "is now situated and exercised at the level of life" rather than sovereignty, and he proposed a now famous bipolar diagram (Rabinow and Rose 2006, 196):

> One pole of biopower focuses on an anatomo-politics of the human body, seeking to maximize its forces and integrate it into efficient systems. The second pole is one of regulatory controls, a biopolitics of the population, focusing on the species body, the body imbued with the mechanisms of life: birth, morbidity, mortality, longevity.

Significantly, power is automatically "built in" and mobile, *embodied* through social practices and norms rather than *invested* in particular individuals or institutions. Neither power nor biopower is in any way exclusively the domain of the state, a mode of external social control, or exclusively punitive—but also positive, productive.

Over the past thirty years, the Foucauldian turn has permeated the sociology, anthropology, and history of medicine, perhaps more deeply in Europe and the United Kingdom, but increasingly in the United States as well. One of the key questions Foucault (1970) raised, following his mentor Canguilhem (1966/1994), was "what is life?" Canguilhem had asserted that what life means changes over time and circumstance. Foucault's own answer to the question "what is life" was specifically concerned with the shift from natural history to the life sciences, from the largely anatomical work of classification and categorization to the study of life processes—physiologies.

As the study of life processes has shifted from cellular to molecular levels over the last half of the twentieth century, the question "What is life?" is again taken up. Important pioneering work by social scientists and historians theorized the new biotechnologies. Yoxen (1981, 1982), for example, discussed life as a productive force and the capitalizing of molecular biology. Duly (1987) was concerned with efforts at "controlling life," and Kay (1993a, 1993b) with "life as technology" and "molecularizing" life. Since our original development of biomedicalization theory (Clarke et al. 2000, 2003), these concerns have increasingly intersected with projects of biomedicine. A new generation of scholarship is coalescing around the assertion that the very grounds of life itself are changing (again). We argue here that it is through a new framing of bios — against this broader theoretical backdrop — that biomedicalization theory and practices now need to be understood.

VITAL POLITICS AND LIFE ITSELF

A number of scholars have been discussing with growing intensity changing conceptions of life, death, birth, and nature. For example, Lock (1993, 48) argued that "nature will be 'operationalized' for the good of society" — remade in the service of "man." Haraway (e.g., 2004, 202) has argued that the better term is "nature-culture," as the two are inextricable — co-constitutive. Rabinow's (1996, 1999) work engaged biopower in the making. Clarke (1998, 275) argued that "the biomedicalization of life itself (human, plant and animal) was the key *social* process." Fischer (1999) asserted that we should attend to "emergent forms of life." Lash (2001) discussed "technological forms of life." Rheinberger (2000) asserted an epochal shift of "rewriting life." And Franklin and Lock (2003) argued that we are now in the midst of "remaking life and death."

In a pathbreaking essay, Sarah Franklin (2000, 188) captured the broad thrust of recent changes:

> There remains the broad Foucauldian question of how [such changes] can be understood as part of an on-going realignment of life, labour and language. . . . This dual imperative, to take evolution in one hand and to govern it with the other, is a defining paradox of global nature, global culture. Nature . . . has been de-traditionalized. It has been antiquated, displaced and superseded . . . nature is in a spin.

For example, across many plant and animal species, natural forms of creating life are being displaced by technoscientific means. Reproduction itself is being "enterprised up" (Strathern 1992). Angling toward grasping biomedical futures and changing them, Thompson (2005, esp. 245–65; in prep.) asserts that this is a new biomedical/biocapitalist mode of pro—duction needful of "good science" and reconsidered ethicality. The centrality of the reproductive sciences to capacities to transform life itself is vivid.

Another major contributor theorizing current "conditions of possibility" is Nikolas Rose (2007, 3, 262; see also 2001), who takes up the politics of life itself:

> The vital politics of our own [twenty-first] century . . . is neither delimited by the poles of illness and health, nor focused on eliminating pathology to protect the destiny of the nation. Rather it is concerned with our growing capacities to control, manage, engineer, reshape, and modulate the very vital capacities of human beings as living creatures.

With multiple parallels to our framing of biomedicalization, Rose (2007, 5–6) argues more broadly that "contemporary biopolitics has not been formed by any single event" but by several "mutations,"[5] including the epistemic shift from the clinical gaze initiated in the eighteenth century to the molecular gaze of today.[6] This signals a fundamental change in how life can be/is to be lived. In sum, Rose (2007, 7) is asserting that "the new world of vital risk and vital susceptibilities, demanding action in the vital present in the name of vital futures to come, is generating an emergent form of life." Biomedicalization is clearly part of this shift.

BIOECONOMY AND BIOCAPITAL

One of the five basic processes of biomedicalization involves changes in the political economy of health, illness, disease, and medicine per se (Clarke et al. 2003, 166–71). Since our initial work, the terms "bioeconomy" and "biocapital" have come to capture new ways in which capital itself is being (re)conceptualized and (re)organized vis-à-vis life itself (e.g., Cooper 2008; Sunder Rajan 2005, 2006). This is occurring through the imbrications of capital with the biological sciences and technologies, biomedicine, megacorporate pharmaceutical and biotechnological industries, (bio)-nanotechnologies, and so on. A significant and growing proportion of capital globally (private, state, and hybrid, etc.) is linked with things biological, biotechnological, and biomedical, including plant and animal agribusiness and "academic capitalism" (e.g., Jasso-Aquilar, Waitzkin, and Landwehr

2004; Slaughter and Rhoades 2004). The term "bioeconomy" refers to this vast set of activities.

Sunder Rajan (2005, 21) defines biocapital as "not only the systems of exchange and circulation involved in the contemporary workings of the life sciences but also a regime of knowledge pertaining to the life sciences as they become increasingly foundational epistemologies for our times." Drawing on comparative ethnographic research on biotech worlds in the United States and India, Sunder Rajan argues that biocapital is becoming transnationally dominant. He sees this "lively capital" as a key part of "speculative capitalism"—investments in projects that have not yet demonstrated their profitability. Thompson (2005, 258–60) uses the term "promissory capital." Much of stem cell, genetics, and regenerative research falls within this domain (e.g., Ganchoff 2004, 2008; Thompson, forthcoming). It is notable that promise and expectations can suffice. Such future-oriented hopefulness and hypefulness (Debate 2008, 12) legitimates biomedicalization as well.

Franklin (2003) framed "ethical biocapital" as a new form of cultural capital—theoretically interesting because it too legitimates biomedicalization. In the United Kingdom, sustained governmental attention to bioethical issues raised by genetics, cloning, and stem cells, and governmental development of detailed regulations of such activities through citizen-expert collaborative processes, have, over decades, generated ethical biocapital. This capital has been put to work, allowing the United Kingdom to proceed rapidly with government-supported research and production of biocapital with Dolly the sheep as iconic (Franklin 2003; see also Pfeffer and Kent 2007). This contrasts sharply with ethical and moral barriers to such research mounted where there is little "ethical biocapital" such as the United States and Germany (e.g., Ganchoff 2004, 2008; Gottweiss 2005; Jasanoff 2007).

The term "biocapital" also refers to the capacities of certain things—such as organs and tissues—to produce surplus value (e.g., Waldby and Mitchell 2006). These may be derived from humans or animals and are constituent elements in some forms of biomedicalization. There is growing transnational traffic in these forms of biocapital, including human tissues (blood, organs, and cell lines), which Hogle (2003) frames as offering the potential for "life/time warranties" in terms of rechargeable cells and extendable lives, and Serlin (2004) frames as the potentially "replaceable you." Waldby and Mitchell (2006) analyze the increasingly transnational traffic in biocapital as "the global tissue economy," and Almeling (2007) as "medical markets."

Another political economic facet of biomedicalization concerns whether and how "tissue" circulating in the global tissue economy is paid for before being transformed into products bought, sold, and used in biomedicine. Waldby and Mitchell (2006) frame this as "biogifts" transformed into "biocommodities." Tissues may be taken from us for "free" in our interactions with biomedicine (such as tests), regardless of our preferences and regardless of their "biovalue." Cohen's (2005, 86) concept of "bioavailability" similarly suggests state involvement, as seen in transnational trafficking in human kidneys, where "legitimate demands of the state are mediated through invasive medical commitment."

Bioavailable zones are currently established through clinical trials, and many are virtual rather than geographic — produced by payments for participation. In the United States, some regular or career volunteers call themselves "professional guinea pigs" and have published "Guinea Pig Zero: A Journal for Human Research Subjects" (Elliott 2008). Such zones may also be literal. For example, Sunder Rajan (2007) reports that in Indian slums, clinical trials organizations (CTOs) set up hospital-like laboratories where experimental subjects live for the full duration of some trials. A nation-state example of a bioavailable zone is Barbados, where the government and health ministries have linked with a CTO to offer the entire island nation's population as a research base. Barbadians are particularly valued "guinea pigs," as the population is distinctively racially homogeneous and of African origin (Whitmarsh 2008; see also Montoya 2007). These examples reveal future directions of the stratification of biomedicalization by race, class, and citizenship.

Thompson (quoted in Franklin and Lock 2003, 7–8; Thompson 2005, 245–76) goes furthest in theorizing these concerns, capturing the profound importance of *re*production (rather than production) to the technical-commercial processes that characterize biocapitalism. Generating biocapital by these means is a form of extraction that involves isolating and mobilizing the "primary reproductive agency" of specific body parts, particularly cells originating from women. This form of nonhuman agency, such as the capacity of stem cells to proliferate, can be harnessed and deployed for both profit and biomedicalizing transformations. In short, biocapital is seizing the increasingly profitable means of reproduction, often without recompense (see also Almeling 2007; Rapp 2001; Thompson 2007, 2008, in prep.).

Other domains of biocapital linked to biomedicalization are bioprospecting (searching for new bioresources) and biopiracy (extracting and removing such resources from their sites of origin without adequate, if any, compensation). Hayden (2003) focuses on the extraction and patenting of life itself—corporate pursuit of patentable biodiversity largely in less-developed and less-regulated nation-states. Such new biologicals are *made* (by technoscientific interventions) as well as *found* (allowing patenting elsewhere) (see also Adams 2002; Wahlberg 2008). Such extraction has, of course, a long imperial history (Schiebinger 2004). Today both animal and plant materials are preserved in what Parry (2004) calls "technoscientific arks" or "biobanks," and debate about modes of preservation, conservation, and classification runs rife (Bowker 2006; Friese 2007). Many new biologicals are developed only because of their potential downstream profits as biomedicalizing treatments, enhancements, preventatives, and even nourishment for plants, animals, and humans. More direct biopiracy is pursued through appropriation of "indigenous" medicines—much more directly usable and also patentable in the United States and elsewhere.[7]

▷ ▷ ▷

In sum, biopolitics, bioeconomy, and life itself can be understood as forming an emergent and already dense theoretical web within which biomedicalization and its five key processes make deeper sense. The trends we identified in 2003 have been not only sustained but elaborated.

Situating Biomedicalization in Practice

Our major statement of biomedicalization theory in 2003 focused of necessity on elaborating its five key processes. There was no space to situate biomedicalization more broadly in even the recent history of American health practices. We therefore take this opportunity to do so especially because biomedicalization can seem much too "tsunami-like," as Virginia Olesen remarked to us years ago. Ambivalences, complications, countertrends, contradictions, and dense negotiations have long weighed in, often quite strongly, against any simple acceptance of medicalizing or biomedicalizing tendencies. Complications abound.

This section has two parts. First we discuss how biomedicalization is being taken up in actual practices individually, collectively, and as populations. Here emphasis is on the embrace of biomedicalizing possibilities and practices along with ambivalences and negotiations. Then we turn to actual countertrends against both biomedicalizing tendencies and practices. Here too there are ambivalences and negotiations, but also refusals, outright rejection, and active opposition. The recurring theme is the complexity faced in engaging biomedicalization.

ENGAGING LIFE ITSELF

The implications of changes in life itself for biomedicalization concern the ways in which people are actually engaging with these new potentialities. Such engagements occur individually (as embodied selves, vis-à-vis enhancements, and as users/consumers), collectively (as health social movements and patient groups), and as populations (especially in terms of biological citizenship and in the United States via health insurance plans).

In the current rise of a "culture of life," the individual is primary, according to Knorr Cetina (2005, S76–S77). Innovations in the biological sciences are encouraging a move away from concerns with humans in society and social salvation and toward ideals of individual perfectibility and enhancement: "It is a person's 'life' that should be improved; the goal of the commitment to oneself is life enhancement—an idea that not only means increased enjoyment but also reflects on biological life as improvable and extendable."

Considerable recent work has focused on the pursuit of enhancement—going "beyond what is necessary to sustain health or repair the body," including cosmetic procedures, cyborg prosthetics, cognitive and genetic enhancements, and regenerative medicine (Hogle 2006, 696). Rothman and Rothman (2003) frame this as "the pursuit of perfection," widely taken up in both scholarly and popular literatures. For example, cosmetic surgery constitutes the cutting edge of commercial medicine in America to the tune of $15 billion per year—"beauty junkies" indeed (Kuczynski 2006). The use of human growth hormone (Conrad and Potter 2004) is taken up popularly as "the short of it" (e.g., Hall 2005). Anti-aging (Binstock and Fishman 2010) is popularized as "the fountain of youth." Age-defying hormone treatments target women and men (Hoberman 2006; Krieger et al. 2005; Mamo and Fosket 2009; Watkins 2007). And new, highly techno-scientific prosthetics create a "replaceable you" (Serlin 2004). People's definitions of what constitutes "enhancement" rather than what is "necessary to sustain health or repair the body" vary tremendously. National debates about steroid drugs and athletics (both professional and amateur) have raised issues of enhancement as normative. The injunction to "be all you can be" clearly drives many athletes in the documentary *Bigger Stronger Faster* (2008).[8] Distinctions between drugs and biological or nutraceutical interventions (via "functional foods") are also problematized (Schroeder 2007).

Recent research on how people are engaging life itself individually centers on the controversial nature of interventions, demonstrating people's ambivalence and anxieties. Even the highly desired and much-joked-about treatments for male sexual dysfunction (Fishman 2004; Fishman and Mamo 2002; Mamo and Fishman 2001; Loe 2004a) have hit snags as some men abandon pharmacological options (Berenson 2005; Potts et al. 2006). Less frequently discussed are the complicated structural issues of stratified access. Singh and Rose (2006, 97) ask: "Can individuals resist/access the pharmaceutically powered drive toward perfection; is their personal agency sufficient to resist/access enhancing drugs, especially if they are very young, or poor, an ethnic minority, a convicted felon—or, for that matter, if they are students at elite competitive universities?" Few of us will find ourselves wholly immune to biomedicalizing temptations and seductions.

Engaging life itself collectively is largely the purview of health social movements and patients' organizations.[9] Here the actions of people vis-à-vis (bio)medicalization have also expanded dramatically, largely as consumers but also at times as providers, scientists, and corporate sponsors. There is a long and deep history of patient-centered movements, commonly involving physicians and other clinicians. But the nature of such movements began to change as the civil rights, feminist, and environmental movements of the 1960s and 1970s brought about an array of activist organizations focused on diseases and health issues with serious policy as well as treatment agendas, such as AIDS activists in the 1980s. Such groups challenged national biomedical research agendas and the organizational principles of clinical trials vis-à-vis access and inclusion. Their efforts led in the early 1990s to the establishment of (separate) NIH offices of minority and women's health. NIH clinical trials guidelines were reformed to require inclusion of women and racial/ethnic minorities in most research, changing the face of biomedical and pharmaceutical research transnationally.[10] Access to and inclusion in biomedicalization is highly valued — and valuable.

Today new and reconstituted health social movements are engaging life itself at molecular levels: working for stem cell research (Ganchoff 2004, 2008), provoking research on environmental illnesses and other emergent illnesses (Brown 2007; Packard et al. 2004), and focusing on diseases treatable and perhaps even preventable using molecular technologies. For example, Shostak (2005) explored how molecular biomarkers and toxicogenomics are taken up by activist movements, especially vis-à-vis environmental exposures (see also Washburn 2009).

In the past, patient groups focused largely on shifting national level research priorities and funding toward particular diseases and conditions and raising money for targeted research. Today such groups are not only still inside political processes (such as testifying before Congress), but also inside research activities as new forms of collaboration emerge around possibilities for biomedicalization. Some organizations are running their own research endeavors (Callon 2003; Akrich et al. 2008). For example, autism activist groups not only support genetics research financially but are also developing autism genetic databanks (through donation of biological materials, family pedigrees, and diagnostic information) and forming research collaborations (with governmental agencies and other organizations) (e.g., Silverman 2008; Singh et al. 2009). There are potential risks as well as benefits for activists and health social movements — especially the appropriation of ideas and practices and the dilution of alternative agendas (Huyard 2009; Thomson 2009).

Franklin and Roberts (2006, xvii) frame such emergent collaborations as transdisciplinary "interliteracies" requisite for understanding certain diseases and conditions: "These interliteracies also connect scientists and clinicians to policy-makers, representatives of government to social scientists, and patient groups to journalists in a host of emergent alliances that together comprise new forms of 'biological citizenship.'" In a number of senses, they are biomedicalization in action.

Biocapital too is attending to patient movement involvement in research (e.g., Ganchoff 2004, 2008). The *Wall Street Journal* ran an ambitious article about how "Patients Become Fund-Raisers," noting that "a wave of wealthy patients [is] bankrolling a new technology: . . . machines that sift through

DNA at lightning speed" to identify genetic markers linked to particular diseases (Naik and Regalado 2006, A1,12). The article directed companies to
potential patient funders who might seek special access to research findings
in return for their investment. This could move the capacity for stratification of biomedicalization far upstream in a single bound.

Engaging life itself through biomedicalization at the level of population
primarily involves "biological citizenship" (e.g., Ginsberg and Rapp 1995;
Nguyen 2005; Novas and Rose 2000; Ong and Collier 2005; Petryna 2002;
Rose 2001, 2007; Rose and Novas 2004). Biological citizenship emerged in
the late twentieth century as extensions of the rights of citizens to the protection and promotion of their health and well-being (Rose 2007, 131). For
example, Petryna (2002) argued that citizens have rights to health services
and social support for the survival and well-being of their biological bodies.
She was affirming claims of Chernobyl victims for both care and protection as biological citizens. In the United States, the profound inadequacy of
governmental response to Hurricane Katrina was assessed through tacit assumptions of biological citizenship framed in the exclamation "How could
this happen in America?"

In sum, people's varied engagements with life itself actively reformulate
possibilities of biomedicalization through new potentialities — individually, collectively, and as biological citizens.

COUNTERTRENDS AND COMPLICATIONS

In addition to anxieties and ambivalences about biomedicalization, there
are active negotiations, refusals of engagement, resistances, and outright
countermovements. Subjugated knowledges (re)appear to have their say
(Foucault 2003). The varied pushbacks include ongoing critiques of medicine, medicalization, and biomedicalization; seeking help from beyond the
boundaries of biomedicine, thereby expanding American medical pluralism; health social movements resistant to (bio)medicalization; a (re)emer-
gent public discourse that "*more* (bio)medicine is not necessarily *better*";
and a growing discourse attempting to articulate "appropriate" levels and
forms of biomedical intervention from the beginnings to the end of life
(somewhat akin to the "appropriate technology movement").

Serious critique of American medicine emerged after World War II as
the medicalization era was getting under way. It emerged both from deep
within medicine and from without, centered on medical reductionism, and

was skeptical of medical efficacy, power, and ethics. Famous among these critiques is the comment of René Dubos (1959) of the Rockefeller Institute that we should not be seduced by a "mirage of health" (Brandt and Gardner 2003, 29). Thomas McKeown (1979) famously demonstrated that increased longevity in the West was due to improved sanitation, nutrition, income, and housing rather than medical interventions. And Ivan Illich (1976) argued that iatrogenic (medically caused) disease constituted a "medical nemesis." The rise of bioethics and patients' rights movements of the 1970s sustained these critiques (Brandt and Gardner 2003, 29, 32). Foucault certainly questions the magnanimity of medicine (e.g., Armstrong 2006). And scholars and others have long attributed blame for many of medicine's problems to corporate capital (e.g., Jasso-Aquilar, Waitzkin, and Landwehr 2004; Navarro 1986, 2007; Waitzkin 2000). Echoes of all these critiques are heard today.

The main way people act on their critique of biomedicine is to seek other ways of addressing health problems — "other" modes of healing also known as "complementary and alternative medicine" (CAM). In our original paper, we discussed the expansion of medical pluralism in the United States and how its impacts on biomedicine were assessed by analyzing consumers' out-of-pocket expenditures on CAM. A 1993 study estimated out-of-pocket expenditures at $10.3 billion a year (Eisenberg et al. 1993, 346). The 1998 follow-up was $27.0 billion, comparable to out-of-pocket costs to patients for *all* physician services (Eisenberg et al. 1998, 1569). By the late 2000s office visits to such providers outnumbered visits to primary care physicians, costing about $40 billion (Groopman 2008, 14).

Alternative medical knowledge systems — especially, but far from only, Asian traditions — have continued to gain legitimacy among users/consumers (e.g., Harrington 2008; Johnson 2004b; Barnes and Sered 2005). Long-standing AMA efforts against many approaches also continue (Johnson 2004b), including testing their "effectiveness" according to Western rather than epistemologically appropriate standards (e.g., Adams 2002). Simultaneously, biomedicine also continues to incorporate elements of CAM, and a National Center for Complementary and Alternative Medicine was established at the NIH in 1999.[11] Ten years ago no U.S. academic medical centers offered "integrative medicine"; today thirty-nine do so along with research and training opportunities for physicians (Catalyst 2008). Health food movements have existed for well over a century (Ackerman

2004) and are expanding dramatically from organic and local products to farmers' markets and slow foods (e.g., Pollan 2007). Vitamins and supplements too have long histories (e.g., Apple 1996). "Functional foods" and "nutraceuticals" are even being considered as public health interventions as public health itself is individualized (Schroeder 2007).

All of these countertrends vividly demonstrate one of our arguments about biomedicalization: that it demands of patients, consumers, and patient groups that we become more knowledgeable and responsible — essentially more "scientized" — vis-à-vis biomedicine. Brown and Webster (2004, 29) describe this as patients "opening the black bag" — going "inside" biomedicine to evaluate it more thoroughly. Recent ethnographies demonstrate the complex messiness of people's ambivalences and changing subject positions in medical decision making (e.g., Becker 2000a; Rapp 1999; Franklin and Roberts 2006; Kerr and Franklin 2006). Landzelius (2006) and others have framed this as "metamorphoses in patienthood,"[12] including new regimens and integration of information from health social movements, the media, and the Internet (e.g., Barbot 2006; Mykhalovzkiy et al. 2004; Akrich et al. 2008).

Ethically charged situations when people (such as a parent or adult child) make medical decisions for another (such as a child or elderly parent) are particularly fraught. These may include surgical modifications (e.g., Parens 2006), cochlear implants (e.g., Komesaroff 2007), and so on. Thompson (2005) found couples performing elaborate "ontological choreographies" vis-à-vis using reproductive technologies to biomedicalize their infertility — or not. Prenatal genetic diagnostics immediately biomedicalize pregnancies, and there is often such profound discomfort about decision making that Rapp (1999) called the pregnant women "moral pioneers." Latimer (2007) extends this analysis, noting that parents of even *potentially* "dysmorphic" children may become actively involved in the very definition of clinical problems in calculating their risk of recurrence in future pregnancies.

End-of-life care is another moment when biomedicalization may be challenged as "moral pioneering" takes on new forms. The long-standing discourse about what constitutes "a good death" initially resulted in the palliative care movement, situating hospice services across much of the country. Yet the problem is not resolved, as demonstrated by Kaufman's (2005) research on how American hospitals shape the end of life — usually by further biomedicalization.

In terms of collective countertrends, some ongoing health social movements such as the women's health and natural birthing movements have deep roots in explicit resistance to medicalization (e.g., Ruzek 1978). In contrast, the Black Power movement provided community-based healthcare in African American communities that were underserved — even undermedicalized — during the 1960s and 1970s (Nelson, forthcoming). Many other such movements, such as those around AIDS, autism, and neglected diseases, have sought further (bio)medicalization (e.g., Epstein 1996; Silverman 2008; Novas 2006). One remarkable accomplishment of segments of the women's health movement has been persistent opposition to "hormone replacement therapies" (HRT) over thirty-five years.[13] Their anti-(bio)medicalization position was validated not only by research revealing HRT as commonly carcinogenic but also by a dramatic drop in new cancer diagnoses following publication of that research and women's ceasing to take such medications (Ravdin et al. 2007; Watkins 2007).

Many other patient and related movements are also organizing against specific forms of biomedicalization (Epstein 2008). One such movement that has had multiple incarnations and fluctuations over the past century is the antivaccination movement, which has multiple arms today (Johnson 2004a). Critiques range from questioning the very principles of particular vaccinations to technical issues in assuring vaccine safety in preparation and distribution (e.g., Leach and Fairhead 2007). A major recent critique has been mounted by families with autistic children and others who believe autism is caused by the vaccine preservative thimerosal (which contains mercury), although there are also growing parents' movements now committed to genetic explanations and research on autism (e.g., Silverman 2008; Singh et al. 2009). Another broad-based set of patients' movements has long sought both prevention of cancer and improved biomedical treatments. Breast cancer activism has been most extensive (e.g., Fishman 2000; Gibbon 2007; Klawiter 2008; Ley 2009; and Parthasarathy 2007). Critiques of biomedical approaches have included their being unnecessarily disabling and disfiguring of women (Montini and Ruzek 1989), failing to address environmental causes (Fishman 2000; Klawiter 2008; Ley 2009), overpromoting chemoprevention (Fosket 2004, this volume), and so on.

An anticircumcision movement questions the value of this surgery usually performed on males in infancy (e.g., Darby and Svoboda 2007). This is further complicated by the relationship between circumcision and the transmission of HIV/AIDS, a much-studied phenomenon given its prevention promise (Dowsett and Couch 2007). Pharmaceutical treatment for ADHD is another major site of critique of biomedical approaches (e.g., Cohen, Lloyd, and Stead 2007; Conrad 2007). New biomedicalizing treatments for post-traumatic stress disorder have some bioethicists asking, "Is it wrong to erase the 'sting' of bad memories by using Propranolol for the prevention of post-traumatic stress disorder?" (Henry, Fishman, and Youngner 2007). Further questioning of the value of biomedicalizing interventions can be anticipated as individuals and movements become more "scientized" and increasingly capable of critiques of the undergirding science as well as of clinical practices.

In addition to critiques of particular interventions, countertrends against biomedicalization include emergent discourses that explicitly assert that "more (bio)medicine is not necessarily better." One example is Brownlee's *Overtreated: Why Too Much Medicine Is Making Us Sicker and Poorer* (2007). The bioethicist Carl Elliott (2003) critiques the American dream of being "better than well." The saturation of media with information and infomercials about diet, health, and diseases even has media health leaders saying "enough already!" (e.g., Kolata 2006). The science that serves as the basis for such public health recommendations is epidemiology, and it too is under attack. From without came Taubes's *New York Times Magazine* essay "Unhealthy Science: Why Can't We Trust Much of What We Hear about Diet, Health, and Behavior-Related Diseases?" (2007). And within epidemiology serious debates rage about the adequacy of its categories (e.g., Krieger 1999, 2003; Shim 2002a, 2002b, 2005). Evidence-based medicine has also been challenged (Mykhalovskiy and Weir 2004) and even asserted to be fascist in its methods (Murray and Holmes 2009).

Last in this litany of countertrends are attempts to construct "appropriate" levels and kinds of biomedicalization, policies seeking to avoid throwing the baby out with the bathwater, parallel to "appropriate" technology use. Here providers, patients, scientists, and policymakers all experience fraught moments of moral pioneering. For example, physicians, insurers, and the FDA are currently weighing the costs of CT scans of the heart—very expensive and of questionable use and need—given other established procedures (Berenson and Abelson 2008). Revisions of mammogram guidelines and ensuing debates have been on the front pages of major newspapers for days (e.g., Kolata 2009). And there are many other examples (e.g., Hedgecoe 2008).

Overall, these countertrends signal that broader ethical questions are being raised. Collier and Lakoff (2005, 22–23), for example, argue that "living" is increasingly ethically problematic vis-à-vis normative configurations of biopolitics and technology. A "bioethics of technoscience" is emerging, arguing that bioethics not only needs to address the "social causation of illness model" (e.g., Taussig et al. 2006) but also new problematizations of embodiment (Armstrong 2006; Armstrong and Rothman 2008; Hacking 2006).[14] Singh and Rose (2006, 98) complicate this, noting deployments of bioethics to legitimate biomedicalization. At the same time, a critical bioethics is emerging (Murray and Holmes 2009).

Larger species-level ethical issues are being articulated vis-à-vis emergent technoscientific possibilities. Franklin's (2000) "genetic imaginary" appears in recurring "dreams of the perfect child" (Rothschild 2005) achievable by "designing our descendants" (Chapman and Frankel 2003; Green 2007) and "enhancing evolution" (Harris 2007), from individual and familial design to species redesign.[15] Yet providers and scientists also express ambivalence and anxiety around tinkering with both human (Rapp 1999; Franklin and Roberts 2006) and plant and animal heredity and reproduction (Haraway 2007; Friese 2007; Thompson, in prep.). Ian Wilmut, one of the scientists involved in cloning Dolly the sheep, argues forcefully that where nature used to set the limits of possibility, biological control is now in human hands (Franklin 2003, 100–101):

> As decades and centuries pass, the science of cloning and the technologies that flow from it will affect all aspects of human life—the things that people can do, the way we live, and even, if we choose, the kinds of people we are. Those future technologies will offer our successors a degree of control over life's processes that will come effectively to seem absolute.... Human ambition will be bound only by the laws of physics, the rules of logic, and our descendants' own sense of right and wrong.

These are the relentless, unavoidable, and even haunting social aspects—human burdens—of technoscientific innovation that *also* constitute biomedicalization that have concerned us. Franklin (2003, 101, italics mine) points to the ways in which "biologically *in* human control" might produce "biologically *out of* human control" and wonders what new kinds of human and biological controls might be effected to deal with them. So do we.

Medicalization and Biomedicalization Theories Today

As we have moved into the twenty-first century, both medicalization and biomedicalization theories have been extended. We discuss these developments here, noting some recent substantive work.

MEDICALIZATION THEORY TODAY

The concept of medicalization was framed by Irving Zola (1972) to signal extensions of medical jurisdiction, authority, and practice into increasingly broader areas of people's lives. In the United States, the concept gradually extended to any and all instances when new phenomena were deemed medical problems under medical jurisdiction—from infertility to hyperactivity and so on.[16] American and British sociologists' conceptualizations of medicalization have tended to differ (e.g., Seale 2008). British work has long focused on doctors as professional figures who are active agents of medicalization, often with inflections of self- or professional aggrandizement (see, e.g., Strong 1979, 1984; Murcott 2006; Williams 2004; Furedi 2006; Dingwall 2006a, 2006b). More recently, scholars in the United Kingdom have been critical of constructions of patients as consumers and of pharmaceutical companies' promotional efforts, which they call "selling sickness" and "disease mongering" to promote medicalization (Ballard and Elston 2005; Furedi 2006; Moynihan and Cassels 2005). British scholars also noted the importance of the media in the commodification of body parts (Seale, Cavers, and Dixon-Woods 2006).

American medical sociologists, in contrast, have long emphasized definitional and institutional aspects of medicalization. What matters most is whether a phenomenon is defined—socially constructed—as falling within medical jurisdiction and how that is elaborated (Conrad and Schneider 1980, esp. 17–38; 1992, 277–279; Conrad 2006, 2007).[17] Issues of medical professional dominance are also taken up, but largely in relation to the actual practice of medicine vis-à-vis patients and other professionals (Freidson 1970, 2001) and to issues such as professional discretion versus clinical practice guidelines (e.g., Timmermans and Kolker 2004). Relations of physicians to third-party payers are also a major focus in the United States (e.g., Light 2004; McKinlay and Marceau 2002; Quadagno 2004; Scott et al. 2000) but remain less important in Britain because of its universal coverage through the National Health Service.[18]

American medical sociologists also look beyond the profession of medicine to entities such as "big pharma," suppliers of hospital equipment and technologies, and patients increasingly acting as lively consumers of biomedical goods and services. Both medicalization and the persistence of specific forms of medical care rest on economic interests and the motivations of multiple and varied actors situated in different social institutions — not only physicians (e.g., Conrad and Schneider 1980a, 1992; Conrad 1992, 2005, 2006, 2007). Recent American work has emphasized insurance companies (e.g., Quadagno 2004), managed-care organizations (Light 2004; Scott et al. 2000), and philanthropic organizations from the great (Carnegie, Rockefeller, Gates) to the persistent, smaller, and local private giving that over the past century so deeply entrenched medical schools and hospitals in every major city in the United States (e.g., Bishop and Green 2008; Pescosolido and Martin 2004; Pescosolido 2006; Schneider 2002).

The major American scholar of medicalization theory is Peter Conrad, who has recently (2006, 2007; Conrad and Leiter 2004) emphasized the increasing salience of both the market and enhancement. In 2006 he asserted that "the shifting engines of medicalization" today are biotechnology (especially the pharmaceutical industry), consumers, and managed care.[19] His latest work (2007) takes up the extension of medicalization to focus on men (via the medicalization of andropause, baldness, and sexual dysfunction) and its expansion across age categories (such as children's hyperactivity morphing into adult ADHD). Conrad also sees new forms of bodily improvement as falling within the traditional framework of medicalization (e.g., use of human growth hormone) and frames the appeal of such interventions as "temptations" of biomedical enhancement.[20]

Few American medical sociologists attend to the centrality of the life sciences or the computer and information sciences to the reorganizing of biomedicine as we do in biomedicalization theory. (Medical technology assessment is a separate and applied domain largely centered on cost containment.) Nor do they address the changes we discussed regarding engagements with life itself. However, in a recent staged debate, Hafferty (2006, 42) argued that there are two "missing witnesses" in terms of understanding medicalization today: the capital market and science. He asks: "What if, over time, medicalization has become less about medicine and more about science?" If biocapital were included, this is indeed our argument.

Recent substantive research on medicalization has focused on risk (Crawford 2004; Halpern 2004), prevention of fetal intersexuality by consumption of hormones during pregnancy (Casper and Muse 2006), "unruly" children (Rafalovich 2005), mania and manic depression (Martin 2007), shyness (Scott 2006), insomnia and snoring (Williams et al. 2008), autism (e.g., Silverman 2008), and compulsive buying (Lee and Mysyk 2004).

BIOMEDICALIZATION THEORY TODAY

The literature on "life itself" discussed earlier has given us the theoretical apparatus and impetus to extend biomedicalization theory in two main directions—placing greater emphasis on biopolitical economy and on enhancements/optimizations. Vis-à-vis biopolitical economy, we originally argued that biomedicalization was in part generated by, and a product of, the Biomedical TechnoService Complex Inc., a concept we created that merged the old "medical industrial complex," coined by HealthPAC (Ehrenreich and Ehrenreich 1971; and see Wohl 1984) and playing on President Eisenhower's cold war "military industrial complex," with the "New World Order, Inc." coined by Haraway (1997). This concept features technoscience and healthcare services in elaborating corporatized forms.

Today we situate this theoretical entity as part of the emergent biopolitical economy of health, illness, life, death, and medicine. This gestures more fully to the penetrating and transformational bioeconomic character of the shift regarding life itself that we have argued effects biomedicalization. In short, we agree with Rose (2007, 16) and others that "an epistemological change . . . perhaps also an ontological change is in process" from the clinical to the molecular gaze. Within this broader epistemic shift of Foucauldian proportions, biomedicalization is a key potential.

Significantly, and we cannot overemphasize this, the potential for the generation of biocapital relies deeply on the *legitimacies* of medicalization (extensions of the jurisdiction of medicine) and biomedicalization (extensions of biomedicine through technoscience). Such legitimacies are the sociocultural infrastructures on and through which biotechnology and biomedicine can be built. Sociocultural infrastructures of legitimation are of fundamental importance. They also are not unchallenged, as our discussion of countertrends demonstrates, and hence are vulnerable and "must be defended" (Foucault 2003).

Situating biomedicalization within this broader epistemic shift gives us more traction to understand how and where biomedicalization operates — and its complications. The molecular gaze and the remaking of life and death encompass in a more thoroughgoing fashion the reach and extent of what eventually — and more often sooner than later — becomes biomedically significant. That is, within the coalescent biopolitical economy, we can see just how distal R&D, social technologies, people, plants, animals, artifacts, et cetera, can be from immediate use on/in humans. Yet they must simultaneously be seen through the lenses of their biopolitical potential (how they can help manipulate/transform bodies) and their bioeconomic potential (potential future contributions to biocommodities and profitability). Promissory capital indeed! The chapters in this volume by Kahn on BiDil, Fosket on chemoprevention, Mamo on infertility medicine, Fishman on Viagra, and Orr on biopsychiatry each and all expand on this point.

Our second site of reconsideration involves placing greater emphasis within biomedicalization theory on enhancements, what Rose (2007, 7) more broadly calls *optimization* — the increasing legitimacy of securing "the best possible futures." Specifically, we are asserting the need for empirical social science research to better understand how users/consumers/patients and providers/scientists/producers engage new technologies so that policies and procedures for their use might be improved to better and more equitably meet people's needs and desires — as individuals, health social movement activists, and biological citizens.

Placing "optimization" into actual practice occurs not only at institutional levels where options for doing so are made available (including insurance coverage for medical tourism, for example) but also at the individual level discussed earlier. Knorr Cetina (2005) described the deepening individualization produced through the emergent "culture of life" increasingly manifest around optimization. Squier (2004) conceptualizes these directions as "liminal lives" at the frontiers of biomedicine. The U.S. Army's recruitment slogan "Be all that you can be" takes on newly biomedicalizable meanings. The chapters in this volume by Boero on bariatric surgery, Fishman on Viagra, and Mamo on infertility medicine are the next generation of scholarship in this area.

Actualizing optimal futures also involves engagements with regimes of risk and surveillance, extended responsibility for knowledge accumulation and consumption, and taking up new kinds of biomedically inflected identities—all described in our original articulation of biomedicalization. There is an ongoing tailoring of the self via transformations of embodiment, enmindment, and enselfment (especially via drugs),[21] and biosociality (Rabinow 1992a)—almost a hyperuse of "technologies of the self" (Foucault 1988). And, as noted, these engagements are often complicated negotiations *with*, rather than simple acceptances *of*, biomedicalizing alternatives (Strauss 1978). Together, these emergent forms of subjectification point to the elaborating duties and burdens of authorizing and making one's own future and often those of others.

This constant anticipatory orientation toward the future, toward potentialities—which socioculturally parallels the biopolitical focus on promissory capital—gives biomedicalization practices a greater temporal dimension (Adams, Murphy, and Clarke 2009). The chapters in this volume by Fosket on chemoprevention, Shim on the imperative of health implemented through cardiovascular epidemiology, and Shostak on the molecularization of epidemiology all feature these concerns and provoke us to think through these concepts further.

BIOMEDICALIZATION THEORY IN ACTION

Since our 2003 publication, biomedicalization theory has traveled widely. In addition to the case studies in this book, this section offers a brief overview of how it has been taken up theoretically and substantively. A number of works use biomedicalization as an overarching process, often tacitly "updating" medicalization theory. Burri and Dumit (2007) use the concept to frame the present situation culturally as the "biomedicalization of living." Related work sustains our argument that health is being biomedicalized. Here the study by Nichter and Thompson (2006) of dietary supplements found they were "for my wellness, not just my illness." Wheatley's (2005) ethnography of heart disease argues for the biomedicalization of fitness and rehabilitation. Preda (2005) asserts that lifestyles per se have been biomedicalized, especially vis-à-vis HIV/AIDS. And Cockerham (2005b) asserts that within this era of biomedicalization, theorization of a health lifestyle stressing structural dimensions is needed.

Biomedicalization has also been viewed as "technogovernance": using the key processes analytically, May and colleagues (2006, 1022) argue that in clinical encounters, information technologies become significant nonhuman actors, nicely addressing the "how" of biomedicalization in practice. At times, biomedicalization is interpreted as an essentially negative phenomenon, implying biological reductionism (Midanik 2006). In contrast, we agree with Bury (2006, 38) that "drawing [overall] inferences about the impact of [bio]medicalization . . . is hazardous."

Biomedicalization has also been taken up in studies of changes at the beginning and end of life (Kaufman and Morgan 2005). Friese and colleagues (2006) found the biomedicalization of infertility occurring to manipulate the "biological clock." The biomedicalization of pregnancy was greeted by expectant Israeli fathers with ambivalence and criticism and provoked complex power negotiations (Ivry and Teman 2008). The regulation of human cloning and stem cell research in Israel is analyzed by Prainsack (2006) as "negotiating life." She found both religious commitments and population issues result in Israeli openness to human cloning as a biomedicalizing reproductive option in the face of "the demographic threat" of Israelis being outnumbered by Palestinians. As Franklin (2006, 86) argues, "Reproductive substance can never be fully de-differentiated from social definitions of obligation, responsibility, equality, or health and well-being."

Studies of aging have incorporated biomedicalization theory as a helpful lens to view aging per se and life extension practices. Joyce and Mamo (2006) assert that new research should "gray the cyborg" to better attend to intersections of age, technology, science, and gender. Targeted consumers become anyone who does not meet youthful standards. The emergent specialty of "anti-aging medicine" is a particularly commercial endeavor; patients primarily pay out of pocket and are often prescribed "off-label" (non-FDA-approved) treatments (Mykytyn 2006; Binstock and Fishman 2010).

Technoscientific innovations are pushing when old age is believed to begin further into a receding future. Late life is transformed into a mutable and reversible state, in stark contrast to normative expectations of age-associated decline (Kaufman 2005; Kaufman, Shim, and Russ 2004). At the same time, through a discourse of health and risk, old age per se is biomedicalized, not only its pathologies (Kaufman et al. 2004; Shim et al. 2006). Finally, aging patients themselves take up a new kind of "clinical life," using biomedical interventions to rejuvenate bodies from the inside out (Shim, Russ, and Kaufman 2007). These researchers found an unprecedented willingness to allow biomedicalizing interventions (Kaufman, Shim, and Russ

2006), and a sense of ethical obligation to promote ever-longer life (Kaufman et al. 2004).

Our concept of technoscientific identities has been taken up in the growing field of the sociology of diagnosis (Jutel 2009), for example in studies of diagnostic imaging which may reveal them — like it or not (e.g., Burri and Dumit 2008; Joyce 2005a, this volume; van Dijck 2005). Genetic diagnostics too may reveal unwanted genetic identities that must be negotiated (Atkinson, Glasner, and Greenslade 2007; Clarke et al. 2009). Sulik (2009) found that women diagnosed with breast cancer developed and maintained technoscientific identities through four processes: immersion in biomedical knowledge; locating themselves in a biomedical framework; receiving support for this identity from both the medical system and support networks; and eventually giving priority to biomedical classification over their suffering. In some senses, then, technoscientific identities may be displacing illness identities (Charmaz 1995; Whyte 2009).

A host of other studies have also taken up biomedicalization — of sexual dysfunction (Loe 2004a; Potts et al. 2006; Fishman, this volume), gender and headache (Kempner 2006), geneticization and ethics (Arnason and Hjörleifsson 2007), disease representation and diagnosis (Rosenberg 2006; Joyce 2005a, this volume) — and it has been used in developing a research agenda on relations among people, other animals, and health knowledges (Rock, Mykhalovskiy, and Schlich 2007).

We hope that the five key processes of biomedicalization described at the beginning of this introduction will be used as analytics — lenses for pursuing research questions. The chapters in this volume by Fishman and Mamo are exemplars. That is, a particular topic of interest can be examined in terms of (1) the dynamics of its biopolitical economy; (2) precisely how it intensifies the focus on health (in addition to illness, disease, injury), on enhancement by technoscientific means, and on the elaboration of risk and surveillance at individual, group, and population levels; (3) how it elaborates the technoscientization of concrete biomedical practices; (4) how it engenders transformations of biomedical knowledge production, information management, distribution, and consumption; and/or (5) the kinds of transformations of bodies, selves, and new individual and collective technoscientific identities that are promoted. Biomedicalization, then, can be conceptualized as a "theory-methods package" offering sensitizing concepts and analytic direction (Clarke and Star 2008).

Gender, Race, Class, and Intersectionality in Medicalization and Biomedicalization Theories

Gender, race, and class have been taken up extensively in social theory and health services research. Here we discuss the relationships of these categories and social positions to medicalization and biomedicalization theories per se.

GENDER

Gender entered medicalization theory with Ruzek's (1978) ambitious study of the nascent women's health movement centered on the overmedicalization of women. This was the fundamental assumption of *Our Bodies, Our Selves* since its first incarnation in 1969. Women's health movements ever since have sustained this analysis (e.g., Ehrenreich and English 1978/2005; Weisman 1998; Ratcliff 2002; Morgen 2002). Social scientists have pursued it as well (e.g., Clarke 1983; Morgan 1998; Riska 2003; Clarke and Olesen 1999). Both Ruzek's (1978) and Riessman's (1983) analyses of women's health began to complicate the gender and medicalization story with the intersectionalities of race and class discussed hereafter (see also Nelson 2003; and Murphy 2003, 2004, forthcoming).

Before 2000, gender-focused work on medicalization was largely on women and/or how the concepts of sex/gender are produced, maintained, and negotiated in health, illness, and biomedicine. That is, much work has conceptualized gender not as forming discrete categories but as being produced in relations — as an effect of power (e.g., Fox and Worts 1999; Malacrida 2004; Moss and Dyck 2002). A fascinating early exception was Pfeffer's "The Hidden Pathology of the Male Reproductive System" (1985), which analyzed the relative absence of such research. A plethora of studies have more recently taken up the (bio)medicalization of men and masculinities. These include work on men's health more generally (e.g., Courtenay 2000; Payne 2006; Robertson 2007; Sabo and Gordon 1995), Type A men and coronary heart disease (Riska 2000, 2002, 2004), baldness (Conrad 2007), hormone treatments (Hoberman 2006), sexuality- and reproduction-related studies (Daniels 2006; Tjornhoj-Thomsen et al. 2005), sexual dysfunction (Mamo and Fishman 2001; Fishman and Mamo 2002; Fishman 2004; Loe 2004a), masculinization of infertility medicine (Thompson 2005), and the male pill (Oudshoorn 2002, 2003; van Kammen and Oudshoorn 2002). Rosenfeld's and Faircloth's *Medicalized Masculinities* (2006) addresses gaps in the study of medicalization of men's bodies and lives — new sites of commodification, consumption, and enhancement. Case studies demonstrate how medicalization increasingly occurs when performances of appropriate masculinities fail in schools (ADHD), in the military (PTSD), and in bed (ED).

Since around 2000, a wave of publications has taken up gender and health in terms of *both* males and females, from more general works (e.g., Bendelow et al. 2001; Lorber and Moore 2002; Krieger 2003; Doyal 2000; Segal et al. 2003) to gender and Prozac (Blum and Stracuzzi 2004) and gender and race on the American health research agenda (Epstein 2007, 2008). Last, as we were about to edit this volume, we found Riska's major paper "Gendering the Medicalization Thesis" (2003). She has extended that work here to include biomedicalization theory, offering the most ambitious theoretical examination of gender, medicalization, and biomedicalization to date.

RACE AND CLASS

Little work has explicitly addressed race, class, and medicalization *theory* aside from the work we cited in our 2003 paper, around stratified (bio)medicalization.[22] For this reason we do not have a separate chapter on race, class, and (bio)medicalization theory in this volume, though there is a chapter on gender and (bio)medicalization theory (Riska), and several of the case studies do take up race and class issues.

More specifically, while there are vast literatures on race and/or class and medicine and science, medicalization is rarely taken up substantively, much less theoretically. Instead, race and class issues have been taken up historically and contemporarily largely as causes of *exclusion from medicine* — as undermedicalization due to limited access to medical care. Not until the Medicare and Medicaid programs were initiated in the 1960s did major segments of populations of color in the United States have such access, especially in terms of hospitalization (e.g., Reynolds 1997). Access has remained highly partial and deeply impaired (e.g., hooks 1994; Wailoo 2001). Racism in American medicine, including serious maltreatment and exploitation of the poor, has also been a serious issue (e.g., Byrd and Clayton 2000–2002; Gilman 2008; Gravlee and Sweet 2008; Jones 1981/1993; Reverby 2000; Washington 2007; Nelson, forthcoming).

One could certainly argue that the substantial and growing literature on racial inequalities (often depoliticized by the term "disparities") deals largely with undermedicalization—unequal access to medicine due to class- and race-related factors. There is widespread (though by no means universal) refusal of medical science and clinical practice to take on social problems of health inequality.[23] Nevertheless, much of this work at least suggests how (bio)medicalization is highly stratified.

The remedial inclusion of women and people of color in clinical trials became a major issue in the late twentieth century. In his important work, Epstein (2007) argues that while the new biopolitical paradigm of "difference and inclusion" on which this is based has produced improvements, it has also produced new problems. One problematic is racial/ethnic categorization: What *is* race? What is ethnicity? (Bowker and Star 1999; Duster 2005). Second, this new paradigm is generating new interpretations of racial/ethnic differences and gender differences as biological rather than social, intensifying research on biological differences rather than on ameliorating health inequalities. In short, inclusion may generate biologization *instead of* (bio)medicalization and/or *as the organizing principle of* (bio)medicalization. This new biopolitics of the state is increasingly influenced by economic concerns (Foucault 2008), making biological "solutions" appealing because of their appeal to biocapital as well (Thompson 2006). Kahn's work (this volume) on BiDil and Lee's (2005) on race and pharmacogenomics sustain Epstein's (2007) argument that treating "difference" biologically often substitutes for addressing health inequalities. How race is addressed in genetics also illustrates this point (Duster 2005; Fujimura, Duster, and Rajagopalan 2008; Fullwiley 2008; Smart et al. 2008).

Here we emphasize the work that our concept of *stratified biomedicalization* can do.[24] Specifically, it emphasizes the selectivity and strategic nature of biomedicalization, its unequal (and sometimes unintended) effects across populations, and how these may exacerbate rather than ameliorate social inequalities along many different dimensions. More broadly, we theorize that exclusion, inclusion, and the embeddedness of race, class, gender, and other sites of inequality dwell in the very structures and processes of biomedicalization—in the very ways that technoscience is itself inherently social. Thus biomedicalization carries *within itself* the ideological, social, and cultural infrastructures that support and maintain racial and class inequalities (see, e.g., Shim, this volume). How biomedicine has sought to address "problems" of racial and socioeconomic inequalities in health—what it considers part of its domain and what it has deemed as beyond its jurisdiction—has been selective, fluid and inadequate.[25]

Recent work on race and medicalization includes attention to BiDil, the first FDA-approved drug targeted for a racial group—African Americans (Kahn 2004, 2005, 2006, 2008, this volume). BiDil served as a race-based model for pharmaceutical design and marketing for companies seeking to extend the market potential of drugs that have had more equivocal effects in the general population (see also Hatch 2009). Another genre of work on race and (bio)medicalization centers on genetics and genomics (e.g., Koenig, Lee, and Richardson 2008; Clarke, Shim, Shostak, and Nelson 2009). Reardon (2004) has done major work on race and the Human Genome Diversity Project. Nelson (2008a, 2008b) examines black family genealogies and DNA testing. Two chapters in this volume, by Shim and Shostak, explore ways in which biomedicalization is stratified in terms of race and class. And Mamo (this volume) explores the construction of racialized masculinities in sperm banks and sperm selection. Much remains to be done.

INTERSECTIONALITY THEORY

Through intersectionality theory, analyses of gender, race, and class have also begun to be brought together over the past two decades. This approach, pioneered by Patricia Hill Collins (1986, 1990/2000, 1999), insists that understanding the lives of people of color requires *simultaneous* consideration of their situatedness vis-à-vis race, class, and gender. These are—in life, in bodies, in practice—nonfungible categories. They cannot meaningfully be understood separately from one another; they are dynamic, changing, and co-constitutive, not merely contextual.

In relation to health, Litt (2000) explored themes of gender, race, class, and ethnic intersectionality in her *Medicalized Motherhood: Perspectives from the Lives of African-American and Jewish Women*. Weber (2006) radically seeks to reform health inequalities research by promoting dialogue and collaboration between feminist intersectional and biomedical paradigms. Dill, McLaughlin, and Nieves (2007) offer guidance for doing intersectional feminist research more broadly. Intersectionality theory is among the most important areas of theoretical development in the social sciences (Clarke and Olesen 1999; Schultz and Mullings 2006). And scholars of biomedicalization are pioneering fresh ways of thinking about it, including, for example, Shim's innovative conceptualization of cultural health capital (2010; see also Shim 2000).

▷▷▷

In sum, it remains interesting that little work has explicitly addressed race, class, and medicalization or biomedicalization *theory* — a situation we hope will soon change.

Focusing on Biomedicalization in the United States

> The life sciences are transforming different societies, locally, nationally, and globally in a multitude of different practices from the clinic to the factory.
> Harrington, Rose, and Singh (2006, 1)

All medicines and modes of healing and maintaining health travel, and they have always done so with the people who produce and use them. Western medicines — the modes addressed here — have been traveling widely for many centuries, especially but not only through imperial colonial and postcolonial projects (e.g., Anderson 2004, 2006; Arnold 1994). As we planned this volume of case studies of biomedicalization, we faced the decision whether or not to focus exclusively on the United States when such processes are certainly burgeoning elsewhere as well. In this section, we discuss our rationales and advantages of our decision to do so. The epilogue to this volume frames possibilities for using our biomedicalization framework transnationally.

We decided to focus on biomedicalization in the United States for multiple reasons. First and foremost, the research projects of all the editors were then U.S.-based. Second, we did not want to claim global reach without having a volume that seriously accomplished that quite daunting task. Third, there are serious historical and institutional advantages of focusing on a single nation-state in terms of "seeing" biomedicalization processes and comparing innovations. That is, our focus on the United States offers distinctive advantages, especially in terms of the comparability of the conditions under which the various case studies of biomedicalization presented here have occurred. Of course, local, state, and regional variations exist, but the similar structuring of the kinds of health services available and relative access to them across the country foregrounds biomedicalization processes across the volume rather than their geopolitical locus. This amplifies those processes rather than complicating them.

Finally, in part because of its unique historical reliance on private corporations (as opposed to the state) to finance, develop, and provide healthcare services and technologies, the United States offers an enormous range of diseases, conditions, and groups that have been highly medicalized and biomedicalized, as well as profoundly neglected. As social scientists, we can see more vividly how the biopolitical economy of health and illness, the discourses and practices of risk and surveillance, technoscientization, transformations of knowledge and of bodies and identities elaborate in relation to one another. These transformations of medical care and of life itself are, of course, increasingly exported and (re)modeled as well as being produced elsewhere, discussed in the epilogue.

Overview of the Book

THEORETICAL AND HISTORICAL FRAMINGS

The volume turns next to our original framing of biomedicalization theory, "Biomedicalization: Technoscientific Transformations of Health, Illness, and U.S. Biomedicine" (Clarke et al. 2003).[26] That paper elaborates in detail the five key processes of biomedicalization noted at the beginning of this introduction. We see these as analytic tools for the study of biomedicalization in specific cases. That chapter is followed by our "Charting of (Bio) Medicine and (Bio)Medicalization in the United States, 1890–Present." This dense chart served as the database for our development of biomedicalization theory and shows the range of things medical we attempted to take into account.

Next is Clarke's chapter on popular, especially visual, cultures of medicine, health, and healing characteristic of three eras in the United States: the rise of medicine from 1890 to 1945, the era of medicalization from 1945 to 1985, and the emergence of biomedicalization from 1985 to the present and beyond. She argues that developments of "things medical" in the past and today are not only accompanied by and reflected in popular cultural iconography—multiple media—but also were and continue to be generated and produced by and through them. She uses the concept of "healthscapes," which in the rise-of-medicine era featured physicians, surgeries, and hospitals as powerful new sites of care. During the medicalization healthscape, drugs, patients, and providers as people dominated. The biomedicalization healthscape spotlights high-tech interventions performed by teams in complex biomedical settings, expanded knowledge and decision-making burdens placed on patients, and the cultural pervasiveness of (bio)medicalization. But there was a lot more going on, and the chapter describes alternative approaches and countertrends always complicating and challenging biomedicalizing tendencies.

Last in this section is Riska's "Gender and Medicalization and Biomedicalization Theories," examining theoretical perspectives and research trends that underlay early medicalization theory and contemporary developments. Initially gender neutral, the medicalization thesis was soon gendered through its adoption by feminist critics of both medicine and medicalization. Specifically, between 1975 and 1985, the medicalization thesis was used to frame feminist critiques that elucidated how medicine itself was a gendered ideology and system of control disproportionately targeting women's bodies. In contrast, gender-focused studies of biomedicalization have since attended to socially constructed boundaries of gendered bodies and increasingly to men and masculinities. Riska demonstrates how biomedicalization theory provides a new framework for grappling with the emergence of the "hormonal body" (Oudshoorn 1994), naturalized for males and females through biomedical interventions that blur treatment and enhancement.

CASE STUDIES OF BIOMEDICALIZATION

The case studies fall into two main groupings. The first focuses on dimensions of "difference" and how biomedicalization processes and practices have engaged in their articulation, redefinition, and elaboration (Mamo, Joyce, Shim, Shostak, Kahn). The second focus is on sites of biomedicalization where implicit and explicit objectives center on enhancement—optimizing selves, families, and making life better and safer (Fishman, Boero, Fosket, Orr).

FOCUS ON DIFFERENCE

Laura Mamo's case study concerns lesbian practices of getting pregnant circa 2000. Drawing on empirical field research, she highlights how lesbian conceptive practices, neoliberal ideologies and policies, and what she calls Fertility, Inc. (a largely market-driven, for-profit healthcare sector) (Kolata

2002) are mutually productive. Analyzing interviews with lesbian consumers of sperm bank services, Mamo elaborates how their consumption practices are shaped by neoliberal discourse. Yet these women also pragmatically and instrumentally negotiate medical labeling and various control loci as they pursue their desire to become pregnant and attain motherhood. Mamo emphasizes that processes of biomedicalization include the *production* of "choice"—a seemingly vast array of offerings that apparently allows and encourages individualized selection. Yet choice is often illusory and directed toward more technoscientific and expensive options for enhancement and risk reduction. For example, low-tech options for conceiving have largely been supplanted by high-tech biomedical infertility interventions.

Elaborating the concept of technoscientific identities, Mamo argues that while processes of subjectification of patients continue, they are no longer driven only by medical classifications and diagnostics. Today they are expanding into consumption-driven cultural scripts central to (re)making bodies, identities, and ways of being through, for example, the organization of sperm selection. Consumers make choices as persons embedded in networks of social relations that constrain technological choices. Here Mamo brings the broader context of neoliberalism forward, provoking analysis of the relations between biomedicalization, consumption, and biopolitical economy. As part of the biomedical landscape of assisted reproduction today, neoliberalism produces new subjectivities, including moral imperatives to be healthy, maximize health, and minimize risk through biomedical "choices."

Kelly Joyce explores a key component of biomedicalization—the use of imaging technologies to construct and redefine individual and collective technoscientific identities. She argues that visualizing the inner body through the use of technologies such as ultrasound, CT, and MRI poses a new variant in knowledge production and identity formation. While MRIS are deployed in different arenas to construct individual, niche, and population identities (Epstein 2007), their legitimacy is not a guaranteed outcome across arenas. For example, studies using MRIS to assert older categories of difference such as race and gender have remained marginal. In contrast, studies using MRIS to forge new technoscientific confirmations of illness identities such as ADHD and schizophrenia have been more successful at achieving legitimization. Psychiatry and psychology textbooks now routinely include MRI-based images to illustrate these identities, and diagnostic MRIs are becoming routine.

Joyce reveals how institutions, policies, technoscience, and popular ideas co-produce new truths and identities, while the outcomes of such initiatives remain unpredictable. In concluding, Joyce explores why organized resistance to the use of such visualizing technologies has not occurred. In particular, cultural ideologies equating mechanically produced images with "reality" and "truth" make it difficult for patient groups to question the need for, and the quality and interpretation of, such technoscientific images.

Janet Shim takes up the subject of racial and class inequality in the stratified biomedicalization of heart disease. She argues that as citizens in a technoscientific era, we are subject to imperatives of health that are simultaneously broader in scope and increasingly minute in their surveillance of behaviors, thoughts, attitudes, and desires. Knowledge produced by epidemiology—the basic science of public health—serves as the basis for much of what constitutes clinical and popular advice about maximizing health, managing risk factors, and minimizing disease. Drawing on her sociological analysis of contemporary epidemiology, she asserts that the impacts of these health mandates are uneven, targeting racialized, classed, and gendered groups in distinctive and disproportionate ways.

Closely examining epidemiological claims about heart disease, Shim traces how constructions of race, class, and sex/gender become embedded in the frameworks, discourses, and practices of research on cardiovascular risk, including its assessment and management. The biomedicalizing of different populations thus reveals linkages between the production of scientific knowledge and its stratified applications. These applications themselves constitute significant new forms of stratification that not only document health inequalities but also help produce them. They deserve considerable attention.

Sara Shostak's case study of the emergence of molecular epidemiology and its consequences for environmental health research and practice asserts a shift in focus from whole populations to individual persons at risk. This is accomplished through the signal technology of molecular epidemiology—molecular biomarkers—and their "translations" into clinical and regulatory settings. Today the modus operandi of environmental epidemiological measurement and intervention is thereby shifting from assessments of the ambient *external* environment (air, water, soil) to assessments of the *interior* molecular levels of individual human bodies.

Shostak's fundamental argument is that molecular epidemiology thus directly contributes to the biomedicalization of environmental health and illness. Specifically, as problems of pollution are redefined and measured as individual rather than social population-level risk phenomena, modes of addressing them will also likely be individualized and biomedicalized. Instead of removing pollutants from the external environment, individuals will be instructed on how to minimize their risks of exposure — and bear the burdens of responsibility if they fail. This individuation of failure eliminates the need for social accountability, including that of the state for its biological citizens.

Jonathan Kahn's chapter traces the racialization of BiDil, the first "ethnic-racial" drug approved by the FDA. BiDil's saga specifically elucidates the power of legal and regulatory regimes in processes of biomedicalization. Touted as a significant step toward the promised era of personalized pharmacogenomic therapies by the FDA and others, BiDil's story is complex. Kahn describes the process. Two drugs already on the market, previously found not to benefit a general population of heart failure patients, were combined into a new race-specific drug. Kahn shows how the legal and regulatory environment shaped where and how this biomedicalization occurred, what ultimate forms it took, and the implications and ramifications it had for health inequalities and commercial interests.

Significantly, this case study demonstrates how a technoscientific entity can exploit the claim to biomedicalize — that it targets a new population and opens a new commercial market — even when its actual ability to do so is far from clear. In fact, BiDil has not cornered a major sector of the market and is not considered a success. However, many other race-specific drugs are in the pipeline.

Together, these papers on the biomedicalization of difference demonstrate the workings of biomedicalization from the anatomo-politics of molecules to the bioburdens of populations.

FOCUS ON ENHANCEMENT

Four chapters are concerned with enhancement and optimization, taking up drugs for sexual dysfunction, bariatric surgery, drugs for preventing cancer, and biopsychiatry (Fishman, Boero, Fosket, Orr).

Jennifer Fishman focuses on Viagra and other drugs in research and development for treating sexual dysfunction. Examining the historical, social, and cultural milieus within which this research developed, she analyzes how these drugs and biomedical research on them contribute to a biomedicalization of sexual dysfunction. These influences include biomedical promotion of "successful aging," our heightened expectations of functionality and normality across the lifespan, and the biomedicalization of health and "lifestyle." This situation simultaneously facilitates and legitimates research for new drugs to treat "lifestyle conditions" such as sexual dysfunction. Using our five key processes of biomedicalization as analytics, Fishman details how sexual dysfunction has become an exemplar case of biomedicalization.

Next, shifting from the desire for more to the desire for less, Natalie Boero takes up the case of bariatric surgical weight loss, increasingly popular in the midst of an "obesity epidemic," but not always successful. Over decades, the medicalization of obesity reconstructed it as a medical phenomenon, largely displacing earlier moral and often religious (sin-based) discourses of stigmatization. Thus in promoting bariatric surgery, physicians often stress that the obese are not themselves "at fault" for their overweight. They simultaneously remove blame from individual fat persons and reframe obesity as a chronic disease.

Boero draws on interviews with bariatric surgeons, pre- and postoperative weight loss surgery patients, participant observation, and textual analysis. She shows that when bariatric surgeries "fail"—when weight is not lost or weight loss is not maintained—specifically gendered notions of individual moral culpability are reinvoked. That is, newer biomedicalized constructions of obesity as disease-based can quickly be displaced by older moral rhetorics of blame and stigmatization. Boero asks how understanding this example of biomedical failure adds to a theory of biomedicalization in an era when definitions of epidemics extend far beyond mass contagion and death.

Primary cancer prevention has historically been aimed at protecting the population by identification and removal of cancer-causing agents "around us," usually in the ambient environment (as Shostak's chapter attests). A new and classically biomedicalizing strategy instead aims to prevent cancer through the individual's ingestion of synthetic compounds that prevent, reduce, or reverse carcinogenesis "within us." Jennifer Fosket's paper explores this use of prevention pills or chemoprevention as it has emerged in the realm of breast cancer to enhance or optimize at-risk women's lives.

The concept of administering daily doses of prescription pharmaceuticals as a health maintenance strategy marks the biomedicalization of health, risk, and prevention. The bodies, selves, and lives of women deemed at high risk for breast cancer and targeted for chemoprevention are thereby marked. Fosket theoretically explores how biomedicalization has both resulted from and also helped give rise to chemoprevention for breast cancer. She examines women's reactions, resistances, and accommodations to that form of biomedicalization, revealing complicated processes of negotiation rather than simple acceptance.

Computer and information technologies are central features of the transformative processes of biomedicalization. Jackie Orr focuses on these in her study of the informatics of diagnosis and the marketing of mental disorders. The emergent field of biopsychiatry—founded on claims to biologically oriented and empirically grounded practices of diagnosing and treating mental disorders—relies increasingly on computer and information technologies. Computerization of psychiatric diagnosis was made possible by systematizing and formalizing changes in the 1980 *Diagnostic and Statistical Manual of Mental Disorders* (*DSM-III*). Today, online diagnostics combine new diagnostic capacities with infomercials about mental disorders and their pharmaceutical "cures" on the Internet and in other electronic media. Formats that state "Take this short assessment test and see if you suffer from X" are flourishing.

As biopsychiatry becomes increasingly embedded in networks of communication and information technologies, these networks simultaneously situate biopsychiatry in ever-tighter feedback relations with the transnational pharmaceutical industry and its circuits of commodification and flows of biocapital. Orr tells the story of this new informatics of psychiatric diagnosis in relation to questions of governmentality and power as they intersect with empirical and conceptual claims of technoscientific sophistication.

The breadth of the topics these case studies engage vividly demonstrates the wide reach of biomedicalization today. Overall, the papers on the elaboration of difference demonstrate some shifts in the locus of causality from the world around us to bodily interiors that can now be revealed by technoscientific means, illustrating what Duster (2006) termed the newly reconfigured reductionist challenge to sociology. Biomedicalizing regimens for prevention and treatment too are increasingly individualized. The chapters on enhancements and optimizations also illustrate individualizing tendencies of biomedicalization in these neoliberal political times. The further development of biosocialities and entitlements of biological citizenship may be anticipated in the future as vital politics elaborate.

The volume concludes with Clarke's epilogue on biomedicalization in its transnational travels. One of our goals is to provoke empirical studies of the rise of medicine, medicalization, and biomedicalization, both as phenomena in and of themselves and in relation to particular drugs, technologies, procedures, and practices, including biocapital. The epilogue aims to support such empirical projects outside the United States. It frames some anticipated complications in translating biomedicalization theory for use in other sites, presents key concepts useful in transnational work, and notes related research. It also offers strategies for researching local sociohistories of the rise of medicine, medicalization, and biomedicalization as healthscapes — traveling assemblages that are adopted, adapted, and transformed by the geopolitics of place.

Conclusions

"Yet we also wish to avoid the breathless futurology that often pervades discussions in this area." (Harrington, Rose and Singh 2006, 3)

One of the most enduring critiques of medical sociology has been that it is atheoretical and merely applied (e.g., Seale 2008, 678–79). Obviously biomedicalization theory has explicitly sought to counter such assertions on several fronts: linkages to Foucauldian notions of biopower and biopolitics; to science, technology, and medicine studies; and to the sociology of knowledge. Situating biomedicalization theory vis-à-vis the new generation of scholarship that takes up vital politics, biocapital, the shift from the clinical to the molecular gaze, and the changing nature of "life itself," as we have done in this introduction, sustains this theoretically driven angle of vision.

In this move, we gesture first to the increasingly dense webs and assemblages being woven among (bio)political economy; the life sciences; technoscientific interventions; human, animal, and plant tissues and parts; and biological citizens who do the actual work of transforming life itself.

Biomedicalization processes operate at core sites of such changes. Biomedicalization theory thus provides greater traction to empirically trace not only how biopolitical economy matters but also how the sites of action and transformation of life itself are remade. Through attending to biopolitical economy as integral to biomedicalization, we also further link medical sociology to both economics and political science.[27] These disciplinary and specialty boundaries seem, in fact, to be melting in the face of vital politics and biomedicalization.

The life sciences matter today both *within* medicine in terms of its technoscientific capacities and *for* medicine in terms of setting "conditions of possibility" vis-à-vis the legitimacy of intervening in life itself and optimizing futures. Biomedicalization theory has pioneered in extending medicalization theory to more broadly grasp and engage market/biocapital elements and the tremendous salience of the life sciences in the new millennium (e.g., Cockerham 2005a; Gottweis 2005; Prainsack 2006; Hafferty 2006; Marzano 2009), especially in terms of "mapping the new genomic era" (Clarke, Shim, Shostak, and Nelson 2009).

Within biomedicalization, the legitimacy and practices of optimization and enhancement to secure "the best possible futures" are becoming increasingly central. Beyond engaging with biocommodities and other biomedical enhancements, optimization also engages regimes of risk and surveillance and heightened responsibilities for knowledge accumulation and consumption. Optimization also involves taking up new kinds of bioinflected individual and collective (including familial) identities — technoscientific identities. Together these new forms of subjectification highlight the duties and burdens of authorizing one's own — and others' — futures.

This constant orientation toward the future — a sociocultural parallel to the bioeconomic/biopolitical focus on speculative and promissory capital — gives a greater temporal dimension to biomedicalization theory. In these senses, then, biomedicalization theory is anticipatory, offering a conceptual frame for analyzing the emergent, the about-to-be, the evanescent (Adams, Murphy, and Clarke 2009). As Rapp (2003, 142–43) notes, although "laboratory life cycles" used to be decades long, this situation is changing. There are "purgatorial dimensions" (Rabinow 1999) — a sense of the almost there — in terms of potential biomedicalizing outcomes. Biopolitical anxieties loom large.

The contributions here examine varied instantiations of biomedicalization in these new lights — case studies of the future as well as the present.

Biomedicalization theory broadly conceived thus demonstrates that domains of health, illness, and medicine are and will continue to be key theoretical as well as substantively interesting sites in the new millennium. The centrality of these domains to life itself is transparent. It is *not just* the biopolitical economy that matters but also the precise sites of action and sociocultural transformations of life itself through the (re)organization and (re)institutionalization of capital into biocapital and biomedicine.

Medicalization (which continues) and biomedicalization are what Williams (1976/1985) called "keywords," carrying significant cultural and historical weight. Their definitions, contours, and deployment therefore require ongoing interrogation and analysis. The cutting-edge research offered here begins to do some of this work. We anticipate that biomedicalization theory will continue to travel productively in many new directions. Speculatively, one fresh direction for future research would use the nicely honed lenses of consumption studies to examine the modes and consequences of patients becoming consumers and biomedicine staging itself through the language and sales strategies of other consumer products.[28] Another emergent topic is the biomedicalization of defense, including the expansion of biomedical approaches to the development of weaponry (Vogel 2008) and during warfare, such as the biological alteration of warriors by technoscientific means (Hogle 2006). Biosecurity is another "growth industry" deserving attention (King 2002, 2005; Lentzos 2006). In terms of biomedicalization and computer and information science, electronic patient records (e.g., Jensen 2005) and implantation of people with bar codes with medical data in part to reduce medical error (Stein 2004) are worthy of study (Mechanic 2002). Biomedicalization studies might explore the potential and actual consequences of evidence-based medicine on the tendency to move directly to highly sophisticated interventions.[29] What happens when the evidence does not support biomedicalizing tendencies but instead indicates historically conservative approaches? What happens when leading physicians themselves challenge the use of technologies of biomedicalization (e.g., Berenson and Abelson 2008)? Again, the five key processes of biomedicalization can be useful analytics with which to approach and interrogate emergent practices.

Critical work like this volume, which itself interrogates biomedicalization, is largely diagnostic (Fosket, this volume; see also Burri and Dumit 2007). But we also need to ask, as Ferguson (2008) enjoins us, "What is to be done after the critical?" What are the policy implications of our diagnoses? To improve biopolitical governance, for example, we need more explicit investigations of how biomedicalization is stratified in its objects, practices, and effects — so that these may be more directly countered (e.g., Cockerham 2007; Navarro 2007). In turn, we need a better grasp on how ideas about race, gender, class, sexuality, disability, and so on propel selective and uneven biomedical efforts to transform, regulate, and optimize bodies and futures. And playing on a science and technology studies dictum, we end our suggestions for future directions with the injunction "Follow biocapital to follow the future" of biomedicalization.

Because biomedicalization in theory and practice stands precisely at the intersection of medical sociology, medical anthropology, medical history, political science, economics, and new science, technology, and medicine studies, we intentionally seek here to produce and sustain cross-disciplinary provocations (e.g., Casper and Berg 1995; Casper and Koenig 1996; Lock 2008). We seek to introduce medical sociologists and anthropologists to creative and exciting science and technology studies perspectives that should be useful as technoscientific innovations increasingly transform biomedicine in the United States and transnationally. In turn, we also seek to introduce science and technology studies scholars to the range and depth of work in medical sociology, medical anthropology, and the history of medicine that is pertinent and too often ignored.[30] We hope this volume performs some of the necessary work to build productive bridges and exciting sites of intersection for future work on biomedicalization.

Notes

Citations here are largely to work done after 2000. For earlier work, please see our 2003 paper, reprinted as chapter 1 in this volume. We thank Susan Bell and the Duke reviewers for valuable comments.

1 The term "technoscience" indicates that science and technology are not easily distinguishable (the one "basic" and the other "applied") but should instead be understood as co-constituted and hybrid (Latour 1987).

2 We thank Donna Haraway (1997) for inspiration for our originally naming this the Biomedical TechnoService Complex Inc. (discussed later). On co-constitution, see Jasanoff 2004.

3 See Hacking 1999 and Epstein 2007 on the shift from assuming a "standard human" in biomedical research and treatment protocols to "niche standardization" based on race, gender, and other markings of "difference" in NIH-funded research since the early 1990s.

4 See, e.g., Armstrong 1995; Lupton 1994/2003, 1995, 1997; Rabinow 1992; Rabinow and Rose 2006; Risse 1999; Shim 2000; and Turner 1984, 1997.

5 Rose's five mutations are *molecularization* as a "style of thought" that is reshaping technoscience per se; *optimization* as the moral responsibility of citizens to secure their "best possible futures"; *subjectification* in terms of recoding the "duties, rights, and expectations of human beings" in terms of health and illness to be ever more individually (rather than socially) responsible; *somatic expertise* as the growing numbers and kinds of subprofessions dedicated to managing aspects of somatic existence with which we are involved as patients/consumers; and *economies of vitality* such that "biopolitics has become inextricably intertwined with bioeconomics."

6 Others have noted the salience of molecularization (e.g., Yoxen 1981, 1982; Kay 1993a, 1993b). Anker and Nelkin (2004) used the term "molecular gaze." Rose makes a broader Foucauldian argument.

7 See, e.g., Adams 2002; Waldram 2004; and the "Comparative Perspectives Symposium: Bioprospecting/Biopiracy," *Signs* 32, no. 2 (2007): 307–46.

8 Dick Scott brought this vividly to our attention. See http://www.bigger strongerfastermovie.com.

9 Health social movements are generally multi-issue (e.g., women's and minority health movements), while patients' organizations or movements focus on particular diseases. Both differ from other social movements in ways too complex to address here. See, e.g., Brown et al. 2004; Brown and Zavetoski 2005; Epstein 2008; Landzelius 2006; and Novas 2006.

10 See, e.g., Epstein 1996, 2007; Lakoff 2006, 2008; and Petryna, Lakoff, and Kleinman 2006. On transnational work, see the epilogue, this volume.

11 See Young 1999 and http://www.nih.gov/about/almanac/organization/NCCAM .htm.

12 See the special issue of *Social Science and Medicine* 62, no. 3 (2006).

13 See, e.g., websites for the Boston Women's Health Book Collective (http://www .ourbodiesourselves.org) and for the National Women's Health Network (http:// www.nwhn.org).

14 See the special issue of *Sociology of Health and Illness* 28, no. 6 (2006).

15 On the intersection of biomedicalization with genetics, see Clarke, Shim, Shostak, and Nelson 2009.

16 See our discussion in Clarke et al. 2003. For historical overviews, see Nye 2003 and Pfohl 1985.

17 For us, classic questions of the sociology of knowledge were central: whose knowledge counts and how lay and professional knowledges are addressed (e.g., Fleck 1935/1979; Wright and Treacher 1982).

18 Debate about national differences due to different health systems is sustained. See *Society* 43, no. 6 (2006): 14–56.

19 Life scientists conventionally use the term "biotechnology" when basic protein structures are altered affecting genetic structure(s) (e.g., recombinant DNA).

Most pharmaceuticals would not fit, nor most medical technologies. We follow this usage.

20 Demedicalization—the ending of medical jurisdiction over a phenomenon—has also been a concern of medicalization theory (e.g., Fox 2001). The classic case was the demedicalization of homosexuality by its exclusion as an illness from the psychiatric *Diagnostic and Statistical Manual.* Conrad (2007, chap. 5) points to its potential remedicalization.

21 The most dominant and portable mechanisms of biomedicalization are drugs. Lakoff (2008) reviews recent social science research. Martin (2006, 2007) describes "the pharmaceutical person," and Davis-Berman and Pestello (2005) "the medicated self." Tone and Watkins (2007) analyze relations among doctors, patients, and prescriptions; Daemmrich (2004) takes up drug regulation as pharmacopolitics. Greene (2006) analyzes (re)definition of disease vis-à-vis congruence with specified pharmaceuticals and insurance coverage categories. Fisher (2009) analyzes clinical trials. Sismondo (2009) examines the long-term planning of pharmaceutical company-sponsored research publication in terms of timing, placement, and recruitment of legitimizing authors. Comparative work examines the social lives of medicines in different nations (Whyte et al. 2002; epilogue, this volume). Big pharma is a major focus (Goodman 2003; Moynihan and Cassels 2005; Bastian 2002).

22 We examined two editions of Lupton's *Medicine as Culture* (1994 and 2003), Petersen's and Bunton's *Foucault, Health, and Medicine* (1997), and White's *Introduction to the Sociology of Health and Illness* (2002). In their indexes, only gender is explicitly related to medicalization. Indexes of the *Handbook of Medical Sociology* (Bird, Conrad, and Fremont 2000) and *The Blackwell Companion to Medical Sociology* (Cockerham 2000) had no listings of medicalization and race or gender.

23 See, e.g., Berkman and Kawachi 2000a; Cockerham 2007; Institute of Medicine 2002; Kawachi and Kennedy 2001; Krieger 1999; and Krieger et al. 1993.

24 See Clarke et al. 2003 (chap 1 in this volume). We drew on Colen's (1995) concept of stratified reproduction, further elaborated by Rapp 2001.

25 There is a vast literature. See, e.g., Cockerham 2007; Duster 2003, 2005; Krieger 1999; Krieger et al. 1993; and Wailoo 2001.

26 For a simplified version of this paper, see Clarke et al. 2004.

27 For other work linking science and technology studies to these fields, see Callon 2003; Gottweis 2005; Mirowski 2007; and Sunder Rajan 2007.

28 See, e.g., Applbaum 2004; Baudrillard 1998; Henderson and Petersen 2002; Miller 1998; and Miller and Rose 1997.

29 See, e.g., Murray and Holmes 2009; May et al. 2006; Mykhalovskiy and Weir 2004; Pope 2003; and Timmermans and Berg 2003.

30 While presentations on medical topics at the Society for Social Studies of Science meetings have gone from 11 percent in 1988 to 29 percent in 2001 (Amsterdamska and Hiddinga 2004), both broader and deeper knowledge across these disciplines remains rare.

▷▷▷PART I / THEORETICAL AND HISTORICAL FRAMINGS

Adele E. Clarke, Janet K. Shim, Laura Mamo,
Jennifer Ruth Fosket, and Jennifer R. Fishman

1 / Biomedicalization

TECHNOSCIENTIFIC TRANSFORMATIONS OF
HEALTH, ILLNESS, AND U.S. BIOMEDICINE

The growth of medicalization—defined as the processes through which aspects of life previously outside the jurisdiction of medicine come to be construed as medical problems—is one of the most potent social transformations of the last half of the twentieth century in the West (Bauer 1998; Clarke and Olesen 1999; Conrad 1992, 2000; Renaud 1995). We argue that major, largely technoscientific changes in biomedicine[1] are now coalescing into what we call *biomedicalization*[2] and are transforming the twenty-first century. Biomedicalization is our term for the increasingly complex, multisited, multidirectional processes of medicalization that today are being both extended and reconstituted through the emergent social forms and practices of a highly and increasingly technoscientific biomedicine. We signal with the "bio-" in biomedicalization the transformations of both the human and nonhuman made possible by technoscientific innovations such as molecular biology, biotechnologies, genomization, transplant medicine, and new medical technologies. That is, medicalization is intensifying, but in new and complex, usually technoscientifically enmeshed ways.

Institutionally, biomedicine is being reorganized not only from the top down or the bottom up but *from the inside out*. This is occurring largely through the remaking of the technical, informational, organizational, and hence the institutional infrastructures of the life sciences and biomedicine via the incorporation of computer and information technologies (Bowker and Star 1999, Cartwright 2000b; Lewis 2000; National Research Council 2000). Such technoscientific innovations are reconstituting the many institutional sites of healthcare knowledge production, distribution, and information management (e.g., medical information technologies/informatics, networked or integrated systems of hospitals, clinics, group practices, insurance organizations, the bioscientific and medical technology and supplies industries, the state, etc.). Such meso-level organizational/institutional changes are cumulative over time and have now reached critical infrastructural mass in the shift to biomedicalization.

Clinical innovations are, of course, at the heart of biomedicalization. Extensive transformations are produced through new diagnostics, treatments and procedures from bioengineering, genomics, proteomics, new computer-based visualization technologies, computer-assisted drug developments, evidence-based medicine, telemedicine/telehealth, and so on. At the beginning of the twenty-first century, such technoscientific innovations are the jewels in the clinical crown of biomedicine and vectors of biomedicalization in the West and beyond.

The extension of medical jurisdiction over health itself (in addition to illness, disease, and injury) and the commodification of health are fundamental to biomedicalization. That is, health itself and proper management of chronic illnesses are becoming individual moral responsibilities to be fulfilled through improved access to knowledge, self-surveillance, prevention, risk assessment, the treatment of risk, and the consumption of appropriate self-help and biomedical goods and services. Standards of embodiment, long influenced by fashion and celebrity, are now transformed by new corporeal possibilities made available through the applications of technoscience. New individual and collective identities are also produced through technoscience (e.g., "high-risk" statuses, DNA profiles, Syndrome X sufferers).

Biomedicalization processes are situated within a dynamic and expanding politico-economic and sociocultural biomedical sector. In this sector, the incorporation of technoscientific innovations is at once so dense, dispersed (from local to global to local), heterogeneous (affecting many different domains simultaneously), and consequential for the very organization and practices of biomedicine broadly conceived that they manifest a recorporation—a reconstitution—of this historically situated sector. We term this new social form the "Biomedical TechnoService Complex Inc."[3] The growth of this complex since World War II is clear. The U.S. health sector has more than tripled in size over the last fifty years from 4 percent to 13 percent of GDP, and it is anticipated to exceed 20 percent by 2040 (Leonhardt 2001). At the same time, Western biomedicine has become a distinctive sociocultural world, ubiquitously webbed throughout mass culture (e.g., Bauer 1998; Lupton 1994). Health has been the site of multiple old and new social movements (e.g., Brown et al. 2001). Biomedicine has become a potent lens through which we culturally interpret, understand, and seek to transform bodies and lives. That is, if the concept of the Biomedical TechnoService Complex Inc. particularly captures some politico-economic dimensions of biomedicalization, the concept of biomedicine as a culture per se, as a regime of truth (Foucault 1980, 133), particularly captures some sociocultural dimensions.

Although we can conceptually tease apart organizational, clinical, and jurisdictional axes of change and their situatedness within a politico-economic and sociocultural sector—however vast—the ways in which these changes are simultaneous, co-constitutive, and nonfungible inform our conceptualization of biomedicalization. That is, a fundamental premise of biomedicalization is that increasingly important sciences and technologies *and* new social forms are *co-produced* within biomedicine and its related domains.[4] Biomedicalization is reciprocally constituted and manifest through five major interactive processes: (1) the political economic constitution of the Biomedical TechnoService Complex Inc.; (2) the focus on health itself and elaboration of risk and surveillance biomedicines; (3) the increasingly technoscientific nature of the practices and innovations of biomedicine; (4) transformations of biomedical knowledge production, information management, distribution, and consumption; and (5) transformations of bodies to include new properties and the production of new individual and collective technoscientific identities. These processes operate at multiple levels as they both engender biomedicalization and are also (re)produced and transformed through biomedicalization over time. Our argument, thus, is historical, not programmatic.

We begin by examining the historical shift from medicalization to biomedicalization. We then elaborate the five key historical processes through which biomedicalization occurs. We conclude by reflecting on the implications of the shift to biomedicalization.

From Medicalization to Biomedicalization

Historically, the rise in the United States of Western (allopathic) medicine as we know it was accomplished clinically, scientifically, technologically, and institutionally from 1890 to 1945. This first "transformation of American medicine" (Starr 1982) was centered not only on the professionalization and specialization of medicine and nursing but also on the creation of allied health professions, new medico-scientific, technological, and pharmaceutical interventions, and the elaboration of new social forms (e.g., hospitals, clinics, and private medical practices) (Abbott 1988; Clarke 1988; Freidson 1970, 2001; Gaudillière and Löwy 1998; Illich 1976; Lock and Gordon 1988; Pauly 1987; Pickstone 1993; Risse 1999; Stevens 1998; and Swan 1990). Then, in the decades after World War II, medicine, as a politico-economic institutional sector and a sociocultural "good," grew dramatically in the United States through major investments, both private (industry and foundations) and public (e.g., the National Institutes of Health [NIH], Medicare, Medicaid) (Kohler 1991; NIH 1976, 2000a, 2000b). The production of medical knowledges and clinical interventions—goods and services—elaborated rapidly.[5]

As medicine grew, sociologists and other social scientists began to attend to its importance, especially as a profession (Abbott 1988; Becker et al. 1961; Bucher 1962; Bucher and Strauss 1961; Freidson 1970; Parsons 1951; Starr 1982; Strauss et al. 1964). The concept of medicalization was framed by Irving Zola (1972, 1991) to theorize the extension of medical jurisdiction, authority, and practices into increasingly broader areas of people's lives. Initially, medicalization was seen to take place when particular social problems deemed morally problematic and often affecting the body (e.g., alcoholism, homosexuality, abortion, and drug abuse) were moved from the professional jurisdiction of the law to that of medicine. Drawing on interactionist labeling theory,[6] Conrad and Schneider (1980) termed this a transformation from "badness to sickness." Simultaneously, some critical theorists viewed medicalization as promoting the capitalist interests of medicine and of the medical industrial complex more broadly (e.g., Ehrenreich and Ehrenreich 1978; McKinlay and Stoeckle 1988; Navarro 1986; Waitzkin 1989, 2001).

Through the theoretical framework of medicalization, medicine came to be understood as a social and cultural enterprise as well as a medico-scientific one, and illness and disease came to be understood not as necessarily inherent in any particular behaviors or conditions but as constructed through human (inter)action (Bury 1986; Lupton 2000). Further, medicalization theory also illuminated the importance of widespread individual and group acceptance of dominant sociocultural conceptualizations of medicine and active participation in its diverse, interrelated macro-, meso-, and micropractices and institutions, however uneven (Morgan 1998).

Gradually the concept of medicalization was extended to include any and all instances of new phenomena deemed medical problems under medical jurisdiction—from initial expansions around childbirth, death, menopause, and contraception in the 1970s to post-traumatic stress disorder (PTSD), premenstrual syndrome (PMS), and attention deficit hyperactivity disorder (ADHD) in the 1980s and 1990s, and so on (Armstrong 2000; Conrad 1975, 2000; Conrad and Potter 2000; Conrad and Schneider 1980a/1992; Figert 1996; Fox 1977, 2001; Halpern 1990; Litt 2000; Lock 1993; Riessman 1983; Ruzek 1978; Schneider and Conrad 1980; Timmermans 1999). Social and cultural aspects and *meanings* of medicalization were elaborated even further and, as we argue next, largely through technoscientific innovations. For example, conditions understood as undesirable or stigmatizable (Goffman 1963) "differences" were medicalized (e.g., unattractiveness through cosmetic surgery; obesity through diet medications), and medical treatment of such conditions was normalized (Armstrong 1995; Crawford 1985). These were the beginnings of the biomedicalization of *health*, in addition to illness and disease—the biomedicalization of phenomena that heretofore were deemed within the range of "normal" (Arney and Bergen 1984; Hedgecoe 2001).

Then, beginning about 1985, we suggest that the nature of medicalization itself began to change as technoscientific innovations and associated new social forms began to transform biomedicine from the inside out. Conceptually, biomedicalization is predicated on what we see as larger shifts-in-progress from the problems of modernity to the problems of late modernity or postmodernity. Within the framework of the industrial revolution, we became accustomed to "big science" and "big technology"—projects such as the Tennessee Valley Authority, the atom bomb, and electrification and transportation grids. In the current technoscientific revolution, "big science" and "big technology" can sit on your desk, reside in a pillbox or inside your body. That is, the shift to biomedicalization is a shift from en-hanced control over external nature (i.e., the world around us) to the harnessing and transformation of internal nature (i.e., biological processes of human and nonhuman life-forms), often transforming "life itself." Thus it can be argued that medicalization was co-constitutive of modernity, while biomedicalization is also co-constitutive of postmodernity (Clarke 1995).

Important to the shift are the ways in which historical innovations of the medicalization era (organizational, scientific, technical, cultural, etc.) became widely elaborated and dispersed material infrastructures, resources and sociocultural discourses, and assumptions of the biomedicalization era (Clarke 1988). Biomedicalization is characterized by its greater organizational and institutional reach through the meso-level innovations made possible by computer and information sciences in clinical and scientific settings, including computer-based research and record keeping. The scope of biomedicalization processes is thus much broader and includes conceptual and clinical expansions through the commodification of health, the elaboration of risk and surveillance, and innovative clinical applications of drugs, diagnostics, and treatment procedures. This includes the production of new social forms through "dividing practices" that specify population segments such as risk groups (Rose 1994). These groups are to be given special attention through new "assemblages" (Deleuze and Guattari 1987) of spaces, persons, and techniques for caregiving. Innovations and interventions are not administered only by medical professionals but are also "technologies of the self," forms of self-governance that people apply to themselves (Foucault 1988; Rose 1996). Such technologies pervade more and more aspects of daily life and the lived experience of health and illness, creating new biomedicalized subjectivities, identities, and biosocialities—new social forms constructed around and through such new identities (Rabinow 1992). We seek to capture these changes in the ordering of health-related activities and the administration of individuals and populations[7]—including self-administration—referred to as governmentality.[8]

The table offers an overview of the shifts from medicalization to biomedicalization cobbled and webbed together through the increasing application of technoscientific innovations. One overarching analytic shift is

MEDICALIZATION CONTROL	BIOMEDICALIZATION TRANSFORMATION
Institutional expansion of professional medical jurisdiction into new domains	Expansion *also* through technoscientific transformations of biomedical organizations, infrastructures, knowledges
Economics: The U.S. Biomedical TechnoService Complex Inc.	
Foundation and state (usually NIH) funded biomedical, scientific, and clinical research with accessible public results	*Also* increasing privatization of research including through university-industry collaborations with increased privatization and commodification of research results as proprietary knowledge
Increased economic organization, rationalization, corporatization, nationalization	*Also* increased economic privatization, devolution, transnationalization-globalization
Physician-dominated organizations	Managed-care-system-dominated organizations
Stratification largely through the dual tendencies of selective medicalization and selective exclusion from healthcare based on ability to pay	Stratification *also* through stratified rationalization, new population-dividing practices and assemblages for surveillance and treatment based on technoscientific identities
The Focus on Health, Risk, and Surveillance	
Works through a paradigm of definition, diagnosis (through screening and testing), classification, and treatment of illness and diseases	Works *also* through a paradigm of treatment of risks and commodification of health and lifestyles
Health policy as problem solving	Health governance as problem defining
Diseases conceptualized at level of organs, cells	Risks and diseases conceptualized at level of genes, molecules, and proteins
The Technoscientization of Biomedicine	
Highly localized infrastructures with idiosyncratic physician and clinic generated paper medical records of patients (photocopy and fax are major innovations)	Increasingly integrated infrastructures with widely dispersed access to highly standardized digitalized patients' medical records, insurance information processing and storage
Individual/case-based medicine with local (usually office-based) control over patient information	Outcomes/evidence-based medicine with use of decision-support technologies and computerized patient data banks in managed-care systems

MEDICALIZATION CONTROL	BIOMEDICALIZATION TRANSFORMATION
Medical science and technological interventions, e.g., antibiotics, chemotherapy, radiation, dialysis, transplantation, reproductive technologies	Biomedical technoscientific transformations, e.g., molecularization, biotechnologies, bioengineering, biomonitoring, chemoprevention, gene therapies, and stem-cell based regenerative medicine and cloning
Medical specialties based on body parts and processes and disease processes (e.g., cardiology, gynecology, oncology) assumed to be universal across populations and practice settings	New medical specialties based on assemblages — loci of practice and knowledge of accompanying distinctive populations and genres of sciences and technologies (e.g., emergency medicine, hospitalists, prison medicine, regenerative medicine)

Transformations of Information, Knowledge Production, and Distribution

Professional control over specialized medical knowledge production and distribution with highly restricted access usually limited to medical professionals	Heterogeneous production of multiple genres of information/knowledges regarding health, illness, disease, and medicine, widely accessible in bookstores and electronically by Internet, etc.
Largely top-down medical-professional-initiated interventions	*Also* heterogeneously initiated interventions and collaborations. New actors include, e.g., health social movements, consumers, Internet users, pharmaceutical corporations, advertisements, websites

Transformations of Bodies and Identities

Normalization	Customization
Universal Taylorized bodies; one-size-fits-all medical devices/technologies and drugs; superficially (including cosmetically) modified bodies	Individualized bodies; niche-marketed and individualized drugs and devices/technologies; customized, tailored, and fundamentally transformed bodies
From badness to sickness; stigmatization of conditions and diseases	*Also* new technoscientifically-based individual and collective identities, and biosocialities

from medicine exerting clinical and social *control over* particular conditions to an increasingly technoscientifically constituted biomedicine also capable of effecting the *transformation of* bodies and lives (Clarke 1995). Such transformations range from life after complete heart failure, to walking in the absence of leg bones, to giving birth a decade or more after menopause, to the capacity to genetically design life itself—vegetable, animal, and human. Of course, many biomedically induced bodily transformations are much less dramatic, such as Botox and laser eye surgery, but these are no less technoscientifically engineered.

The rest of the table describes shifts from medicalization to biomedicalization within the five key processes that co-constitute biomedicalization. Analytically, the shift from medicalization to biomedicalization occurs unevenly across micro, meso, and macro levels. Significantly, biomedicalization theory emphasizes organizational/institutional meso-level changes, and these are highlighted here in order to describe the processes and mechanisms of action and change in concrete—if widespread—practices. Biomedicalization is constituted through the transformation of the organization of biomedicine as a knowledge- and technology-producing domain as well as one of clinical application. Computer and information technologies and the new social forms co-produced through their design and implementation are the key infrastructural devices of the new genres of meso-institutionalization (Bowker and Star 1999). The techno-organizational innovations of one era become the (often invisible) infrastructures of the next (Clarke 1988, 1991).

The following points are at the core of our argument about the shift from medicalization to biomedicalization. We offer an alternative understanding of historical change beyond that of technological determinism (e.g., Jasanoff 2000; Rose 1994). While we see sciences and technologies as powerful, we do not see them as *determining* futures. With other science, technology, and medicine studies scholars, we start with the assumption that sciences and technologies are made by people and things working together (e.g., Latour 1987; Clarke 1987). Human action and technoscience are *co-constitutive*, thereby refuting technoscientific determinisms (Smith and Marx 1994). Although the changes wrought by biomedicalization are often imaged as juggernauts of technological imperatives (Koenig 1988), bearing distinctive Western biomedical assumptions (Lock and Gordon 1988; Tesh 1990), the new social/cultural/economic/organizational/institutional forms routinely produced as part and parcel of technoscientific innovations are usually analytically ignored (Vaughan 1996, 1999). That is, the realms and dynamics of the social *inside* scientific, technological, and biomedical domains are too often rendered invisible. At the heart of our project lie the tasks of revealing these new social forms and opening up critical spaces to allow greater democratic participation in shaping human futures *with* technosciences.

Therefore, central to our argument is the point that in daily material practices, biomedicalization processes are not predetermined but quite contingent (Freidson 2001; Olesen 2002; Olesen and Bone 1998). In laboratories, schools, homes, and hospitals today, workers and people as patients and as providers/health system workers are responding to and negotiating biomedicalization processes, attempting to shape new techno-scientific innovations and organizational forms to meet their own needs (Strauss, Schatzman, et al. 1964; Wiener 2000). In practice, the forces of biomedicalization are at once furthered, resisted, mediated, and ignored as varying levels of personnel respond to their constraints and make their own pragmatic negotiations within the institutions and in the situations in which they must act (Lock and Kaufert 1998; Morgan 1998; Olesen 2000; Smith 1997). As a result, the larger forces of biomedicalization are shaped, deflected, transformed, and even contradicted.

Many of the themes we develop here are not new, but their synthesis within an argument for technoscientifically based biomedicalization is. Further, the shifts are shifts of emphasis—these trends are historical and historically *cumulative* from left to right across the table, not separate and parallel. Traditional medicalization processes can and do continue temporally and spatially at the same time as more technoscientifically based biomedicalization processes are also occurring. Innovations accumulate over time such that older, often "low(er)" technologically based approaches are usually simultaneously available somewhere, while emergent, often "high(er)" technoscientifically based approaches also tend over time to drive out the old. There is no particular event or moment or phenomenon that signals this shift, but rather a cumulative momentum of increasingly technoscientific interventions throughout biomedicine since roughly 1985. The unevenness of biomedicalization persists and will continue to persist historically and geographically in the United States and elsewhere.

We turn next to an elucidation of the concrete practices and processes of biomedicalization.

Key Processes of Biomedicalization

Biomedicalization is co-constituted through five central (and overlapping) processes: major political economic shifts; a new focusing on health and risk and surveillance biomedicines; the technoscientization of biomedicine; transformations of the production, distribution, and consumption of biomedical knowledges; and transformations of bodies and identities. We emphasize historical developments in the transitional and current biomedicalization era.

ECONOMICS: THE U.S. BIOMEDICAL TECHNOSERVICE COMPLEX INC.

One theoretical tool for understanding the shift from medicalization to biomedicalization is the concept of the "medical industrial complex" put forward in the 1970s in the midst of the medicalization era. Changes in medicine in that era were critically theorized as reflecting the politico-economic development of a "medical industrial complex" (taking off from President Eisenhower's naming in the 1950s of "the military industrial complex" consolidated through World War II). This concept was coined by a progressive health activist group, HealthPAC (Ehrenreich and Ehrenreich 1971), and was subsequently taken up inside mainstream medicine by Relman (1980), then editor of the *New England Journal of Medicine* (also see Estes, Harrington, and Pellow 2000). For the current biomedicalization era, we offer a parallel concept: the Biomedical TechnoService Complex Inc. This concept emphasizes the *corporatized* and *privatized* (rather than state-funded) research, products, and services made possible by technoscientific innovations that further biomedicalization. The corporations and related institutions that constitute this complex are increasingly multinational and are rapidly globalizing both the Western biomedical model and biomedicalization processes per se.

The size and influence of the Biomedical TechnoService Complex Inc. are significant and growing. The healthcare industry is now 13 percent of the $10 trillion annual U.S. economy. In the economic downturn of late 2001, the healthcare sector was even viewed by some as the main engine of the U.S. economy, offering a steadying growth. Pharmaceutical-sector growth is estimated at about 8 percent per year (Leonhardt 2001). Americans spent more than $100 billion on drugs in 2000, double the amount spent in 1990 (Wayne and Petersen 2001). The emergence of a global economy dominated by flexible accumulation by interdependent multinational corporations (Harvey 1989), streamlined production arrangements, new management technologies (Smith 1997), and increased specialization enables many of the biomedicalization processes discussed here.[9]

Through its sheer economic power, the Biomedical TechnoService Complex Inc. shapes how we think about social life and problems in ways that constitute biomedicalization. The most notable socioeconomic changes indicative of and facilitating biomedicalization are, as indicated in the table, (1) corporatization and commodification; (2) centralization, rationalization, and devolution of services; and (3) stratified biomedicalization.

CORPORATIZATION AND COMMODIFICATION

Trends in corporatization and commodification are embodied in the moves by private corporate entities to appropriate increasing areas of the healthcare sector under private management and ownership. In biomedicalization, not only are the jurisdictional boundaries of medicine and medical work expanding and being reconfigured, but so too are the frontiers of what is legitimately defined as private versus public medicine, and corporatized versus nonprofit medicine. For example, in the United States, federal and state governments have been instrumental in expanding the private healthcare sector by inviting corporations to provide services to federally insured beneficiaries. Historically, since the Social Security Act established the government as a direct provider of medical insurance coverage through the Medicaid and Medicare programs in 1965, most recipients have been treated in public/not-for-profit clinics, hospitals, and emergency rooms. As healthcare costs and competitive pressures for personnel and revenues escalated, however, many of these facilities closed or were bought out and consolidated by for-profit corporations. By the late 1990s, efforts were under way to move such patients into private HMOs, effectively privatizing social healthcare programs (e.g., Estes, Harrington, and Pellow 2000).

Second, under pressure from powerful biomedical conglomerates, the state is increasingly socializing the costs of medical research by underwriting startup expenses of research and development yet allowing commodifiable products and processes that emerge to be privatized—that is, patented, distributed, and profited from by private interests (Gaudillière and Löwy 1998; Swan 1990). The Human Genome Project is one high-profile example. What began as a federally based and funded research effort culminated in the shared success of sequencing the genome between Celera Genomics and government-funded scientists. In related developments, genetic and tissue samples collected from the bodies of individuals and communities have become patented commodities of corporate entities that offered no patient or community reimbursement (Adams 2002; Landecker 1999; Rabinow 1996). Another striking example is the patenting of the BRCA1 genes (breast cancer markers) by Myriad Genetics. The company not only receives royalties each time a genetic test for breast cancer is given but also holds sole-proprietor rights over research conducted on those genes (Zones 2000), though ownership of such rights is being challenged in the company's own country (Canada) and in France (Bagnall 2001).

Further, as suggested in table 1, industry-academy collaborations are also becoming routine sources of funding for universities, including academic medical centers (combinations of medical schools, hospitals, clinics, and research units) that had been federally funded for thirty years. The U.S. Balanced Budget Act of 1997 cut an estimated $227 billion, with large cuts of hospital budgets, while federal indirect medical education payments were also trimmed (Fishman and Bentley 1997). Strapped academic medical centers are filling this gap in part by conducting extensive clinical trials for pharmaceutical companies, requisite to bringing new products to market. Special contracts units, a new social form, have been established at major medical centers, often within their "offices of industry and research development," to negotiate blanket contract overhead rates with pharmaceutical companies.

Trends toward increased pharmaceutical company sponsorship of research have become highly problematic, however. The current and former editors of thirteen major medical journals stated in an editorial in *Journal of the American Medical Association* that they would reject any study that does not ensure that the sponsor gave researchers complete access to data and freedom to report on findings (Davidoff et al. 2001). Further, a new study found that industry-sponsored research is 3.6 times more likely to produce results favorable to the sponsoring company, implicating both universities and individual scientists (Bekelman, Li, and Gross 2003).

CENTRALIZATION, RATIONALIZATION, AND DEVOLUTION OF SERVICES

Centralization of facilities, healthcare services, and corporate healthcare coverage has been on the rise through the merger and acquisition of hospital facilities, insurers, physician groups, and pharmaceutical companies.

This has resulted in the loss of many community, public, and not-for-profit facilities that either could not compete or were acquired expressly for closure. The underlying objectives are to boost the efficiency and uniformity of services, to centralize and rationalize decision making about service provision, to capture more markets and arenas of health for profit, and to exert greater economic control within these arenas. In practice, Foucauldian panoptical patterns of physical decentralization with administrative centralization are common (Foucault 1973, 1991a). These patterns are greatly facilitated by meso-level computer and information science practices and programs that automatically monitor highly dispersed developments for centralized management operations.

Although such healthcare consolidations bring some efficiency, they also pose numerous dangers as a result of corporate concentration. Such dangers include, for example, inflationary tendencies from the concentration of pricing power, new administrative burdens, and the enhanced political power of conglomerates. Such consolidations now exert significant leverage over political and regulatory processes, as well as decision making that affects provider groups, patient care, and service options in highly stratified ways (Waitzkin 2001; Waitzkin and Fishman 1997). For example, in northern California recently, Blue Cross (a health insurance company) and Sutter Health (a for-profit corporatized provider network) were locked in contractual conflicts over reimbursement rates. Because of Sutter's acquisition of large numbers of healthcare facilities in the area, it was able to effectively deny services to many Blue Cross subscribers by not accepting Blue Cross insurance, eventually compelling the insurer to agree to higher rates.

Devolution of healthcare services also demonstrates the trend toward rationalization. That is, there are attempts to routinize and standardize health services while also shifting increasing proportions of the expensive labor of hands-on care to families and individuals (Timmermans and Berg 1997). Outpatient surgery, home healthcare, and elaborating subacute care facilities (e.g., skilled nursing facilities, nursing homes) are a few examples of devolution. Devolution itself also contributes to the fragmentation of healthcare and its geographic dispersal, making rationalizing more difficult.

STRATIFIED BIOMEDICALIZATION

Morgan (1998) recently reasserted the unevenness and instabilities of medicalization processes, reminding us that medicalization was not monolithic and unidirectional but heterogeneous and fraught with paradoxical problems of exclusion, inclusion, participation, and resistances. Such arguments were initially elaborated in Ehrenreich's and Ehrenreich's (1978) critical elucidation of the dual tendencies of medicalization. The first tendency, *co-optative medicalization*, refers to the jurisdictional expansion of modern medicine — extending into areas of life previously not deemed medical. The second tendency, *exclusionary disciplining*, refers to the *simultaneous* exclusionary actions of medicine that erect barriers to access to medical institutions and resources that target and affect particular individuals and segments of populations. Historically, these dual strategies have stratified the U.S. medical market by race, class, gender, and other attributes. For example, co-optative tendencies have long predominated for white middle- and upper-class groups, especially women, while exclusionary tendencies or particular kinds of co-optative medicalization (such as provision or imposition of birth control and sterilization) have prevailed for peoples of color and the poor (Riessman 1983; Ruzek 1980; Ruzek, Olesen, and Clarke 1997). Medicalization was stratified, and so too is biomedicalization.

We term the reformulation and reconstitution of such processes in the biomedicalization era *stratified biomedicalization*.[10] The co-optative and exclusionary tendencies we have noted persist and become increasingly complex, and new modes of stratification are also produced. Even as technoscientific interventions extend their reach into ever more spaces, many people are completely bypassed, others impacted unevenly, and while some protest excessive biomedical intervention into their lives, others lack basic care. Such innovations are far from the goal of universally accessible and sustainable healthcare promoted by some bioethicists and others (e.g., Callahan 1998).

Even rationalization itself is stratified, producing fragmentation. For example, the availability of routine preventive care, screening services, pharmaceutical coverage, and elective services such as bone-marrow transplants or infertility treatments are differentially available depending on one's health insurance plan, or lack thereof. There are still over one thousand different insurers in the United States, all providing different kinds of coverage, and thus, as a whole, the system is highly uncentralized, inefficient, and uncertain — the very things that, in theory, rationalization attempts to eliminate.

In 2001 the share of the population wholly uninsured for the entire year rose to 14.6 percent, or 41.2 million people, up from 14.2 percent in 2000 and an increase of 1.4 million people (Mills 2002, 1). In 2001 and 2002, about 75 million people under age sixty-five went without health insurance for at least one month; nearly three in four were in working families, and more than half were white (Meckler 2003, A4).

Cutbacks in government coverage of medical care are also widespread and are being made in concert with reductions in a range of social services that affect the health status of individuals and groups downstream. There has even been research on the efficacy of group medical appointments for the poor instead of (or with) short individual examinations (McInaney 2000). Such gatekeeping becomes ever more imperative in efforts to eke out economic profits from increasingly expensive and highly technological procedures, and from providing services to less desirable but financially still necessary markets and population groups.

At the same time, there are dramatic increases in stratifying fee-for-service options for those who can afford them. The most common and affordable alternatives are choosing high-end preferred providers through such an insurance plan. Here providers to whom you pay a higher co-payment are often more available (within weeks rather than months) and may have better reputations. Some plans offer high-end hospital options — you pay more to go to certain "better" hospitals. Out-of-pocket boutique medicine options usually range from cosmetic surgeries to new reproductive/conceptive technologies to some organ transplants. In addition, there are emerging options for "boutique or concierge primary care" based on privately paid annual fees to individual physicians in private practice. Here individuals pay providers an annual amount (from a few to many thousands of dollars) in return for which they get appointments within twenty-four hours for longer and even unlimited durations, cell phone and e-mail access to their physicians, house calls, and so on. High-end versions (at about $13,000 per year) are located in chic, spalike offices with marble baths, terry robes, and complete privacy and are being organized through franchises. This "concierge" model is popular with wealthy seniors, people with chronic illnesses, and the youthful rich (Heimer 2002). In short, even "good" medical insurance no longer ensures good primary care.

In sum, the politico-economic transformations of the biomedical sector are massive and ongoing, ranging from macro structural moves by industries and corporations to meso- and microlevel changes in the concrete practices of health and medicine. Not only do such transformations produce new and elaborated mechanisms through which biomedicalization can occur, but also biomedicalization, in turn, drives and motivates many of these economic and organizational changes.

THE FOCUS ON HEALTH, RISK, AND SURVEILLANCE

In the biomedicalization era, what is perhaps most radical is the biomedicalization of health itself. In commodity cultures, health becomes another commodity, and the biomedically (re)engineered body becomes a prized possession. Health matters have taken on a "life of their own" (Radley, Lupton, and Ritter 1997, 8).

HEALTH AS MORAL OBLIGATION

Specifically, health becomes an individual goal, a social and moral responsibility, and a site for routine biomedical intervention.[11] Increasingly what is being articulated is the individual moral responsibility to be and remain healthy (e.g., Crawford 1985) or to properly manage one's chronic illness(es) (Strauss et al. 1984), rather than merely attempt to recover from illness or disease when it "strikes" (Parsons 1951). In the biomedicalization era, the focus is no longer on illness, disability, and disease as matters of fate but on health as a matter of ongoing moral self-transformation.

Health cannot be assumed to be merely a base or default state. Instead health becomes something to work toward (Conrad 1992; Edgley and Brissett 1990), an ongoing project composed of public and private performances (Williams 1998, 1999), and an accomplishment in and of itself (Crawford 1994, 1999). Terms such as "health maintenance," "health promotion," and "healthy living" highlight the mandate for work and attention toward attaining and maintaining health. There has been a steady increase in mandates for self-regulation until, with biomedicalization, there is a shift in the general cultural expectations of whole populations. In this constant self- and other public disciplining, there is no rest for the weary.

RISK FACTORS AND SELF-SURVEILLANCE

In the biomedicalization era, risk and surveillance practices have emerged in new and increasingly consequential ways in terms of achieving and maintaining health. Risk and surveillance concerns shape both the technologies and discourses of biomedicalization as well as the spaces within which biomedicalization processes occur (Bud, Finn, and Trischler 1999; Fosket 2002). Risk and surveillance mutually construct one another.

Risks are calculated and assessed in order to rationalize surveillance, and through surveillance risks are conceptualized and standardized into ever-more-precise calculations and algorithms (Howson 1998b; Lupton 1995, 1999).

Risk and surveillance are aspects of the medical gaze that is disciplining bodies. They are aspects of biomedicalization that, in a quintessential Foucauldian sense, are no longer contained in the hospital, clinic, or even within the doctor-patient relationship (Armstrong 1995; Waitzkin 1991). Rather, they implicate each of us *and* whole populations through constructions of risk factors, elaborated daily life techniques of self-surveillance, and the management of complicated regimens around risk and chronic conditions.[12]

It is no longer necessary to manifest symptoms to be considered ill or "at risk." With the "problematization of the normal" and the rise of what Armstrong (1995) calls "surveillance medicine," everyone is implicated in the process of eventually "becoming ill" (Petersen 1997). Both individually and collectively, we inhabit tenuous and liminal spaces between illness and health, leading to the emergence of the "worried well" (Williams and Calnan 1994), rendering us ready subjects for health-related discourses, commodities, services, procedures, and technologies. It is impossible not to be "at risk."

Instead individuals and populations are judged for *degrees* of risk — "low," "moderate," or "high" — vis-à-vis different conditions and diseases, and this then determines what is prescribed to manage or reduce that risk. Thus biomedicalization elaborates through daily lived experiences and practices of "health" designed to minimize, manage, and treat "risk" as well as through the specific interactions associated with illness (Fosket 2002; Press, Fishman, and Koenig 2000). Risk technologies are therefore "normalizing," not in the sense that they produce bodies or objects that conform to a particular type, but more that they create standard models against which objects and actions are judged (Ewald 1990).

Of particular salience in the biomedicalization era is the elaboration of standardized risk-assessment tools (e.g., to assess risk of breast cancer, heart disease, diabetes, hypertension, etc.) that take epidemiological risk statistics, ostensibly meaningful only at the population level, and transform them into risk factors that are deemed meaningful at the individual level (Gifford 1986; Rockhill et al. 2001). For instance, current breast cancer risk-assessment technologies construct a standardized category of "high risk" for breast cancer in the United States. Women classified as "high risk" are given the option of taking chemotherapy—pharmaceuticals usually used only to treat cancer because of their toxicity and other negative side effects—to "treat" the *risk* of cancer (Fosket 2002). Genomic technologies and profiling techniques mark the next wave in such risk assignments (Fujimura 1999; Shostak 2001).

Further, with the institutionalization of the assumption that everyone is potentially ill, the health research task becomes an increasingly refined elaboration of risk factors that might lead to future illnesses. Such research and knowledge production—as well as their active consumption by patients/consumers as well as providers—are primary and fast-growing components of biomedicalization and will continue to be major contributors to the development of "surveillance medicine" (Armstrong 1995) and new forms of public health in the twenty-first century (Shim 2000, 2002a, 2002b). Health is thus paradoxically both more biomedicalized through processes such as surveillance, screening, and routine measurements of health indicators done in the home, and seemingly less medicalized as the key site of responsibility shifts from the professional physician/provider to include collaboration with or reliance upon the individual patient/user/consumer.

THE TECHNOSCIENTIZATION OF BIOMEDICINE

The increasingly technoscientific nature of the practices and innovations of biomedicine is, of course, a key feature of biomedicalization. While science and technology became increasingly constitutive of medicine across the twentieth century, in its final decades, technoscientific transformations gained significant momentum. These changes are part of major shifts in the social organization of biomedicine itself, the objects of biomedical knowledge production, the ways in which biomedicine intervenes, and the objectives with which it does so. Moreover, innovations are increasingly likely to be hybrid ones that are generated simultaneously through sciences *and* technologies *and* new social forms—most often computer and information technologies and the organizational structures developed to articulate them into the flows of biomedical and related work (Berg 1997, 2000; Star 1995; Wiener 2000). These changes, we argue, have spurred biomedicalization and are also manifest in how it is effected.

We describe three overlapping areas in which the technoscientization of biomedicine is manifest: (1) computerization and data banking; (2) molecularization and geneticization of biomedicine and drug design; and (3) medical technology design, development, and distribution.

COMPUTERIZATION AND DATA BANKING

Fundamental to biomedicalization is the power (past, present, and especially future) of computerization and data banking. These technoscientific advances are pivotal to the meso-level (re)organization of biomedicine. That is, many of the biomedical innovations of the twenty-first century are situated in *organizations* that are themselves increasingly computer dependent in heterogeneous ways that in turn are increasingly *constitutive* of those organizations. The application of computer technologies within multiple biomedical domains and their organizational infrastructures are thereby mutually constructed, creating new social forms for orchestrating and performing the full range of biomedically related work.[13]

One important computer-based organizational innovation involves the reorganization of, and much wider access to, individual medical records. Centralized storage and access to patient records have been hopes of doctors, hospitals, and insurers since at least the nineteenth century (Blois 1984). Recent technological breakthroughs in hardware, software, and data-processing and storage technologies have allowed the integration of medical data into heterogeneous and widely dispersed databases to become routine in systemic and ubiquitous ways. Considerable pressure is being brought to bear to computerize *all* medical records according to standardized formats that can be webbed across multiple domains. Thus, as noted in table 1, from paper versions of medical records dwelling in individual physicians' offices, clinics, and hospitals, common during the era of medicalization, patient information can now be uploaded and accessed via websites managed by HMOs, pharmacies, and other third-party entities in faraway places for multiple purposes. Also, new companies are engineering "doctor-friendly" formats (Lewis 2000; National Research Council 2000).

These new and elaborating meso-level infrastructures are facilitating many of the downstream processes requisite for biomedicalization, not only enabling the expansion of medical jurisdiction but also producing infrastructures for greater public-private linkages and new iterations of biomedical governmentality. Computerization allows more aspects of life to be scrutinized, quantified, and analyzed for their relationships to health and disease. Integration and compatibility of data across various sites are articulated via specialized software that increasingly imposes standardized categories and forms of information (Bowker and Star 1999). Such formats make it all but impossible to enter certain kinds of data in the medical record, especially highly individualized information common to medical practice on unique individual bodies. At the same time, these data formats render it all but impossible *not* to record other kinds of data, such as the information required to comply with "clinical decision-support technologies" (Berg 1997) and highly detailed diagnostic and treatment regimens. These are the very meso-level techno-organizational transformative "devices" that biomedicalization demands and *is*.

Decision-support technologies are generated through outcomes research and evidence-based medicine that depend on major computerized databases, as noted in table 1 (Ellrodt et al. 1997; Traynor 2000). Here the safety and efficacy of specific protocols and treatments are assessed based on data from very large populations of patients and providers across time and space. The geographic variations in "conventional" treatments and the different "community standards" revealed by regional health statistics have long irked segments of the American medical profession (Reverby 1981). As the production of biomedical knowledge is accelerated through the use of computer technologies, both behavioral and outcomes research are increasingly defining new biostatistical criteria for what counts as "scientific." Such research allows for the "objective" statistical identification of "industry standards" (Porter 1995), and insurance companies are already moving toward covering only those procedures demonstrated as "valid" through such standardizing research. Such developments will likely cut in many different and even paradoxical directions simultaneously. For example, vis-à-vis women's health, "unnecessary" yet costly hysterectomies and Cesarean sections, so long criticized by feminists (e.g., Ruzek and Hill 1986), will be highlighted for deletion. Other highly vaunted treatments, such as bone-marrow transplants for breast cancer and estrogen replacement therapy for menopausal symptoms, have already been challenged by such outcomes studies (Weiss et al. 2000; Writing Group 2002).[14]

Further, such protocols are being developed in concert with the spread of another new social form, the specialty of "hospitalists" — physicians who practice only in hospitals and to whose care medical responsibility is almost completely shifted from the patient's own primary physician upon hospitalization (Pantilat, Alpers, and Wachter 1999). A major rationale here is that the technoscientific infrastructure of hospital medicine is so complex and rapidly changing that only a localized specialist can keep up with its applications in acute patient care.

Finally, error in medicine—mistakes at work—is a recent focus of research using the new massive computer databases (Institute of Medicine 1999). Prevention of errors and the knowledge thought to be gleaned from analyses of centralized data will likely drive the rhetoric that justifies the dramatic losses of privacy and the creation of new vulnerabilities caused by the computerization of medical records. Thus the potential generated by the compilation, storage, analysis, and control of computerized patient data furthers the possibilities of biomedicalization processes in new and important ways.

The guiding assumptions common to these developments are that care and treatment services can and should be better rationalized such that variations are indicative of up-to-date scientific decision making rather than "unnecessary" or "discretionary" treatment. However, provider discretion about individual-case treatment, continuity of care, doctor-patient relationships, situationally appropriate care, privacy of treatment, and patient involvement in treatment decision making will likely be drastically, though unevenly, limited and stratified.

MOLECULARIZATION AND GENETICIZATION

Second, the biomedical sciences of the new millennium are being transformed by molecular biologies. Molecular biological approaches initiated in the 1930s yielded in the 1950s the discovery of DNA structure. This and related developments in basic science and research technologies are now propelling attempts to understand diseases at the (sub)molecular levels of proteins, individual genes, and genomes (proteomics, genetics, and genomics), partially displacing previous emphases on germs, enzymes, and biochemical compounds (Chadarevian and Kamminga 1998). The study of differences among humans is also devolving to the level of the gene—called "geneticization" (Lippman 1992; Hedgecoe 2001).

In current treatment and drug development, these innovations have generated a shift from "discovery" of healing properties of "natural" entities to computer-generated molecular and genetic "design," or what Jacques

Loeb would have called "engineering" (Pauly 1987), that can be targeted precisely at diseases and conditions likely to generate high profits (e.g., baldness, obesity). Pharmacogenomics — the field that examines the interaction of genomic differences with drug function and metabolism — offers the promise that pharmaceutical therapies can be customized for groups and individuals. Such gene therapies (including the just patented "genepill") and related innovations are beginning to hit the market (Genteric 2001). Further, reengineering human germ lines through choosing and assembling genetic traits for offspring will become possible and desired by some, a "do-it-yourself evolution" (Buchanan et al. 2000), while strongly opposed by others as further stratifying reproduction (Rapp 1999).

These applications of molecular biology and genomics to medicine are themselves highly dependent on computer and information sciences, and the convergence of these two domains was further fueled by the announcement in 2001 of the completion of the first rough map of the human genome. For example, software to analyze and predict how genome interactions might promote health or cause disease, developed by scientists at the National Human Genome Research Institute, is being scaled up to run on supercomputers. Such large-scale information technologies are being enlisted by biotechnology and pharmaceutical groups to crunch through hundreds of such genome interactions to find potential intervention points (Abate 2000a). In the process, novel meso-level organizational partnerships are being forged among government entities, information technology companies, and biotechnology firms. The mutual constitution and dependency of computerization and molecularization trends are reflected in new hybrid sciences like bioinformatics, which pairs biology with computer science. Called "the career choice of the decade" (Wells 2001), bioinformatics is giving rise to new, well-funded training programs to produce a workforce able to sort through and translate the findings of genomic and proteomics research into information eventually usable for medical purposes.

Biotechnological pursuits of genomic manipulations today stand at the pinnacle of technoscience. While computerization is standardizing patient data, it paradoxically also enables the further tailoring and customization of bodies (Conrad 2000), central to processes of biomedicalization. The basic medical assumption about intervention in the United States and other highly/over developed countries will be that it is "better" (faster and more effective, though likely not cheaper) to redesign and reconstitute the problematic body than to diagnose and treat specific problems in that body.[15] Molecular biologies and genomics will make such redesign possible "from the inside out," or transformatively, rather than operating externally as most prosthetics traditionally do (Clarke 1995).

MEDICAL TECHNOLOGY DEVELOPMENT

Third, medical technology developments of all kinds are being transformed through digitization, miniaturization, and hybridization with other innovations to create new genres of technologies. These extend the reach of biomedical interventions and applications in fundamentally novel ways. For instance, recent advances in material sciences make possible hybrid and bionic devices. Examples from corneal implants to computer-driven limbs, continuously injecting insulin packs for diabetics, electronic bone growth stimulation devices, and heart and brain pacemakers (the latter initially used for treatment of depression) are becoming routine in boutique Western medicine. Hybridization is also apparent in the next generation of transplant medicine, termed "tissue engineering," which will include new kinds of implants: body parts custom grown through molecular means, modified through materials science, and triggered by "biological switches" (Hogle 2000).

Digitization has also transformed medical technologies in ways that further their gaze and reach into both the interior of the body and its behaviors. In addition to the computerization of patient data, including genomic, behavioral, and physiological information, visual diagnostic technologies are also elaborating rapidly with technical innovations, at times outpacing local organizational capacities to use them safely and effectively (Kevles 1997). Imaging technologies are increasingly digitized, facilitating their resolution, storage, and mobility among multiple providers, distributed sites of care such as telemedicine, and agencies or entities interested in centralizing such information (Cartwright 2000b). The costly reading of cytological and pathological specimens such as Pap smears and biopsies is also being computerized after decades of effort (Bishop, Marshall, and Bentz 2000). Finally, transplant medicine has shifted from a local medical charity to a transnational web of organizations made possible through computer and information sciences, ranging from local hospitals to cutting-edge biotechnology firms to multinational distribution organizations (Hogle 1999).

But this is also intensifying the stratification of biomedicalization globally through organ purchasing by the rich from the poor, largely arranged on-line (Cohen 1999; Delmonico et al. 2002; Organs Watch 2001; Scheper-Hughes 2000).

Biomedicine is increasingly part of what Schiller (1999) calls digital capitalism. The Internet is a key reorganizing/transforming device and hence a key technology of biomedicalization. The Internet has recently been called "the first global colony," in part because its economics and individualist culture "feel awfully American" (Lohr 2000, 1). The National Research Council (2000) published recommendations and guidelines for extending health applications of the Internet, from virtual (remotely guided) surgery to education, consumer health, clinical care, financial and administrative transactions, public health, and research. An important digital aspect over the coming decades is likely to be the application of distance-learning techniques and technologies to professional education for all kinds of health-care careers, also easily globalized.

In sum, the ongoing technoscientization of biomedicine lies at the heart of biomedicalization. Theorizing these technoscientific transformations of biomedicine requires that their meanings *and* their material forms and practices, including embodied corporeal transformations and manifestations, be conjointly studied and analyzed as co-constitutive (Casper and Koenig 1996; Gray, Figueroa-Sarriera, and Mentor 1995; Haraway 1991, 1997; Hayles 1999).

TRANSFORMATIONS OF INFORMATION AND THE PRODUCTION AND DISTRIBUTION OF KNOWLEDGES

Information on health and illness is proliferating through all kinds of media, especially newspapers, magazines, the Internet, and direct-to-consumer prescription and over-the-counter drug advertising. In fact, biomedicine, more than being a subculture, is today so much a fundamental element of mass culture that Bauer (1998, 747, 744; also see Hodgetts and Chamberlain 1999) suggests that its constant presence in popular media points to the medicalization of science news and of society generally: "Medicine is the current core of popular representations of science. . . . Our evidence of the dominance of health news is an empirical indicator of the advent of a medicalized society. . . . [The] medicalization of science news is a correlate of these larger changes in society, celebrating the successes of medical sciences, anticipating breakthroughs on the health front, and mobilizing an ever greater demand for medication and services." The cultural imaginary of biomedicine travels widely and is locally and flexibly accessed and (re)-interpreted.

Thus the production and transmission of health and medical knowledges are key sites of biomedicalization in terms of both the transformation of their sources and distribution channels and the reformulation of who is responsible for grasping and applying such knowledges. Biomedicalization also works through the co-optation of competing knowledge systems, including alternative medicine and patient-based social movements (Adams 2002; Belkin 1996). Finally, techniques for the legitimation of biomedical knowledge claims are also changing.

HETEROGENEITY OF PRODUCTION, DISTRIBUTION, AND ACCESS TO BIOMEDICAL KNOWLEDGES

First, the sources contributing to the production of health-related information have both increased and diversified. In cyberspace, for example, federally sponsored websites target not only researchers and healthcare providers but also Internet-savvy healthcare consumers. On one such site, http://www.clinicaltrials.gov, potential human subjects can find clinical trials for which they may be eligible. Numerous private companies also provide medical information. The information on these websites comes from a variety of sources. Although there is still a reliance on medical professionals for answers to health questions, sites often have discussion boards where users exchange their own knowledges and experiences. Another rapidly growing source of medical knowledges is patient advocacy groups that have their own organizations, newsletters, websites, and serious stakes in knowledge production and dissemination (Brown 1995; Brown et al. 2001).

In principle, these changes democratize production and access to medical and health knowledges in new ways. In practice, the waters are muddy (e.g., Kolko, Nakamura, and Rodman 2000; National Research Council 2000; Yates and Van Maanen 2001). First, it is often difficult to know whether the seemingly objective information located on the Internet is produced by medical experts holding professional credentials and what kinds of financial and scientific stakes they might have in presenting information in a particular way. Potential profits rise every time someone logs onto the growing number of healthcare websites on the Internet that couple the provision of information with the marketing of products (including alternative-medicine products and dietary supplements). In addition, corporate agreements with search engine companies have found ways to limit the access of Internet consumers to the diversity of information sites available on the Web. Companies can purchase "prime-time" and "sole-supplier" status from search engines, thereby preempting access to their competition, and consumers are often unaware of such agreements (Rogers 2000). Last, it is unknown whether do-it-yourself sites are more or less common or more or less likely to be hot linked (National Research Council 2000). However, the heterogeneity of knowledge sources can also be interpreted as disrupting the division of "expert" versus "lay" knowledges and enabling new social linkages. For many, these new modes of access to health information are a welcome change; for others, they confound more than they clarify. For yet others, the digital divide is all too real, and access remains elusive and stratified.

Second, biomedical knowledges have been transformed in terms of access, distribution, and the allocation of *responsibility* for grasping such information. Historically in the United States, nonexperts' ability to obtain biomedical information was severely limited, as such knowledges dwelled almost exclusively in medical libraries and schools that were closed to the public, creating what amounted to a professional monopoly on access to information. Popularized "lay" health information was also scarce. Health sections in bookstores were rare and small until the 1970s, when women's health and consumer health movements began producing self-help books. Activists in such movements were instrumental in altering the self-help landscape, including the Boston Women's Health Book Collective's first *Our Bodies, Ourselves* in 1970.[16] A breast cancer patients' movement challenged the use of radical mastectomies as the de facto treatment, advocating greater patient involvement in surgical decisions (Montini 1996), and AIDS activists successfully challenged the NIH's clinical trial practices (Epstein 1996). In each case, activists challenged the professional monopoly over the production of medical knowledges by insisting on their own participation as they acquired and disseminated scientific information and demanded immediate access to innovative healthcare. Today, enabled by computer technologies, individuals are organizing to articulate new research interests, fund research studies, and, at times, open up new research frontiers (Brown 1995; Brown et al. 2001; Fishman 2000; Kroll-Smith and Floyd 1997). Some groups are even starting to fund their own science directly (Rabeharisoa and Callon 1998). Because of increasing congressional responsiveness to their demands, some supposed patients' groups are now started by scientists, pharmaceutical companies, and professional medical organizations (Zola 1991; Zones 2000), known among health NGOs as "AstroTurf" rather than grassroots based.

In the biomedicalization era, while knowledge sources proliferate and access is streamlined in ways purportedly in the interests of democratizing knowledge, the interests of corporate biomedicine predominate. This point is highlighted by the loosening, in 1997, of the criteria under which direct-to-consumer advertising of prescription pharmaceuticals is allowed by the Food and Drug Administration, a profound shift in social policy on the proper relationship between the public and biomedical knowledge. Previously, provider-patient relationships were based on a notion of protecting laypeople from knowledge best left to professionals. Now, pharmaceutical companies encourage potential consumers to first acquire drug information and then proactively ask their providers about the drugs by brand name. In 2001 the industry spent about $2.5 billion on consumer advertising (Freudenheim and Petersen 2001, 1,13). One recent survey (Kaiser Family Foundation 2001, 18–20) found that 30 percent of Americans surveyed who viewed direct-to-consumer advertising said they talked to their doctor about a specific medication they saw advertised, and 44 percent of those reported that their doctors provided them with the prescription medicine they asked about. While direct-to-consumer advertisements do help to educate the public about potential treatment options, such marketing undeniably boosts pharmaceutical revenues: Prescriptions for the top twenty-five drugs directly marketed to consumers rose by 34 percent from 1998 to 1999, compared with a 5.1 percent increase for other prescription drugs (Charatan 2000, 783). This both transforms doctor-patient relationships and increases the power and profit of the pharmaceutical industry, furthering biomedicalization (Woloshin et al. 2001).[17]

But all is not *new* knowledge and information. Within these new techno-scientifically based knowledge sources, there is also amped up access to older *cultural* discourses of stratification. Through what are called "remediations," new visual technologies such as computer graphics and the World Wide Web "are doing exactly what their predecessors [film, television, photography] have done in (re)enacting similar inequities . . . yet they present themselves as refashioned and improved versions of other media" (Bolter and Grusin 1999, 14–15). The continuities are significant, as the media often import historic cultural stratifications regarding sex, race, sexuality, and gender—and patienthood as well—that usually remain unquestioned. For example, Forsythe (1996) studied a patient information system for migraine sufferers that was intended to provide information distinct from that provided by physicians. She found the system "in fact offers information characterized by the same assumptions and deletions as that provided by neurologists" (Forsythe 1996, 551). Intended to empower migraine patients, the system may instead reinforce rather than reduce power differentials between doctor and patient.

CO-OPTATION OF COMPETING
KNOWLEDGE SYSTEMS

Another transformation of knowledge constitutive of biomedicalization is the co-optation of competing knowledge systems and the reconfiguration of healthcare provision and organizations in ways originally proposed and implemented by social movements.

The last decades of the twentieth century in the United States saw a profound rise in the use of alternative and complementary medicines. In 1993 one study estimated that $10.3 billion consumer dollars a year were spent on alternative medicines in the United States (Eisenberg, Kessler, et al. 1993, 346). In 1998 a follow-up study conservatively estimated out-of-pocket patient expenditures for alternative medicines at $27 billion, which is comparable to the out-of-pocket costs to patients for all physician services (Eisenberg, Davis, et al. 1998, 1569). These findings, perceived as an economic threat to Western biomedicine, clearly repositioned alternative medical knowledge systems as legitimate (at least to users/consumers), shifting them from the margins of healthcare to the center. The response from deep within the structures of Western biomedicine has been a marked increase of interest in such approaches. At the beginning of the twentieth century, Western biomedicine dealt with such approaches by organizing antiquackery committees and recruiting the state to make such practices illegal (Gevitz 1988), and similar efforts continue today (Adams 2002). Additionally, at the beginning of the twenty-first century, Western biomedicine is attempting to co-opt and incorporate many elements of alternative medicines. As understandings of health and healing systems from other cultures have spread, and as people knowing such systems have migrated globally, there have been interesting nomenclature shifts in Western medical fields, from considering "other" people's health/life/healing systems as "superstitions" to "culture based healing systems" to "alternative medicines" (Anderson 2002; Arnold 1988). Numerous large-scale clinical trials are testing the "effectiveness" of alternative medical practices and therapies (Adams 2002).[18] Major pharmaceutical companies now market their own brands of herbal and nutritional supplements and vitamins.

Similarly, biomedicalization includes co-optation of organizational and ideological shifts and innovations brought about by grassroots social movements such as women's health movements, disability rights, AIDS activism, and other disease-specific movements (Belkin 1996; Worcester and Whatley 1988). For example, early feminist consumer activism centered on expanding patient access to drug information via "patient package inserts" and medical information via readable materials on health and illness (e.g., Boston Women's Health Book Collective 1971) and feminist women's health centers (Ruzek 1978). Displacing feminist centers, biomedicine now offers "sleeker" versions of women's health (Worcester and Whatley 1988). Building on decades of efforts by women's health movements, AIDS activists in the 1980s and 1990s provoked major changes in the testing and approval of new drugs. Rapid patient access to experimental therapies for AIDS and many other conditions through innovative clinical programs is now administered by the FDA (Epstein 1996) with participation information accessible over the Internet.

TECHNIQUES OF LEGITIMATION OF BIOMEDICAL CLAIMS

A final shift regarding knowledges within biomedicalization concerns techniques used for the legitimation of biomedical claims—the standards by which the innovations offered by biomedical sciences are tested and deemed acceptable. As noted in table 1, early standards of care and quality control over various drugs and technologies from about 1890 to 1940 were established through the classic individual case-observation method. Reform efforts and a series of U.S. policies passed in the early twentieth century created a federal "pure food and drugs" infrastructure for oversight and regulation, acting through institutional medicine and public health.

New standards required drug manufacturers to submit evidence from "adequate tests" to demonstrate that a drug was "safe" before it could be licensed for sale.

The development of the randomized clinical trial as the "gold standard" for the legitimation of biomedical claims soon followed. In 1962, after the Thalidomide crisis in which many children were born with birth defects, the FDA began requiring pharmaceutical companies to also obtain evidence of drug "efficacy" through "adequate and well-controlled investigations incorporating 'appropriate statistical methods'" (Marks 1997, 129). The randomized controlled trial consisting of three phases of testing in human subjects has become the ideal instrument for producing scientific knowledges and evidence for the therapeutic appropriateness of releasing any drug or medical device onto the market. With the rise of biostatistics, methods of drug evaluation have achieved a distinctive form of scientific and bureaucratic standardization (MacKenzie 2001; Marks 1997; Porter 1995). Major policy events indicative of this shift in the science of legitimation include the 1993 NIH guidelines requiring the inclusion of women and racial minorities in NIH-funded clinical studies and the 1998 FDA requirement that clinical trials produce explicit data on women and minorities (Epstein 2007a). Today clinical trials are big business, offering new careers in clinical trial management to nurses and others (Mueller 1997; Mueller and Mamo 2000). However, serious ethical problems, including patient deaths attributed to conflicts of providers' interest, have led the NIH to close down *all* NIH-sponsored research temporarily at several major university medical centers in the past few years.[19] Informed consent and other trial protocols were typically found inadequate, and there was serious underreporting of safety problems to the FDA, along with inadequate record keeping.

These emergent forms of legitimation contribute to a biomedicalization of clinical trials not only through a scientization of the FDA's approval process but also through new linkages created among government agencies (e.g., the FDA), private industry (e.g., pharmaceutical companies), and academic research institutions. These new assemblages, which often give rise to different criteria for drug approval, also create new structural and infrastructural ties between what were formerly known as the "public" and the "private" (Fishman 2003).

TRANSFORMATIONS OF BODIES AND IDENTITIES

The fifth and last basic process of biomedicalization, as noted in table 1, is the transformation of bodies and the production of new individual and collective identities. There is an extension of the modes of operation of medical research and clinical practice from attaining "control over" bodies through medicalization techniques (e.g., labeling disease and concomitant medical interventions) to enabling the "transformation of" bodies to include desired new properties and identities (Clarke 1995). As a Foucauldian technique, regulation through biomedicalization works "from the inside out" as a type of biomedical governance. It is achieved through alterations of biomedicalized subjectivities and desires for transformed bodies and selves. The body is no longer viewed as relatively static, immutable, and the focus of control, but instead as flexible, capable of being reconfigured and transformed (Martin 1994). Thus opportunities for biomedicalization extend beyond merely regulating and controlling what bodies can (and cannot) or should (and should not) do to also focus on assessing, shifting, reshaping, reconstituting, and ultimately transforming bodies for varying purposes, including new identities. Such opportunities and imperatives, however, are stratified in their availability—imposed, made accessible, and promoted differentially to different populations and groups.

FROM NORMALIZATION TO CUSTOMIZATION

Where medicalization practices seemed driven by desires for normalization and rationalization through homogeneity, techniques of stratified biomedicalization additionally accomplish desired tailor-made differences. New technoscientific practices offer "niche marketing" of "boutique medicine" (Hannerz 1996) to selected healthcare consumers, usually on a fee-for-service basis. Institutionally, customization has been increasingly incorporated into biomedicine through projects including computer-generated images of the possible results of cosmetic surgery, the proliferation of conceptive technologies promoting "rhetorics of choice" (Rothman 1998), and the promise of individualized gene therapies and pharmacogenetics. Such customization is often part of the commodification and fetishization of health products and services common in the biomedicalization era, wherein health products and services become revered, valued, and imbued with social import that has little to do with their use value or physical properties.

Such desires are concomitant with another trend in stratified biomedi-calization: "lifestyle" improvement. The pharmaceutical industry's atten-tion to developing "lifestyle drugs" such as Viagra exemplifies this move-ment toward enhancement and the concern with "treating" the signs of aging (Mamo and Fishman 2001), targeting the fastest-growing U.S. popu-lation segment. For another example, "Better Bodies" was the name of a 2000 conference focusing on innovations in cosmetic surgeries, sponsored by the UCSF Foundation and promoted to major campus donors.

Such attention to customization applies not only to bodily improve-ment and enhancement, including anti-aging strategies, but also to "health promotion" through obtaining enhanced knowledge about individualized susceptibilities and potential pathologies. One of the newest incarnations of this phenomenon is the public availability of "total body scans"—high-resolution CAT scans of the body billed as *preventive* in that they may de-tect early signs of disease or verify the healthiness of various parts of the body, including the brain, heart, lungs, colon, ovaries, abdomen, and kid-neys. These imaging services are available on demand in many U.S. cities and suburban malls in stand-alone offices and are generally paid for out of pocket. The biomedical governmentality to "know thyself" that is asso-ciated with such bodily techniques often relies on a neoliberal consumer discourse that promotes being "proactive" and "taking charge" of one's health.

In the move from universalizing bodies to customizing them, biomedi-cine has also allowed for some destabilization of differences. Human bodies are no longer expected to adhere to a single universal norm. Rather, a multiplicity of norms is increasingly deemed medically expected and ac-ceptable. Technoscience is seen as providing the methods and resources through which differences of race/ethnicity, sex/gender, body habitus, age, and so on can be specified, measured, and their roots ascertained. Sig-nificantly, biomedicalization processes are appropriating both the defini-tion and management of bodily differences as within the proper jurisdic-tion of biomedical scientific research and technologies. This new regime of biomedical governance allows the further stratified customization of medical services, technologies, and pharmaceuticals to "manage" such dif-ferences (Lock and Gordon 1988), thus further biomedicalizing them. Ex-amples of such stratified biomedicalization include "culturally competent care," pharmacogenetics, and new social forms—new systems of service provision designed to render increasingly customized care, ranging from high-end birthing clinics to AIDS nursing care delivered in satellite offices located in single-room-occupancy hotels to avoid costly hospitalization.

How the body is conceived of and treated by biomedicine has also changed over time and constitutes another important site of biomedicalization. In the early twentieth century, conventional medical treatments focused on the ill body, emphasizing surgery (as technologies of anesthesia and asepsis were refined) and control of acute infectious diseases (such as tuberculosis, through quarantine and isolation). Over the course of the twentieth century, improved living conditions, the advent of antibiotics around World War II, and successful interventions into acute diseases gradually shifted the focus to management of chronic illnesses such as some cancers, heart disease, and AIDS (Strauss et al. 1984; Strauss and Corbin 1988; Strauss and Glaser 1975). In biomedicalization, the focus shifts to behavioral and lifestyle modifications (e.g., exercise, smoking, eating habits, etc.) literally promoted by the government, among others. Such techniques have become part of conventional treatments, with an enormous contiguous industry that has grown up around stress-management regimens, wellness programs, the diet industry, and extensive direct-to-consumer advertising of both prescription and over-the-counter pharmaceutical and nutraceutical technologies for "maintaining" health and "controlling" chronic illness. Thus, although in some respects no less normalizing or disciplining, biomedicalization enacts its regulation of bodies through offering not just "control over" one's body through medical intervention (such as contraception) but also "transformation of" one's body, self, health. Thereby new selves and identities (mother, father, walker, hearer; beautiful, sexually potent person) become possible. Some such identities are sought out, while others are not.

TECHNOSCIENTIFIC IDENTITIES

Technoscientific identities is our generic term for the new genres of risk-based, genomics-based, epidemiology-based, and other technoscience-based identities. The core criterion is that such identities are constructed through technoscientific means. That is, technoscientific identities are produced through the application of sciences and technologies to our bodies directly and/or to our histories or bodily products, including images (Dumit 1997). These new genres of identities are frequently inscribed on us, whether we like them or not. For example, individuals today may unexpect-edly learn they are genetic carriers of inherited diseases (Karlberg 2000) or may seek out such information about themselves. The new subjectivities that arise through the availability of these technosciences do so through a biomedical governmentality that encourages such desire, demand, and need to inscribe ourselves with technoscientific identities (Novas and Rose 2000). Of course, people negotiate the meanings of such identities in heterogeneous ways.

This is not to say that the identities themselves are all new but that technoscientific applications to bodies allow for new ways to access and perform existing (and still social) identities. There are at least four ways that biomedical technoscience engages in processes of identity formation. First, technoscientific applications can be used to attain a previously un-available but highly desired social identity. For example, infertility treat-ments allow one to become a "mother" or "father," while the identity of "infertile" can be strategically taken on by lesbians and single women to achieve pregnancy through technoscientific means (Mamo 2002). Second, biomedicalization imposes new mandates and performances that become incorporated into one's sense of self. The subjectivities that arise out of these performances of what it is to be healthy (e.g., proactive, prevention-conscious, neorational) suggest how biomedical technoscience indicates a type of governmentality that can enact itself at the level of subjective identities and social relations. Third, biomedical technosciences create new categories of health-related identities and redefine old ones. For example, through use of a risk-assessment technique, one's identity can shift from being "healthy" to "sick," or to "low risk" or "high risk" (Fosket 2002).

Fourth, biomedicalization also enables the acquisition and perfor-mance of identities as patients and communities through new technoscien-tific modes of interaction, such as telemedicine. As new computer-based technologies allow cosmopolitan providers to "reach out and heal" people whom Cartwright (2000b) has called "remote locals" in their communities, new social identities and social formations are created. Telemedicine "is a method of reordering geography and identity through new styles of health management that involve new configurations of population and different ways of imagining what global health is and will be . . . unhinged from local practices" (348–49). One wonders what will happen, through such technoscientific interventions, to what Lock (1998, 182) has called "local biologies," often centuries-long established cultural differences in meaning making associated with what we today term biomedical issues.

In discussing the relations between medicalization and disease concepts, Lock (1998, 180) has noted the tendency to "streamline and normalize" specific conditions/diseases into entities wholly (or at least normally) treatable by an available or soon-to-be-available drug, device, or procedure. The classic case she examines is menopause, which was transformed in the West from a complex and unevenly symptomatic syndrome into a standardized "estrogen deficiency disease" treatable by hormone replacement therapies (now deemed dangerous after sixty years of increasingly intense use). Here we see how the meaningful identities of disorders and diseases as well as of persons and groups are also being redefined at this historical moment and also through technoscientific means (also see Fishman and Mamo 2002). Fleck (1935/1979) was among the earliest to alert us to such possibilities.

The major framing of technoscientific identities to date is Rabinow's concept of biosocial identities and biosocialities that "underline[s] . . . the certain formation of new group and individual identities and practices arising out of these new truths" (1992, 241–42) (e.g., neurofibromatosis groups). "These [biosocial] groups will have medical specialists, laboratories, narratives, traditions, and a heavy panoply of pastoral keepers to help them experience, share, intervene in and 'understand' their fate" (242). However, attribution of identity does not equal acceptance of it (Novas and Rose 2000). Interactionist labeling theory again becomes relevant, raising questions of power—who gets to label whom, with what consequences, and what "responses" may occur. Technoscientific identities' origin stories usually lie in sites where technoscience successfully dwells: in research/medical/insurance/governmental/legal domains which are often socially and culturally highly privileged and potent. Yet on an individual basis, technoscientific identities are selectively taken on, especially when accepting such identities seems worthwhile, including access to what can be experienced as "medical miracles." Such an identity can be handled as a "strategic" identity,[20] seemingly accepted to achieve particular goals, but (typically in other situations) it may also be refused. Such identities may also be ignored in favor of alternatives. Negotiations with biomedicalization processes are ongoing.

Conclusions

We have offered an analysis of the historical shift from medicalization to a synthesizing framework of biomedicalization that works through, and is mutually constituted by, economic transformations that together constitute (1) the Biomedical TechnoService Complex Inc.; (2) a new focus on health, risk, and surveillance; (3) the technoscientization of biomedicine; (4) transformations of knowledge production, distribution, and consumption; and (5) transformations of bodies and identities. We have argued that biomedicalization describes the key processes occurring in the domains of health, illness, medicine, and bodies, especially but not only in the West. We have asserted that the shifts are shifts of emphasis: medicalization processes can and do continue temporally and spatially, if unevenly. Innovations thus are cumulative over time such that older approaches are usually available simultaneously somewhere, while new approaches and technoscientifically based alternatives also tend to drive out the old over time.

In addition to being temporally uneven, we have argued that biomedicalization is *stratified*, ranging from the selective corporatization of "boutique" biomedical services and commodities directed toward elite markets, to the increasingly exclusionary gatekeeping made possible by new technologies of risk and surveillance, to the stratification of rationalized medical care. Through emergent "dividing practices," some individuals, bodies, and populations are perceived to need the more disciplinary and invasive technologies of biomedicalization, as defined by their "risky" genetics, demographics, and behaviors; others are seen as especially deserving of the customizable benefits of biomedicine provided through innovative assemblages, as defined by their "good" genetics, valued demographics (e.g., insurance and income status), and "compliant" behaviors.

Stratified biomedicalization both exacerbates and reshapes the contours and consequences of what is called "the medical divide"—the widening gap between biomedical "haves" and "have-nots" (Abate 2000b). Surveillance, health maintenance, increased knowledge, and extended health and biomedical responsibilities for self and others are, however, promoted for all. This imperative to "know and take care of thyself," and the multiple technoscientific means through which to do so currently, have given rise to new genres of identities, captured in our concept of *technoscientific identi-*

ties. The ubiquity of the culture of biomedicine renders it almost impossible (and perhaps not even desirable) to avoid such inscriptions.

We believe the concept of biomedicalization offers a bridging framework for new conversations across specialty divides within sociology and more broadly across disciplinary divides within the social sciences. Biomedicalization engages the concepts of structure and agency, stratification, and the complex intersectionalities of culture, political economy, organization, and technoscience. The transformations of biomedicalization are manifest in large, macrostructural changes as well as in new personal identities and subjectivities, but especially at the meso-level of new social forms and organizational infrastructures. Further, we assert that the processes and experiences of biomedicalization illustrate the importance of interaction and contingency in social life. Finally, biomedicalization demonstrates the mutual constitution of political economic, cultural, organizational, and technoscientific trends and processes. Our view of the complex transformations we are witnessing in Western biomedicine is that their roots, manifestations, and consequences are most often co-produced and reciprocally (re)constructed and (re)generated continuously over time.

Those of us who dwell in the sociology of health, illness, medicine, and related areas tend to vividly see the increasing pervasiveness of biomedicine in everyday life. Although not all-encompassing, its ubiquity must be negotiated by each of us on a daily basis. We are awash in a sea of biomedicalizing discourses. And we agree, however anxiously, with Abir-Am (1985) that in the sense that any advertising is good advertising, our project here cannot help but constitute and promote biomedicalization. (Re)naming *is* creating; representing *is* intervening (Hacking 1983).

Yet biomedicalization is punctuated—in fact, rife—with contradictions and unanticipated outcomes that complicate this trend relentlessly. The power-knowledges produced by social sciences of, in, and for biomedicine transgress those boundaries, percolate widely, and are potentially disruptive. There are no one-way arrows of causation, no unchallenged asymmetries of power, no simple good versus bad. In fact, the blurrings of certain boundaries in the creation of new social forms—public/private, government/corporation, expert/lay, patient/consumer, physician/insurer, university/industry/state, among others—are unleashing new and sometimes unpredictable energies. Thus we refuse interpretations that cast biomedicalization as a technoscientific tsunami that will obliterate prior practices and cultures. Instead we see new forms of agency, empowerment, confusion, resistance, responsibility, docility, subjugation, citizenship, subjectivity, and morality. There are infinite new sites of negotiation, percolations of power, alleviations as well as instigations ,of suffering, and the emergence of heretofore subjugated knowledges and new social and cultural forms. Such instabilities always cut in multiple and unpredictable directions (Strauss 1993). Thus we end by calling for case studies that attend to the heterogeneities of biomedicalization practices and effects in different lived situations.[21] We have attempted to elucidate some rich contradictions here in hopes of provoking more democratizing interventions.

Notes

This chapter originally appeared in somewhat different form in the *American Sociological Review* (2003).

We thank our generous colleagues who read and commented on the paper: Isabelle Baszanger, Simone Bateman, Ilana Löwy, Jean-Paul Gaudillière, Phil Brown, Monica Casper, Peter Conrad, Eliot Freidson, Donald Light, Virginia Olesen, Guenter Risse, Sara Shostak, and especially Leigh Star, Herbert Gottweis, Vincanne Adams, and the ASR editors and anonymous reviewers. This paper is part of an ongoing collaboration initiated by Clarke; coauthors are listed in random order.

1 Following Latour (1987), we use the term "technoscience" to indicate an explicit move past scholarly traditions that separated science and technology conceptually and analytically. We argue that these two domains should be regarded as co-constitutive; we thus challenge the notion that there are "pure forms" of scientific or technological research totally distinguishable from their practical applications. Similarly, the term "biomedical" features the increasingly biological scientific aspects of the practices of clinical medicine. That is, the technoscientific practices of the basic life sciences ("bio") are increasingly also part of applied clinical medicine — now biomedicine.

2 Other scholars have used the term "biomedicalization" (Cohen 1991, 1993; Estes and Binney 1989; Lyman 1989; Weinstein and Weinstein 1999). They were not, however, concerned with technoscience. See Clarke and Olesen 1999 and Clarke et al. 2000 for earlier formulations of these ideas.

3 This concept merges the "medical industrial complex," a term coined by Health-PAC (Ehrenreich and Ehrenreich 1971), with the "New World Order, Inc." coined by Haraway (1997).

4 For reviews of the history and sociology of medical technologies and related practices, see Marks 1993 and Timmermans 2000. Co-constitution is defined as the mutual and simultaneous production of a social phenomenon; for a discussion, see Jasanoff 2000.

5 We use the plural "knowledges" to signal that knowledges are heterogeneous and may be incommensurate and contested. On the production of "situated knowledges," see Haraway 1991; for an exemplar, see Clarke and Montini 1993.

6 For a review and extended citations to this theoretical approach, see Pfohl 1985.

7 The term "population health" is increasingly used to refer to studies of particular population groups (the aged, women, ethnic groups, adolescents, etc.).

8 Governmentality is a Foucauldian concept used to refer to particular kinds of power often guided by expert knowledges that seek to monitor, observe, measure, and normalize individuals and populations (Foucault 1973, 1980, 1988, 1991a). This kind of power relies not on brute coercion but on diffuse mechanisms such as discourses that promote the pursuit of happiness and healthiness through certain modes of personal conduct including self-surveillance and self-regulation. We use "governmentality" to connote various governing rationalities based in disciplining and surveillance, biopower, and technologies of the self (also see Rose 1996; Turner 1997).

9 For discussions of trends in the political economy of healthcare, see, for example, Bond and Weissman 1997; Estes 1991; Estes, Harrington, and Pellow 2000; Estes and Linkins 1997; Light 2000a, 2000b; Navarro 1999; Robinson 1999; Salmon 1990; Whiteis and Salmon 1990.

10 We borrow aspects of Ginsburg's and Rapp's (1995) framing of stratified reproduction.

11 For more on the links between health and morality, see, for example, Bunton, Nettleton, and Burrows 1995b; Crawford 1985, 1994, 1999; Edgley and Brissett 1990; Howson 1998a; Illich 1976; Lupton 1993, 1995; Tesh 1990; Williams 1998, 1999; and Zola 1972.

12 On risk factors, see Armstrong 1995; Castel 1991; and Petersen 1997. On techniques of self-surveillance, see Crawford 1994; Edgley and Brissett 1990; Featherstone 1991; and Turner 1984, 1992. On chronic conditions, see Charmaz 1991; Hunt and Arar 2001; Strauss and Corbin 1988; Strauss and Glaser 1975; and Strauss et al. 1984.

13 The consequences of organizations per se on scientific and technical work are only recently being addressed beyond traditional concerns about productivity (e.g., Vaughan 1996, 1999). On work organization, see Mechanic 2002 and Smith 1997.

14 Bastian (2002) notes that one pharmaceutical company attempted to stem its losses from hormone replacement therapy reductions by promoting an alternative product via a campaign to hairdressers with free salon capes bearing the product logo, "scripted messages" to insert in conversations, and fact sheets to hand out to clients.

15 This is already the situation in infertility medicine, where the notion of a sequential ladder of appropriate care from less to more intervention has largely been abandoned in favor of immediate application of high-tech approaches that are more certain to produce babies regardless of cost (Becker 2000b). For lesbians using assisted reproductive technologies to get pregnant, the social category "les-

bian" often serves as the basis for high-tech infertility interventions, regardless of the complete absence of infertility diagnoses (Mamo 2002).

16 This book has been adapted and translated into nineteen languages and has sold over four million copies. http://www.ourbodiesourselves.org/jamwa1.htm.

17 The birth control pill was an early event in this shift (Oudshoorn 2002). "The pill" was the first serious pharmaceutical designed to be taken by healthy asymptomatic people (women). Grave doubts that people would take powerful drugs in the absence of illness were quickly erased by its immediate success.

18 Researchers at the University of California, San Francisco, for instance, are conducting major clinical trials to assess the impacts of traditional Chinese herbs and acupuncture on negative side effects arising from cancer treatment. The Osher Center at UCSF received a $5 million grant for this work.

19 These university medical centers include the University of Illinois at Chicago, University of Pennsylvania, and Johns Hopkins University, which receives the highest amount of federal NIH research dollars (Riccardi and Monmaney 2000; Russell and Abate 2001).

20 Spivak's (1988) concept of "strategic essentialism" asserts the legitimacy of using essentialist-realist epistemological assertions when they may be more effective politically than assertions of multiplicity or diversity.

21 See Fosket 2002 for a study of chemoprevention as the biomedicalization of breast cancer risk; see Fishman (2003) for a study of the biomedicalization of sexuality; see Mamo 2002 for a study of the biomedicalization of lesbian reproduction; and see Shim 2002a, 2002b, for a study of the biomedicalization of race, socioeconomic status, and sex through epidemiology.

Adele E. Clarke, Jennifer Ruth Fosket, Laura Mamo,
Jennifer R. Fishman, and Janet K. Shim

2 / Charting (Bio)Medicine and (Bio)Medicalization in the United States, 1890–Present

An earlier incarnation of this table was developed by all the editors and essentially served as our "data bank" for our theoretical papers on biomedicalization around which this book is built (Clarke et al. 2000, 2003).[1] Over time, we worked back and forth abductively (Reichertz 2007) in terms of generating the table and theorizing it to empirically ground our analysis of changes in the organization and constitution of medicine, health, and illness since around 1890. The original chronological version of the table was especially useful in determining the historical time frames and break points or turning points associated with the three different historical eras: the rise of medicine as a political economic institutional sector from around 1890 to 1945; the emergence, consolidation, and normalization of medicalization processes from around 1940 to 1990; and the later shift to highly interventive technoscientific biomedicalization with its own normalization and expansion patterns from around 1985 to the present and beyond.[2]

After we had analytically generated the five key processes of biomedicalization, we then reorganized the table according to them. The five areas are (1) the political economic sector of (bio)medicine; (2) the focus on health, risk, and surveillance; (3) the increasingly technological and scientific nature of biomedicine; (4) transformations in how biomedical knowledges are produced, distributed, and consumed, and in medical information management; and (5) transformations of bodies to include new properties and the production of new individual and collective technoscientific identities. Thus the table presented here is a combination of chronological facts and theorization of historical eras and key processes. To reflect our increasing level of concern with countertrends and resistances to biomedicalization (discussed in the introduction), we have also expanded references to alternative medicine, medical pluralism, and grassroots health social movements in this iteration.

There is, of course, a riskiness to the ways in which we retrospectively organized historical facts according to the five key processes, some of which only came to full flower in the last era — biomedicalization. We are both aware of and accept these risks as necessary for the table to be coherent. As discussed in the introduction, we see the five processes of biomedicalization as "analytics" and tables like this as "data." In sum, the table should be understood as a working document, a kind of methodological tool that may be of use to other scholars of health and medicine as well, discussed further in the epilogue.

Notes

1 Our original inspiration for creating such a table was an ambitious one by Donna Haraway, "Twentieth Century U.S. Biological Kinship Categories" (Haraway 1991, 219–29). Hers was a different project and had different periodization. Guenter Risse, then chair of the Department of the History of Health Sciences at the University of California, San Francisco, was an invaluable resource. We thank them both.

2 The complex issues of periodization of the chart and of our analysis of the eras of the rise of medicine, medicalization, and biomedicalization are discussed in chapter 3 of this volume. Special thanks for illuminating discussion of these issues with Clarke go to Keith Wailoo, Michelle Murphy, Laura Harkewicz, April Huff, and others who attended the "Healthscapes and Body States" workshop held at the University of California, San Diego, in 2007, organized by Steve Epstein and Lisa Cartwright.

THE RISE OF MEDICINE CA. 1890–1945	MEDICALIZATION CA. 1940–90	BIOMEDICALIZATION CA. 1985–PRESENT
I. Economic Transformations		
A. Organizational Structure		
Small businesses	Local and national corporatization	Transnational corporatization
Cottage industry	Vertical and horizontal integration	Increasing concentration and transnationalization/ globalization
Religious hospitals and dispensaries	Privatization of hospitals	
Emergent pharmaceutical industry	Proliferating forms and heightened competition Elaborating megapharmaceutical corporations	Elaboration of private hospitals into multi-service medical centers Globalizing megapharmaceutical corporations
Institutionalization of medical sciences and medical profession	Rise of the medical industrial complex producing products, technologies, and services	Rise of the Biomedical Technoservice Complex Inc. Devolution of medical services
Specialization	Increasing specialization	Ongoing subspecialization
Public health services limited	Public health programs and clinics	Decrease in and biomedicalization of public health
Medicalization of birth and childhood (obstetrics and pediatrics)	Medicalization of menopause, aging (gerontology) and conditions once deemed social problems (e.g., alcoholism, addiction)	Selective biomedicalization and formation of niche markets (e.g., transplantation, enhancements)
B. Forms of Payment		
Fee-for-service	Fee-for-service	Fee-for-service
Charity care common	Charity care limits	Decrease in charity care
Initiation of private health insurance	Elaboration of private health insurance with eligibility limits	Private insurance with elaborating eligibility limits Managed care

THE RISE OF MEDICINE CA. 1890–1945	MEDICALIZATION CA. 1940–90	BIOMEDICALIZATION CA. 1985–PRESENT
I. Economic Transformations (continued)		
B. Forms of Payment (continued)		
	Initiation of state/public insurance coverage of poor and elderly under Medicaid and Medicare	Prospective payment to physicians Budgets and caps under managed care Limits on state/public coverage under Medicaid and Medicare
C. Sites of Care, Spatial Cartographies of (Bio)Medicine		
General practice	General practice/family practice	Primary care providers
Public sanitation as prevention	Screening as prevention Expansion in focus to tertiary treatment and cure	Revived focus on primary and secondary prevention, including risk treatment
Minimal medical care for most	Proliferation of forms of medical care and services	Limits on some clinical care Prisons as new site of specialization
Informal and formal caregiving at home predominates Emergent hospitalization Emergent public clinics	Increased hospitalization and formal caregiving Public health clinics	Curtailed hospitalization with new specialties of hospitalists, discharge planners, etc. Reduced public health clinics
Limited public health/ visiting nurses in home	Limited public health/ visiting nurses in home	Increasing informal (unpaid) and formal home care
Limited emergency services	Emergency room a primary care site Rise of clinical trials (mostly local)	Emergency room a primary care site Expansion, nationalization and transnationalization of clinical trials

THE RISE OF MEDICINE CA. 1890–1945	MEDICALIZATION CA. 1940–90	BIOMEDICALIZATION CA. 1985–PRESENT

I. Economic Transformations (continued)

D. Key Economic Phenomena

THE RISE OF MEDICINE CA. 1890–1945	MEDICALIZATION CA. 1940–90	BIOMEDICALIZATION CA. 1985–PRESENT
1912[a] creation of U.S. Public Health Service	Some local, state, and federal coverage of public health services and programs	Coverage of public health care decreasing but expansion of surveillance via epidemiology and biostatistics
Some local, state, and federal coverage of public health	Expanding and globalizing private philanthropic investment in medical research	Combined private, government and philanthropic investments in increasingly transnational medical research
Initial private philanthropic investments in medical research		
Minimal governmental investment in medical research	Major expansion of government investment in medical research and shifts of patents to private corporations (federal subsidization of biomedical industrial complex)	Vastly elaborated government investment in medical research with increased privatization of patent rights; contested intellectual property rights
	Increased government payment for medical care at all levels	Expanding public-private and university-industry research and development ventures
Minimal local, state, and federal government coverage of medical care	1946 Hill-Burton Act provides federal support for hospital construction	Reduction and devolution of federal medical coverage to state and local governments
	Corporate value measured in stock values and net worth	Corporate value measured in global site rights; venture capital central
Corporate value measured in revenues and profits	Rise of health statistics and health services research	Expansion of health statistics and health services research
Initial collection of statistics and actuarial data for health and life insurance		Outcomes-based and evidence-based medicine elaborate

THE RISE OF MEDICINE CA. 1890–1945	MEDICALIZATION CA. 1940–90	BIOMEDICALIZATION CA. 1985–PRESENT

I. Economic Transformations (continued)

E. Key Socioeconomic Processes of (Bio)Medicalization

Organization	Corporatization	Transnationalization
Laissez-faire administration	Rational administration	Privatization and devolution
Institutional elaboration	Rise of the welfare state	Neoliberalism
New technologies of examination and normalization (e.g., institution of regular health examinations, screening)	Routinized monitoring, surveillance, and regulation	Panoptic surveillance Biomedical knowledge production industry Transnationalization of biomedicine
Private payment for healthcare	Private payment for healthcare	Private payment for healthcare
Healthcare as charity	Healthcare as business and charity	Purchase of health and healthcare as commodity

F. Biomedical Research Sponsorship

Rise of private foundations as research sponsors	Expanded private foundations as research sponsors	Expanded private foundations as research sponsors
Beginnings of industrial clinical research	Expansion of industrial clinical research	Highly selective industrial clinical research with transnational expansions
1930 Creation of the National Institutes of Health	Growth of NIH	Major expansions of NIH
1944 Public Health Service Act grants NIH legislative authority to conduct research	1965 Total appropriations for NIH top $1 billion[b]	2007 Total appropriations for NIH top $28.9 billion in FY 2007

THE RISE OF MEDICINE CA. 1890–1945	MEDICALIZATION CA. 1940–90	BIOMEDICALIZATION CA. 1985–PRESENT

I. Economic Transformations (continued)

G. State and Federal Involvement in Medical Practice

THE RISE OF MEDICINE CA. 1890–1945	MEDICALIZATION CA. 1940–90	BIOMEDICALIZATION CA. 1985–PRESENT
State regulation and elaboration of medical education and licensure	State regulation and elaboration of medical education and licensure	State regulation and elaboration of medical education and licensure
Minimal federal government investment in medical research	Increasing federal government investment in medical research	Increasing (federal) socialization of costs of medical research and privatization of product development
1890 predecessor to Association of American Medical Colleges organized	Increased coverage of medical care by states	Decreasing coverage of medical care by states
Limited coverage of medical care by states	Increased federal government coverage of medical care	Decreasing federal government coverage of medical care
1902 Biologics Control Act		
1906 Pure Food and Drug Act	1965 Social Security Act creates Medicare and Medicaid	1996 Personal Responsibility and Work Opportunity Reconciliation Act (welfare reform)
1921-1929 Sheppard-Towner Act	1950s–1970s Regulatory jurisdiction and standards of FDA expand to include areas such as additives, and marketing and advertising of health claims	1990s–present Debates on Medicare reform regarding drugs
1927 Food, Drug, and Insecticide Administration organized		1997 FDA regulations allow direct-to-consumer advertising of prescription drugs by pharmaceutical companies
1938 Food, Drug, and Cosmetic Act		

H. Corporate Foundation Involvement in Structure of Medicine

THE RISE OF MEDICINE CA. 1890–1945	MEDICALIZATION CA. 1940–90	BIOMEDICALIZATION CA. 1985–PRESENT
Rockefeller Foundation	Also	Also
Carnegie Foundation	Robert Wood Johnson Foundation	Bill & Melinda Gates Foundation
1910 Flexner Report of Carnegie Foundation	Pew Foundation	
Ford Foundation	Hewlett Foundation	
Macy Foundation	Packard Foundation	
	Rand Corporation	

THE RISE OF MEDICINE CA. 1890–1945	MEDICALIZATION CA. 1940–90	BIOMEDICALIZATION CA. 1985–PRESENT
	II. Risk and Surveillance	
	A. Key Agents of (Bio)Medicalization	
Public health practitioners	Public health practitioners	The new public health
General practitioners	Family physicians	Primary care providers
Medical specialists	Medical subspecialists	Medical subspecialists
Eugenicists	Medical geneticists	Genetic counselors/
Initial media coverage of medicine as scientific	Growing media coverage of medicine and health issues	geneticists
		Expanding media & Internet coverage of health
		Self-surveillance
	B. Key Discourses of Regulation	
Social hygiene	Individual hygiene	Fetishization/commodification of health and life enhancement
	Healthism, fitness	
	Health promotion and wellness	
	Population-based risk surveillance	Health promotion
	1979 publication of first of Healthy People series by PHS	Personalized, individualized risk surveillance and management
		Rhetorics of choice and knowledge
	1986 World Health Organization releases Ottawa Charter for Health Promotion	**1997** FDA regulations allow direct-to-consumer advertising of prescription drugs
		2000 *Healthy People 2000*
	C. Regulatory Discourses of Heredity and Reproduction	
Social Darwinism and eugenics movement	Rise of medical genetics	Rise of prenatal genetic testing, genomics and back door to eugenics[c]
Eugenic sterilization of institutionalized mentally ill and retarded	Population control of targeted groups in United States and elsewhere via contraception	Differential access to reproductive technologies, population control, stratified reproduction

THE RISE OF MEDICINE CA. 1890–1945	MEDICALIZATION CA. 1940–90	BIOMEDICALIZATION CA. 1985–PRESENT
II. Risk and Surveillance (continued)		
D. Technologies of the (Bio)Medical Gaze		
New concept of annual checkups and physical examinations	Routinization of examinations and screening programs	Further elaboration and computerization of imaging technologies (e.g., digitalized x-ray imaging) and testing and screening programs (e.g., Pap smear)
Discovery of x-rays	Elaboration of diagnostic imaging technologies (x-ray, MRI, PET, CAT)	
Development of clinical laboratories	Automated laboratory testing	Genetic profiling
		Medical informatics, data banking
		Standardized risk assessment tools
		Self-surveillance
III. Technoscientization of Biomedicine		
A. Key Objects of (Bio)Medical Knowledge		
Diseases caused by germs	1953 Discovery of DNA structure	Genes & suceptibilities
Hormones	Intracellular bases of disease (e.g., enzymes)	Risk and behavioral lifestyle factors
Disease classification systems	Medical *control* over bodily processes and properties	Molecular basis of disease
Organ systems and functional medicine	Medicalization of crimes without victims (e.g., alcoholism, homosexuality, and abortion)	Biomedical *transformation* of bodily processes and properties
Physiological and cellular bases of disease		Biomedicalization based on technoscientific identities, subjectivities
		1988 Launching of U.S. Human Genome Project

THE RISE OF MEDICINE CA. 1890–1945	MEDICALIZATION CA. 1940–90	BIOMEDICALIZATION CA. 1985–PRESENT

III. Technoscientization of Biomedicine (continued)

B. Dominant Disciplines of (Bio)Medicine

Bacteriology	Pharmaceutical chemistry	Pharmacogenetics/geno-
Human anatomy and physi-	Microsurgery	mics
ology	Epidemiology	Molecular biology
Human pathophysiology	Psychiatry	Molecular epidemiology &
Biochemistry and endocrin-	Genetics	biostatistics
ology		Health economics and
Surgery		health services research
		Individual & population
		risk assessments
		Genetics/genomics

C. Disciplines Mediating Medicine's Relations with Patients

General practice	Psychiatric treatment and	Health economics and
Psychiatry	drugs	health services research
Mental hygiene	Psychology and psychoso-	Outcomes research/
History of medicine	matic medicine	evidence-based medicine
	Medical sociology/doctor-	Medical anthropology/
	patient interaction	culturally competent care
		Science, technology, and
		medicine studies as
		translational research

D. Conventional Treatments of Disease

Emphasis on acute infec-	Emphasis on acute diseases	Prophylaxis/preventive care
tious diseases	Antibiotics and other new	Elaborate attention to
Isolation and quarantine	drugs	chronic illnesses
Surgery	Initial attention to chronic	Lifestyle modification
Limited use of drugs	illnesses	Risk prevention, reduction
	Lifestyle modification and	and treatment
	risk reduction	Tissue banking
	Technological intervention	Pharmaceutical regimens
		Customizable pharmaceu-
		ticals

THE RISE OF MEDICINE CA. 1890–1945	MEDICALIZATION CA. 1940–90	BIOMEDICALIZATION CA. 1985–PRESENT
III. Technoscientization of Biomedicine (continued)		
E. Reproductive Medicine		
Simple contraceptives: spermicides and mechanical barriers Rise of functional obstetrics and gynecology	Scientific contraceptives: Pill, IUD, Depo-Provera Family planning programs Emergence of new conceptive reproductive technologies	More scientific contraceptives: implants, microbicides (new male method unreleased) Linkage of contraception to HIV prevention Major expansion and transnationalization of infertility medicine Linkage of infertility services to stem cell research (e.g., co-location)[d]
F. Dialysis and Transplantation Medicine		
Research on transplantation at Rockefeller Institute	Transplantation and dialysis as cottage industries supported by local charities 1972 Federal government ensures dialysis medical care for end-stage renal disease Fox and Swazey, *Courage to Fail: A Social View of Organ Transplants and Dialysis*	For-profit corporatized and transnationalized companies, markets, and organ and tissue exchange organizations Replacement part donation promoted Regenerative medicine Legal and illegal payment for organ donation in a transnational black market Fox and Swazey, *Spare Parts: Organ Replacement in American Society*
G. Mental Health		
Institutionalization Psychiatry Lobotomy, shock treatment	Deinstitutionalization Psychiatry New psychopharmaceutical drugs	Prozac Nation Short-term therapies Extensive pharmaceutical armamentarium Neuropsychiatry

THE RISE OF MEDICINE CA. 1890–1945	MEDICALIZATION CA. 1940–90	BIOMEDICALIZATION CA. 1985–PRESENT

III. Technoscientization of Biomedicine (continued)

H. Approaches to Death and Dying

THE RISE OF MEDICINE CA. 1890–1945	MEDICALIZATION CA. 1940–90	BIOMEDICALIZATION CA. 1985–PRESENT
Most deaths at home	Most deaths in hospitals	Most deaths in hospitals
Patients often not infomed of fatal diagnoses	Technological war against death (e.g., CPR, pharma- ceuticals, mechanical life support)	Ongoing technological war against death
Routine denial of fatal diagnoses	Initial public discussion of death and dying	Extensive public discussion of death and dying
	Hospice movement initiated	Right-to-die movements
		Hospice movement institu- tionalized
		Palliative care movement gains momentum

I. Medical Records

THE RISE OF MEDICINE CA. 1890–1945	MEDICALIZATION CA. 1940–90	BIOMEDICALIZATION CA. 1985–PRESENT
Paper patient records	Paper patient records	Paper patient records disappearing
Paper test results	Computerization of test results	Online patient and insur- ance records accessible via multiuser computer networks
X-rays maintained by radiology labs	Medical imaging and other tests maintained by labs	Medical imaging, surgery videos, and other records given to patients on CDs
		Extensive medical informatics

IV. Transformations of Knowledge Production and Distribution

A. Doctor-Patient Discourses

THE RISE OF MEDICINE CA. 1890–1945	MEDICALIZATION CA. 1940–90	BIOMEDICALIZATION CA. 1985–PRESENT
Physician as benevolent (Norman Rockwell image)	Physician linked to ser- vice and technology and "doctor knows best"	Physician as corporate cog, stockholder, venture capitalist "Know thyself, heal thyself"
Ideal patient as trusting and deferential	Ideal patient as emergent consumer	Ideal patient as self- surveilling, informed & scientized consumer
"Sacred" relationship be- tween doctor and patient	Emerging patient and con- sumer group distrust	Often fraught doctor/ patient relations compli- cated by insurance regu- lations and limitations of coverage
	Challenges to medical paternalism especially among women and African Americans	

THE RISE OF MEDICINE CA. 1890–1945	MEDICALIZATION CA. 1940–90	BIOMEDICALIZATION CA. 1985–PRESENT

IV. Transformations of Knowledge Production and Distribution (continued)

B. Key Sources of (Bio)Medical Information

General practitioners	General practitioners	HMO information programs
Family members & friends	Medical specialists and subspecialists	Medical specialists and sub-specialists
Domestic household management texts	Family members & friends	Family members & friends
Alternative healers	Newspapers, radio, and TV	Newspapers, radio, and TV
Newspapers and radio	Self-help publications	Self-help publications
	Massive advertising in various media of over-the-counter drugs	Internet
		Direct-to-consumer prescription drug advertising
		Investing and business news

C. Alternative Medicine

"Irregular" medicine	Alternative medicine goes professional or underground	Alternative Medicine Inc.
Multiple medical systems/ medical pluralism in practice	Shakeout and elaboration of approaches	Clinical trials and co-optation by biomedicalization of selected approaches
Many alternative forms branded as "quackery" and attacked by AMA	State licensure of some forms	Some insurance coverage
	Regulation of some forms by FDA	State licensure of some forms
		Regulation of some forms by FDA

D. Techniques of Legitimation of Biomedical Claims

Case/observation-based medicine	Case/observation-based medicine	Outcomes measurement, evidence-based medicine, cost effectiveness
Scientific determination of causes of infectious and other diseases by laboratory testing	Elaboration of scientific knowledge of disease processes	Clinical protocols and treatment algorithms
	Elaboration of laboratory testing	Elaboration of laboratory testing
	Emergence of clinical trials	Randomized controlled clinical trials
	Medicine-focused media coverage	Rise of the Internet
		Biomedicine-focused media coverage

THE RISE OF MEDICINE CA. 1890–1945	MEDICALIZATION CA. 1940–90	BIOMEDICALIZATION CA. 1985–PRESENT

IV. Transformations of Knowledge Production and Distribution (continued)

E. Social Movements Focused on Health, Illnesses, and (Bio)Medicine

THE RISE OF MEDICINE	MEDICALIZATION	BIOMEDICALIZATION
Alternative health social movements	Alternative health social movements	Alternative health social movements
Quackery, underground subcultures	Abortion and reproductive rights movements	Abortion and reproductive rights
Birth control movement	Women's health movements	Co-optation of women's health, consumer health, and patient movements, alternative medicine
Mental hygiene movement	Consumer health/patients rights movements	
Social hygiene movement		
Eugenics movement	1958 AARP founded	1980s Emergence of HIV/ AIDS pandemic, HIV/ AIDS activism (e.g., ACT UP)
1921-1929 Sheppard-Towner Act for maternal and child health services	1970 *Our Bodies, Ourselves*	
	1971 Federation of Feminist Women's Health Centers founded	
1924 Immigration Restriction Act	1974 Establishment of the National Institute on Aging	1993 NIH Revitalization Act creates Office of Research on Women's Health and Office of Research on Minority Health, requires inclusion of women and minorities in all government-funded clinical studies
1933 White House conference on maternal and child health		

F. Biomedical Ethics

THE RISE OF MEDICINE	MEDICALIZATION	BIOMEDICALIZATION
Physicians as ethical authorities	Concerns about power in doctor-patient relationship	Proliferation of uncertainties in provider-patient relationships
Peer pressure and review	Beginning of professionalized biomedical ethics	Institutionalization of professionalized ethics
Early hospital committees for case review, morbidity and mortality reports	Formalization of informed consent processes to protect patients from unnecessary risk	Informed-consent purposes shift from protection from risk to also include protection of autonomy and personal dignity as subjective dimensions of risks recognized
State licensing boards as watchdog mechanisms		

THE RISE OF MEDICINE CA. 1890–1945	MEDICALIZATION CA. 1940–90	BIOMEDICALIZATION CA. 1985–PRESENT
IV. Transformations of Knowledge Production and Distribution (continued)		
F. Biomedical Ethics (continued)		
Percival's Medical Ethics[e]	**1974** U.S. National Commission for the Protection of Human Subjects of Biomedical and Behavioral Research begins; NIH requires all research be approved by Institutional Review Boards	Emergence of consumer advisory boards Access to clinical trials as a social good requiring equitable distribution **1980s** Emergence of HIV/AIDS activism seeking changes in NIH drug trial protocols **1980s–1990s** Revised FDA regulations allow for accelerated approval and access to drugs for diseases with no alternative therapies **1993** NIH Revitalization Act requires inclusion of women and minorities in all government-funded clinical studies
V. Transformations of Bodies		
A. Discursive Constructions of the Body		
Lived/endured/blatantly controlled Site of care/discipline	Elaborated/subtly controlled Site of discipline/care of the self	Transformable Fetishized consumer object Site of discipline/care of the self
B. Discursive Constructions of Sex		
Little sex discourse Homosexuality as disease	Sex education/SEICUS Homosexuality *not* a disease	Sexualized mass media Pharmaceutical treatment of sexual dysfunction for men (e.g., Viagra) and emerging for women

THE RISE OF MEDICINE CA. 1890–1945	MEDICALIZATION CA. 1940–90	BIOMEDICALIZATION CA. 1985–PRESENT

V. Transformations of Bodies (continued)

C. Targets of Co-optative Medicalization[f]

White, middle- and upper-class women	White, middle- and upper-class women	White, middle- and upper-class women
"Deviants"	Postmenopausal women	Postmenopausal women
Homosexuals	Children	Working/lower/underclass and people of color
Addicts (hospital prisons)	Working/lower/underclass and people of color (e.g., forced sterilization, coerced contraception)	**1991** Launching of Human Genome Diversity Project
Alcoholics		**1993** NIH issues guidelines requiring inclusion of women and minorities in all NIH-funded clinical studies
		1998 FDA requires clinical trial data on women and minorities to ensure approval

D. Objects of Exclusionary Disciplining[g]

Working/lower classes	Working/lower/underclass	Working/lower/underclass
People of color	People of color	People of color
Immigrants	Immigrants	Immigrants
	People with "prior conditions"	People with "prior conditions"

a. Dates in boldface represent marker events. Chart updated 2009.

b. For 1965, see National Institutes of Health 1976. For 2007, see http://www.hrsonline.org/Policy/LegislationTakeAction/nih_approp_index.cfm.

c. Duster 2003.

d. Leake 1975. Thompson 2007, 2008, forthcoming.

e. Leake 1975.

f. Ehrenreich and Ehrenreich 1978.

g. Ehrenreich and Ehrenreich 1978.

Adele E. Clarke

3 / From the Rise of Medicine
to Biomedicalization

U.S. HEALTHSCAPES AND ICONOGRAPHY,
CIRCA 1890–PRESENT

> What these things have in common is loquaciousness: they give rise to an
> astonishing amount of talk. We are interested in how talk and thingness
> hang together.—LORRAINE DASTON, "Speechless"

This chapter centers on popular, especially visual,
cultures of medicine, health, and healing characteristic of three eras in the United States: the rise of medicine from around 1890 to 1945, the era of medicalization
from around 1940 to 1990, and the emergence of biomedicalization around 1985 to the present and beyond. I
argue that developments of "things medical" in the past
and today are not only accompanied by and reflected in
popular cultural iconography—multiple media—but
also were and continue to be in part generated and produced by and through them.

Visual objects of "things medical" (my generic term)
include paintings, photographs, advertisements, packaging materials, books, pamphlets, comic books, magazine images, plays, movies, television, and so on (e.g.,
Lederer and Rogers 2003; Reagan, Tomes, and Treichler
2007). The social sciences have dealt with these aspects
of social and cultural life inadequately.[1] Visual materials
are nonhuman actants writ large; images and objects do
many kinds of work (Latour and Woolgar 1979). "The
argument that 'representation' always carries a double
meaning—as depiction (a picture of) and as proxy
(an act of speaking for)—has a long lineage in social
theory" (Hayden 2005, 187). Identifying the social, cul-

tural, political, economic, and other interests represented by the artifacts as well as their semiotic interpretation is our scholarly task (e.g., Joyce, this volume).[2]

My argument is that in deeply significant but largely ignored ways, contemporary American biomedicalization itself is imbricated with popular and visual cultural materials, representations, and media coverage of things medical. The quantity, pace, and extent of exposure of major segments of the American populace to such popular and visual representations began to expand in the late nineteenth century, the era of the rise of institutionalized medicine. Since then, via new media, it has relentlessly gained momentum. In turn, such images and objects themselves have become constitutive of medical knowledge and practices. As Cartwright (1998, 220) asserts, "Medicine is a cultural field whose meanings are created not only by the elite managers of technoscientific laboratories and research centers, but also through the intervening forces of popular media and by media activist countercultures." All of this has an important history that I explore here as *healthscapes*.

Healthscapes

Inspired by Appadurai (1996), I present the three eras of the rise of medicine, medicalization, and biomedicalization as healthscapes. Healthscapes are ways of grasping, through words, images, and material cultural objects, patterned changes that have occurred in the many and varied sites where health and medicine are performed, who is involved, sciences and technologies in use, media coverage, political and economic elements, and changing ideological and cultural framings of health, illness, healthcare, and medicine. Appadurai (1996, 33–35) saw five main global cultural flows: ethnoscapes, mediascapes, technoscapes, financescapes, and ideoscapes. I am arguing for another, healthscapes, which both cut across and synthesize elements of the others.[3]

In healthscapes, health and medicine are viewed as far more than particular institutions in their conventional sociological framings, separable from others such as family, religion, economy, polity, media. Instead healthscapes attempt to capture all of these together, however inadequately, in their material, semiotic, and symbolic dimensions (e.g., Hall 1997). Healthscapes are kinds of assemblages (Marcus and Saka 2006), infrastructures of assumptions as well as people, things, places, images. Healthscape is a framing concept that provokes both "thick description" (Geertz 1979) of a particular place and era and simultaneously captures the traveling potentials of "things medical" across transnational/global flows.

Healthscapes attempt to capture the temporality and ethicality of "regimes of practices" — "practices being understood here as places where what is said and what is done, rules imposed and reasons given, the planned and the taken-for-granted meet and intersect" (Foucault 1991b, 75). Collier and Lakoff (2006) call these "regimes of living" that posit ethics of how life is to be lived, fraught with "oughts." Laqueur (1989) asserted that the "humanitarian narratives" that appeared in popular culture of the eighteenth and nineteenth centuries helped to establish humanitarianism as a cultural good. I argue similarly that healthscapes/regimes/practices helped to establish the rise of medicine, medicalization, and biomedicalization as cultural goods.

Establishing and sustaining medicine as a cultural good links directly to the cultural authority of medicine so vividly engaged by Starr's *The Social Transformation of American Medicine: The Rise of a Sovereign Profession and the Making of a Vast Industry* and by its many critics (1982).[4] Healthscapes address a gap in Starr's analysis regarding *how* medicine gained cultural authority. I am asserting that there are economic *and* cultural, material *and* semiotic, institutional, collective, *and* individual dynamics that produce, promote, and challenge medical authority. Healthscapes in this sense approximate Foucauldian discursive regimes, appropriating and integrating all kinds of imagery and media into legitimation work — from globe-hopping TV doctor shows to pharmaceutical packaging materials. The visual cultures of things medical do the fundamental work of linking medicine to modernity — a cultural "good" indeed.[5]

Next I briefly frame the importance of taking visual imaging and iconography seriously. I then offer a chart that compares the three healthscapes. The rest of the chapter is devoted to the three healthscapes. For each, there is an overview and discussion of the featured iconography.

Taking Visual Cultures of (Bio)Medicine Seriously

> Like seeds around which an elaborate crystal can suddenly congeal, things in a supersaturated cultural solution can crystallize ways of thinking, feeling, and acting. These thickenings of significance are one way that things can be made to talk.
> —LORRAINE DASTON, "Speechless"

The work done by visual imagery goes beyond words in ways still not well understood. Jordanova (1998, 216) asserts three ranges of meaning of the word "display": (1) how things look or appear, (2) exhibition/demonstration, and (3) ostentation/showing off, occurring simultaneously. Different media also "afford" different kinds of meanings and engender different kinds of meaning-making processes (e.g., Dicks, Soyinka, and Coffey 2006). Moreover, and more complicated, "display concerns what is hidden as well as what is revealed, for these are two sides of the same coin" (Jordanova 1998, 210). All this makes imagery even more important in terms of understanding other historical eras — and other sites. Imagery offers openings through which to see.

Second, (bio)medical framings of "regimes of living" (Collier and Lakoff 2005) have become deeply naturalized. "Popular images of medicine, whether in literature, the cinema, television, or print media, draw upon these visual conventions as iconographies to construct meanings and metaphors that are readily understandable, that translate complex theories for general consumption" (Marchessault and Sawchuk 2000, 3). Significantly, I do not consider the medico-scientific images as "translating" science directly or simply into more popular terms, nor does the process end there (e.g., Irwin and Wynne 1996). Rather, as Fleck (1935/1979) argued, when scientific concepts and facts are (re)represented, some things are "lost" while others are "found" in translation, often including a new "vividness" in more popular and accessible incarnations. These popular versions may loop back, influencing experts, among others, folding new forms into inner circles of technoscience. Meaning making and its consequences ramify endlessly.

A third reason to take visual cultures of medicine seriously is that precisely because illness, suffering, caring, and healing are to many at the heart of the human condition, they have quite densely elaborated iconographies. "The aura of power surrounding medicine and its practitioners has made healers and physicians favorite subjects of dramatists, novelists, and other artists from ancient times to the present" (Jones 1981, 183).[6] Fourth, "Medical images of the interior body have come to dominate our understanding and experience of health and illness at the same time and by the same means as they promote their own primacy. Medical imaging technologies have rendered the body seemingly transparent" (van Dijck 2005, x). Over time, van Dijck argues, "the ideal of transparency" is extending to ever "deeper" levels: "From the pen of the anatomical illustrator to the surgeon's advanced endoscopic techniques, instruments of visualization and observation have mediated our perception of the interior body through an intricate mixture of scientific investigation, artistic observation, and public understanding" (4). As Rose (2007) asserts, we now participate in the molecular gaze, seeking both truth of one's self and future imaginaries.

My fifth point is that in grasping visual cultures of biomedicine we see possible futures—"imagining (bio)medicalization" (e.g., Squier 2004). Media coverage of medicine, especially visual imagery, creates "imagined medicines" of the near and/or distant future (Anderson 1983). In the supposedly democratic United States, this would be an "imagined medicine" available to all (e.g., Tomes 2001). What is at best obscured and at worst made fully invisible are the stratifications, inequalities, and lack of healthcare in the United States and for most of the world. Last, I agree with Marchessault and Sawchuk (2000, 2) that "research which encourages media literacy around popular representations of medicine also makes possible a public discourse around ethics."

The three healthscapes have a wide range of old and extremely new visibilities and meanings in popular cultures. These include whatever "counts" as framings of health, healing, illness, disease, dying, death, medicines, surgeries, drugs, devices, procedures, enhancements, visualizing diagnostic technologies (from x-rays to MRIs to brain scans to ultrasounds), all kinds of medical technologies and devices (including reproductive technologies), and so on. Of course, they also include the full range of human actors as patients, physicians, nurses, technicians, scientists, receptionists, pharmacists, alternative providers, and so on. Healthscapes include whatever is imaged and imagined as things medical.

Overview of the Three Eras

> All these things threaten to overflow their outlines.
> —LORRAINE DASTON, "Speechless"

The table, "From the Rise of Medicine to Biomedicalization," offers an overview of the three eras and their key processes.[7] Several caveats obtain. First, the eras are fuzzily bounded and bleed into one another. Second, processes or phenomena in different eras are matters of emphasis, not exclusion. Innovations are cumulative over time. Older approaches are usually available somewhere.

THE RISE OF MEDICINE TO BIOMEDICALIZATION, 1890–PRESENT

	RISE OF MEDICINE CA. 1890–1945	MEDICALIZATION CA. 1940–85	BIOMEDICALIZATION CA. 1980–PRESENT
Infrastructure	Organizational	Material	Digital
Basic social processes	Specification and legitimation	Control and elaboration	Enhancement and transformation
Focus of clinical gaze	Treatment of acute illnesses and communicable diseases	Medicalization of chronic illnesses and diseases	(Bio)medicalization of health and risk factors
Main mode of clinical action	Surgical success and clinical skills	Routinization of medical care and drugs	Technoscientific interventions and drugs
Main focus of biomedical sciences	Germ theory and disease classification	Biochemistry and pharmaceutical sciences	Molecular biology, genetics, genomics, and nanotechnologies
Main focus of biomedical technology	Amplifying bodily indicators, imaging, and sedation	Imaging, tests, procedures, and treatments	Imaging, devices, biotechnologies, and nanotechnologies
Medical constructions of patients and identities	Patienthood as a privilege / Illness and disease identities in known biographies	Passive patients / doctor knows best / Diagnostic identities	Responsible consumers / Technoscientific identities and biosocialities
Main media construction of (bio)medicine	Great doctors and the need for medicine	Great doctors and hospitals	Great technoscientific innovations
Key rhetorics	Medicine "for the benefit of all mankind"	"Healthcare for all" via Medicare, Medicaid, and private insurance	Direct-to-consumer advertising and infomercials
Major "other" medicines and health social movements	Homeopathy / Maternal and child health movement	Chiropractic / New Age healing and women's health movements	Acupuncture / AIDS and other patient movements

This framework also raises historiographic concerns that I attempt to address. Healthscapes offer "grand narratives" that, however valuable, risk oversimplifying (e.g., Warner 2004). To complicate such pictures, we are urged to go "beyond the great doctors" and "top-down" stories of elites (e.g., Reverby and Rosner 1979, 2004; Cooter 2004). Doing social history of health and medicine "from the bottom up" demands that we attend to self-help strategies (e.g., Risse, Numbers, and Leavitt 1977), patients (e.g., Porter 1985) and, more recently, patient group movements (e.g., Epstein 2008).

Further, doing social history "from the outside in" includes practices in the margins of medicine. We must take medical pluralism — simultaneous availability of multiple kinds of medicines in the same places — very seriously indeed as "other" medicines. This challenges the naturalization of Western, scientific, allopathic or "regular" (bio)medicine as the only real form. Instead it asserts that this approach is indeed "cultured" — neither transcultural nor a transcendent "culture of no culture" (Taylor 2003; Traweek 1988). In the United States, "other" medicines were deemed "irregular" and "quackery" by the AMA in 1849, just two years after its founding, initiating efforts against their legitimacy that continue today.[8]

Therefore, to complicate the grand narratives of the healthscapes, in addition to the five iconographic images of "regular" medicine for each era, I offer two "other" images. One represents the alternative form of medicine that seems to have most distinctively challenged the (bio)medicine of that era. The second portrays some form of resistance to, or rejection of, conventional biomedical practice that occurred during that era. While there is an overarching pattern of (bio)medicalization, *always* there are many other things also going on.

The Rise of Medicine Healthscape

> ... the most stolidly functional things — buildings, soap, newspapers — radiate an aura of the symbolic — LORRAINE DASTON, "Speechless"

The rise of medicine healthscape extends roughly from around 1890 to 1945. I emphasize this healthscape here as our original work addressed only medicalization and biomedicalization eras (Clarke et al. 2003). While the AMA was founded in 1847, its Western scientific/allopathic form of medicine did not achieve dominance in the United States until the late nine-teenth century. By then, through becoming state-legitimated medicine via state licensure laws, generating state-based infrastructures (often the first state medical schools), developing a national journal, sponsoring research, and attacking other forms of healing as quackery, allopathic or regular medicine was succeeding. Infrastructure building was deeply community and state based, often taken for granted (Pescosolido and Martin 2004). Profound sponsorship of regular medicine by major foundations (Rocke-feller and Carnegie) in the early twentieth century then sealed this success by enabling its science and further professionalization. Most competing medical schools closed, including many that admitted women and people of color. "Allopathic" or "regular" medicine became just plain "medicine" in the United States, a generally unmarked category, rendering other forms beyond the pale.[9]

In fact, this period was termed "the golden age of medicine" (Brandt and Gardener 2003) as it became more effective and specialized (e.g., Weisz 2006). The tipping point when it could be assumed that a clinical interven-tion by a regular American physician would be more likely to be positive than negative in outcome has been estimated by historians to be around 1910 (Hansen 1999, 630–31). As a sociologist, I also foreground medicine's sustained institutionalization (Starr 1982).

The elaboration of consumption along with production occurred in health and medicine as well as elsewhere during this era, particularly mani-fest in visual cultures (e.g., Hansen 1999; Brown 2005; Laird 1998; Seale 2004). Working from "a more patient-oriented perspective," Tomes (2001, 521) found many and varied "merchants of health" actively integrating medicine and consumer culture deeply during the "health century." By 1930, healthcare or medicine *was* a major industry in the United States. Even during the Great Depression, Americans spent over $3.5 billion an-nually on medical services and commodities, and the healthcare sector al-ready ranked sixth among American industries, above the automobile, iron and steel, and oil and coal sectors, with 4 to 5 percent of the GDP (Tomes 2001, 526).

Tomes further asserts that medicine was not as "exceptional" a political economic sector—did not historically operate largely *outside* relations of capital in the domains of charity and minor cottage industry—as has com-monly been assumed: "The habits of mind cultivated in a consumer culture organized to sell toothpaste and refrigerators did indeed spill over, if only indirectly, into views of health-related products and services" (524–25). Further, "health in a bottle and a book" was much more widely accessible, while "new forms of media and marketing . . . multiplied the commercial opportunities for self-medication" (Tomes 2001, 531, 534).[10]

During the rise of medicine healthscape, mass production and consump- tion were accompanied by, "indeed instituted by, new visual technologies . . . the visual order of modernity" (Brown 2005, 6). As we argued earlier (Clarke et al. 2003, this volume), in the United States as elsewhere, partak- ing of "the miracles of modern medicine" is a sign of modernity. Signifi- cantly, medical iconography was specifically constitutive of the *standard- ization* of American medicine in concept if not in actual practice—what ought to be, if not what was. This emergent standardization of assumptions was accomplished in part by a key change in popular culture captured in 1923 by Edwin Slosson, head of the Science Service in Washington: "This seeing of the same pictures and reading the same magazines at the same time by the greater part of the one hundred million people of the United States tends towards conformity in taste and ideas, toward the standardiza- tion of the American" (Brown 2005, 1).

Advertising is perhaps the mass-produced visual discourse most "seen" in "modern social life." Raymond Williams (1980, 179) argued, "The half- century between 1880–1930 . . . saw the full development of an organized system of commercial information and persuasion, as part of the modern distributive system in conditions of large-scale capitalism." Drug and toi- letries advertising led the way, becoming the dominant sector in all adver- tising today.[11] The combined use of visual and narrative discourses, and the ways in which commercial information and persuasion cannot meaning- fully be separated, also deeply influenced nonadvertising modes of visual representation (e.g., Goffman 1976).

Imaging in the healthscape of the rise of medicine centered on new kinds of things medical: physicians, the profession of medicine, and the life sciences.[12] Practitioners sought "cultural products that could act as per- manent testimonies to the goodness of medicine. . . . The big historical change is perhaps the use of the mass media to cultivate a sense of medical benevolence, with the result that professionals themselves no longer have to work so hard at it" (Jordanova 1998, 213–15). "Regular" physicians dif- ferentiated themselves from "quacks" by displaying their safeness through "gentlemanly attributes and appeals to science"—education, institutional affiliations, scientific knowledge, humanity, and public spirit (206–9).

After 1890, "America's daily newspapers and weekly magazines began to give high visibility to medical discoveries . . . initiat[ing] a pattern of proclaiming medical breakthroughs in the media that continues to the present" (Hansen 1999, 629–30). This media coverage "helped to establish in mass culture two new intertwined notions: 'medicine is scientific' and 'medicine makes progress.'" By 1895, "Research itself had become visible; medical innovation was now a public thrill" (Hansen 1999, 676; Wailoo 2004).

Public interest in scientific research linked the rise of the medical profession to its growing social and cultural authority in the United States in significant if complicated ways (Tomes 2001, 524; Cohen and Shafer 2004; Starr 1982). "With each new success came new pictures of medicine and new expectations—and the American public never saw medicine in the same old way again" (Hansen 1999, 677–78). If modernity was scientific and medicine was scientific, then (*pace* Latour 1993) we had "modern medicine" by the start of the twentieth century. Hansen (1999, 630–31) further asserts "possible feedback loops from ordinary citizens' concerns and enthusiasms to the social status of the profession and to the willingness of elected officials and philanthropists to support expensive new institutions like the medical laboratory and the rapid growth of health departments and hospitals across the United States."

Along with "reel nature" (Mitman 1999), by the early twentieth century we had "reel medicine." Doctors, surgeons, and dentists became "the first recognizable professionals in the picture palace," and "mass media made familiar the image of the white-coated male physician, stethoscope in hand, and the worlds of the hospital, laboratory and waiting room" (Lederer and Rogers 2003, 487). Nurses, then as now, did double symbolic duty as handmaidens to medicine and sex objects. In 1936, *The White Angel* featured Florence Nightingale and nursing during the Crimean War (Jordanova 1998, 214).

By the 1920s, the AMA was actively shaping media representations of medicine in positive and uncritical directions, often accomplished by dissolving boundaries between entertainment and education (Lederer and Rogers 2003, 487–8). Foreshadowing the infomercial, "Print media and later radio began to cover health issues more widely; they also became increasingly dependent on revenue from health-related advertising . . . its own powerful form of commercialized advice" (Tomes 2001, 531).

But it was not only newspapers, magazines, and movies that took up great doctors and miracles of medical science. Popular books also did so: Thomas Mann's *The Magic Mountain* in 1924 (van Dijck 2005, 85–99), Sinclair Lewis's *Arrowsmith* in 1925 (Rosenberg 1963) (made into a film in 1931), Paul de Kruif's *The Microbe Hunters* in 1926 (Lederer and Rogers 2003, 491–92), and Proust's *Swann's Way* translated in 1929 (van Dijck 2005, 83). On Broadway, medicine was featured hagiographically in plays such as *Yellow Jack*, the story of Walter Reed's work against yellow fever (Jones 1981, 193). In contrast, in "Frankenflicks," doctors gone bad and mad were dominant (Frayling 2005). The first *Frankenstein* and *Dr. Jekyll and Mr. Hyde* appeared in 1910 along with Dr. Moreau in *Island of Lost Souls* (Clark 2004, 129), and *The Bride of Frankenstein* in 1935 (Lederer and Rogers 2003, 493).

Film also became a popular medium for new state-sponsored public health messages seeking to change attitudes and behaviors, initially called "health propaganda" before propaganda was only what "others" did. More than one thousand health-related films were made between 1905 and 1927, often reflecting the cultural stereotyping and ethnocentrism of their makers (Lederer and Rogers 2003, 489). Radio, newspapers, magazines, and films became common sources of health information. "Weekly medical advice programs . . . habituated listeners to the sounds, if not the sights, of modern medicine" (Lederer and Rogers 2003, 491), including conceptions of germs, viruses, and cells (Tomes 1998).

Medical enhancements were in sharp focus. In *The Monkey Gland Affair*, monkey testes were surgically inserted into human males for "rejuvenation," an early-twentieth-century practice later extended to prized stallions and other animals (Hamilton 1986). In a common American version, goat gland transplants were advertised on the radio in the Midwest (Hamilton 1986, 136–37). Following a similar logic, Steinach offered vasectomies to make testes return sperm to the body to increase circulation and testosterone levels (Sengoopta 2006).

ICONOGRAPHY OF THE RISE OF MEDICINE ERA

The main mode of clinical action of allopathic medicine during this era significant for its success was the development of safe and effective surgery. Thomas Eakins's painting *The Agnew Clinic* (1889) is both iconic and hagiographic here (figure 1; see also Doyle 1999). This image often appears

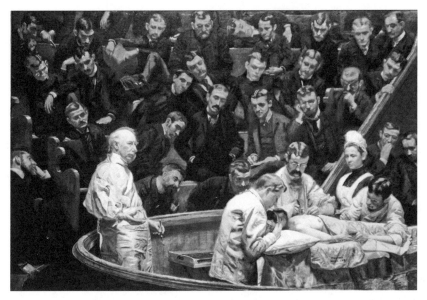

1 Thomas Eakins, *The Agnew Clinic*, 1889. Courtesy of the University of Pennsylvania Art Collection, Philadelphia.

linked to the rise of medicine, brilliantly capturing the era of the "great man" (Leavitt and Numbers 1978). It appears often on book covers such as Porter's *The Greatest Benefit to Mankind: A Medical History of Humanity* (1999), capturing the missionary rhetoric of medicine formulated then and still lively today (Harvey and Abrams 1986), positioning medicine closer to religion than to capitalism.

Though a contemporary of Monet, Eakins is more aptly termed "anti-impressionist," seeking more Rembrandt-like pictorial drama and senti-ment (Solomon 2006, 18). Dr. Agnew is "shown here in dual roles of sur-geon [scalpel in hand] and teacher. . . . By [visually] emphasizing Agnew more than the operating field, Eakins conveys that the enlightenment to be gained in this scene clearly comes from what the professor, not the patient, has to offer" (Cohen and Shafer 2004, 201). Eakins's painting was criticized for its "vivid realism" and "unpleasant subject" (Carmichael and Ratzan 1991, 269–71). He appears to be portraying a mastectomy, allowing a view of a bared female breast, holding down medicine's "soft-core" visual posi-tioning. The Halsted mastectomy was a major new surgery, a radical and often unnecessarily disabling version actively protested by women's health movements a century later (Montini and Ruzek 1989).

Among the most famous and most reproduced medical paintings of all

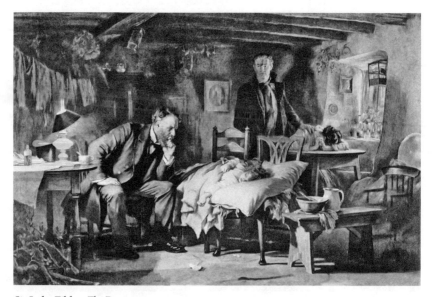

2 Sir Luke Fildes, *The Doctor*, 1891.

time is *The Doctor*, by Sir Luke Fildes (1891) (figure 2), painted after the death of his son. It was the subject of a film by Edison in 1911, and part of a major exhibit at the Century of Progress International Exposition in Chicago in 1933. The painting traveled to eighteen American cities and was viewed by about five million people. In 1947 it became the image on the U.S. postage stamp commemorating the centennial of the founding of the AMA (Lederer and Rogers 2003, 496–97).

It is interesting that this lonely figure was promoted by the AMA itself as iconic during this infrastructure-building era. The physician, Dr. Murray, is seen waiting in silent vigil, without diagnostic aids, with a bottle of something perhaps therapeutic, but unable to prevent the death (Cohen and Shafer 2004, 199).[13] He reflects the "moral mission" of medicine of the era (Sloane and Sloane 2003, 6), rather than the technical innovation captured by *The Agnew Clinic*. The iconographic success of *The Doctor* inspired a subsequent series of eighty-five paintings known as *Great Moments in [the History of] Medicine*, sponsored in the 1930s and 1940s by Parke, Davis and Company and distributed widely to hang in physicians' offices and other medical venues (Bender and Thom 1966; Duffin and Li 1995). *The Agnew Clinic* and *The Doctor* in some senses triggered an elaborating physician-centered and physician-promoted popular iconography still with us today.

3 John French Sloan, *X-rays (or Fluoroscope: Department of Interior)*, 1926. Philadelphia Museum of Art: Gift of Mrs. John Sloan, 1975.

Many new medical technologies were invented during the rise-of-medicine era, often focused on visualizing bodies and amplifying bodily indicators. Probably the most important and continuingly so was the x-ray, which supplemented sight, sound (stethoscope), and touch (palpation) (e.g., Cartwright 1995; Kevles 1997; Pasveer 1989). Figure 3 shows an etching by the American artist John French Sloan titled *The Fluoroscope* (1926).[14] Combining x-rays with a fluorescent screen allowed the interior to be seen immediately by the doctor—without awaiting "development."

As the physicians observe, the patient (Sloan himself) looks on. Cohen and Shafer (2004, 206) note how the patient is literally "screened off," seen now by the physicians as "only an image . . . not as a human being," signaling a new kind of physician-patient relation, partialized and mediated by technology.

Later in this era, advertising industries were influenced by Edward Bernays, nephew of Sigmund Freud, who "translated" elements of Freudian theory (especially vis-à-vis the unconscious and sexuality) into what might be called "capitalist public relations theory."[15] Advertising then shifted from an educational orientation focused on product merits to creating desire through emotions. This was accomplished "through a heavy emphasis on illustration" linked to "visual clichés" and "abstract types" of people with whom consumers were to identify. Initially, painting and drawing were preferred media, offering idealized reflections rather than "images of literal reality" (Brown 2005, 168, 216). One subgenre was called "goodwill advertising." Figure 4, *The Light That Shines for All*, by E. R. Squibb and Sons (1936), is an exemplar. It "promoted positive views of medical science and the medical profession" (Tomes 2001, 532) in the very ways that Jordanova (1998; cf. Helfand 2002) noted became normative.

But the rise of medicine was not only about physicians, patients, and technologies; it was also about place. The photograph in figure 5 is of the Los Angeles County–University of Southern California Medical Center, which opened in 1933.[16] Such edifices were characteristic of the rise of medicine healthscape, vividly capturing its grandiosity and the "emergence of the hospital as a citadel of health care, a medical skyscraper" (Sloane and Sloane 2003, 29).[17]

Pictures of new hospitals routinely appeared as they became, for the first time, places from which one could emerge not only alive but also on the road to recovery, places where middle- and even upper-class people now went for help. From tiny community hospitals to urban behemoths, these images chronicled the major change in locus of medical care delivery in human history. Hospitals became "the new cathedrals" of urban America. However, deep class- and race-based inequalities meant that some Americans could not afford such "white palaces."[18]

During the rise of medicine healthscape, the potent "other" form of medicine was homeopathy. Developed by Samuel Hahnemann in Germany in the eighteenth century, homeopathy is based on ingesting highly

THE LIGHT THAT SHINES FOR ALL

Six centuries ago, the Black Death reached Europe in its march around the world. It killed one-quarter of the population. Into the same wide grave went saint and sinner, rich and poor, alike.

Today, the Black Death slumbers. Its legions are as terrible as ever, but they march no more through lands where medical science stands on guard.

Man's ancient enemy, disease, is everywhere. But now his ranks are thinning, before the light of medical science —the light that shines for all.

For saint or sinner, rich or poor—this bright white light

burns on. Brighter and brighter it grows, piercing the farther darkness, lighting the way to health and happiness.

In its radiance, workers in chemical and biological laboratories have learned to make products for the preservation of health and the relief of suffering, products of uniform effectiveness and safety; and found the way to bring these products, through modern medical service, within the reach of all.

The House of Squibb is dedicated to the service of scientific medicine. We shall go on working with the medical profession, guardian of the public health.

E · R · SQUIBB & SONS

MANUFACTURING CHEMISTS TO THE MEDICAL PROFESSION SINCE 1858

THE PRICELESS INGREDIENT OF EVERY PRODUCT IS THE HONOR AND INTEGRITY OF ITS MAKER

4 *The Light That Shines for All.* Advertisement for E. R. Squibb and Sons, 1936. Medicine and Madison Avenue On-line Project ad MM0215. John W. Hartman Center for Sales, Advertising, and Marketing History. Rare Book, Manuscript, and Special Collections Library, Duke University, Durham, N.C.

5 Los Angeles County–University of Southern California Medical Center. Courtesy of the
University of Southern California Archives.

diluted remedies for diseases that produce, in a healthy person, symptoms
similar to those of the disease (e.g., Rogers 1998). A photo of a homeopathic pharmacy in Philadelphia around 1890 shows the great range of
herbal remedies used in this approach (figure 6). Practitioners, clinics, and
hospitals were found throughout Europe and North America (Porter 1985,
114). Kirschmann (2004) describes the rise, fall, and more recent second
rise and transformation of this approach. Today many such remedies are
found in health food stores rather than drugstores.

Last, in figure 7 we see an image related to an early American effort
of resistance against medicine as solely a private, fee-for-service institution and against governmental refusal to help citizens to be healthy—the
Sheppard-Towner Act of 1921. Riding international war relief efforts aimed
at children and U.S. feminists' success in obtaining the vote for women in
1920, the act framed new initiatives for federal support of the health of children and mothers. Consistent with "maternalist" goals of women's movements, reducing high U.S. infant and maternal mortality rates was a particular focus. Here Mrs. Anna Grosser of the U.S. Children's Bureau stands
with a chart showing the relationship of infant mortality rates to father's

6 *Homeopathic Pharmacy, Philadelphia*, ca. 1890. Archives and Special Collections on Women in Medicine and Homeopathy. Hahnemann Collection, Drexel University College of Medicine.

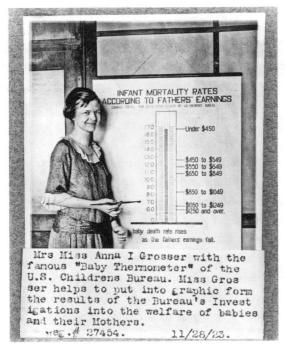

INFANT MORTALITY RATES
ACCORDING TO FATHERS' EARNINGS

170
160 — Under $450
150
140
130 — $450 to $549
120 — $550 to $649
110 — $650 to $849
100
90
80 — $850 to $1049
70 — $1050 to $1249
60 — $1250 and over.

baby death rate rises
as the fathers' earnings fall.

Mrs Miss Anna I Grosser with the famous "Baby Thermoneter" of the U.S. Childrens Bureau. Miss Grosser helps to put into graphic form the results of the Bureau's Investigations into the welfare of babies and their Mothers.
neg.# 27454.   11/26/23.

7 Mrs. Anna Grosser of the U.S. Children's Bureau. U.S. Library of Congress Prints and Photographs Division, Washington.

earnings—that is, social class. Sheppard-Towner authorized federal aid to states for maternal and child health and welfare programs starting in 1921. However, the AMA and other medical groups opposed the act vigorously, and funding ended in 1929. Instead the AMA promoted routine private pay care. Skocpol (1992, 512–19) characterizes this as vital to its antifeminist and anti-state-sponsored medicine strategies. No further federal programs were implemented until the New Deal, and they are still challenged today despite continued high rates of infant mortality.[19]

The Medicalization Healthscape

> Once circumscribed and concretized, the new thing becomes a magnet
> for intense interest. . . . It is richly evocative; it is eloquent.
> —LORRAINE DASTON, "Speechless"

The medicalization era got under way as World War II ended. Medicine as networks of heterogeneous institutions was coalescing and integrating, especially economically and infrastructurally, expanding from the provision of "emergency services" for illness or trauma to include the normalization of medical care such as annual checkups (e.g., Davis 1981). Having a "family doctor" and "going to the doctor" became routine for those who could afford it. The tremendous success of penicillin and an explosion of other pill-based treatments quickly placed "medicine in a container" in people's hands over the next decades (Goodman 2003). Prescriptions from physicians were not necessary for access to most new drugs until Congress passed legislation in 1951 (Tone and Watkins 2007, 3).

Federal funding through the GI Bill and the Hill-Burton Act of 1946 supported the expansion of medical schools and the building of new hospitals. Employer-supported health insurance and the Social Security Act of 1965 creating Medicare and Medicaid provided many Americans with access to professional healthcare for the first time by "covering" it. During this era, medical jurisdiction also extended into new domains of daily life. Simultaneously, the major new medium of visual culture came into common use in the U.S.—television. Both television and medicine then became part and parcel of everyday living—integrated, naturalized, co-constitutive—during the deeply transformative medicalization healthscape.

Extending earlier patterns, popular books on medical topics were often made into movies and television programs, key sites of imaging things medical.[20] The first *Dr. Kildare* book appeared in 1937, and by 1970 there were fifteen films, seven books, one radio program, and two television programs featuring him; clothing, jewelry, children's games and toys, comic books, comic strips, and paperbacks extended the Kildare mystique (Lederer and Rogers 2003, 498–99). The M*A*S*H novels came out in the 1960s, followed by Robert Altman's film and the television series, militarizing medical media. Robin Cook's *Coma* and *Women in White* also appeared (Jones 1981, 184–85, 188). In terms of sites of practice, both medical offices and hospitals were commonly imaged (e.g., Feldman and Douplitzky 2002; Karpf 1988).

Interestingly, as medical sociology discovered the patient in the 1960s and 1970s (e.g., Strauss and Glaser 1964), so too did Broadway. In what Gussow (1979) called "The Time of the Wounded Hero," successful plays featured deformity, disability, serious illness, and prolonged dying. Jones (1981, 193) asserted that "ironically, the technological progress of medicine has shifted the dramatic focus from the doctor to the patient who is the new hero." Heroic or not, patients were certainly increasingly medicalized before our very eyes. What were the cultural consequences of all this? Basala (1976, 276) argued for a "correlation between the image of science in popular culture and public understanding of science or public willingness to support scientific education and research." He found that by the 1970s, while science and scientists were still commonly imaged as isolated and antisocial, clinical medicine was often imaged as positive and humane (see also Frayling 2005; Turow 1989)

New forms of medical enhancement also appeared. After the improvement of plastic surgery through war wound repair, it began to be used for cosmetic purposes. A popular autobiography titled *Doctor Pygmalion* (1954) hailed the remaking of noses, bodies, and lives (Gilman 1995). In one of the most gendered domains of medicine, then as now, most of the surgeons were male and patients female. Perhaps the main arena of action on the ever-moving boundary between treatment and enhancement was female hormone replacement therapy (HRT). Initiated and advertised in the 1930s, HRT blossomed in visual and other media for the rest of the twentieth century, including a book titled *Feminine Forever* (Wilson 1966), which went through multiple editions (Houck 2006; Watkins 2007).

In sharp contrast with medicine generally, psychiatry and mental hospitals did not fare well in the media. The book *The Snake Pit* (1945) became a movie in 1948 portraying straitjackets and shock therapies, and was sometimes banned (Lederer and Rogers 2003, 497). The now classic and still horrifying *King of Hearts* and Ken Kesey's *One Flew over the Cuckoo's Nest* appeared in the 1960s (Jones 1981, 193). And Frederick Wiseman made an important muckraking documentary on a Massachusetts hospital for the criminally insane, *Titticut Follies*.[21] Across all of these, institutionalization was portrayed as a very undesirable form of medicalization indeed, as a patient-led antipsychiatry movement was simultaneously demonstrating (Morrison 2005). Newspaper coverage of mental illness continued to link it to violence and criminality (Wahl 2004).

In the medicalization era, "the ideal of bodily transparency" was further elaborated. In 1966 the film *Fantastic Voyage* offered an early incarnation of what Sawchuk (2000, 10) intriguingly calls "biotourism of sublime inner space" — "the persistent cultural fantasy that one can travel through the inner body, a bodyscape which is 'spatialized' and given definable geographic contours." It offers a new subjectivity for the medicalized subject — the biotourist. A parallel and fascinating change during this era appeared in ads for psychotropic medications. Metzl's (2004, 18) sophisticated analysis portrays a shift from the dominance of images of psychiatrists and a language of psychoanalysis and psychotherapy to dominant images of patients and of physically much "larger than life medications" — giant pills — between 1965 and 1985. Big pharma indeed! This accompanied a shift from imaging predominantly female to mixed-sex patients, a move toward diversity more common in the next era.

A final theme in popular visual cultures of things medical during the medicalization healthscape is fear. Referring to animated cartoons focused on things medical, Cantor (2007, 1) noted that "most of these movies present a comforting message about the capacity of medical science to detect and combat dread disease. But, it is often a message built on fear." This tension is sustained into the present.

ICONOGRAPHY OF THE MEDICALIZATION ERA

As the physician's house call disappeared in the 1950s and 1960s, the major new medium of television entered the homes of most Americans, delivering images of physicians instead. Among the most famous of the doc-

8 *Marcus Welby, M.D.*, with Robert Young
and James Brolin. Photo by Gene Trindl,
Motion Picture and Television Photo
Archive.

tor shows was *Marcus Welby, M.D.*, which ran from 1969 to 1976 (figure 8;
Jones 1981, 184–85, 188). Welby's irreverent sidekick, Dr. Kiley, linked the
program to buddy Westerns. Karpf (1988, 189–91) asserts that the younger
Kiley worked as a heartthrob, the older Welby as solacing father figure,
and Welby's nurse Consuelo mothered doctors and patients alike. While
females were largely missing from the media as physicians and patients,
Karpf notes that male patients were encouraged to be more like women,
foreshadowing key arguments in the current men's health movement and
gendered health analyses (e.g., Riska 2004, this volume). At the height of
the show's popularity, "Dr. Welby" received approximately five thousand
letters per week from people seeking medical assistance; the show received
over thirty awards, including one from the AMA, to which scripts were submitted for approval in advance (Jones 1981, 183).

Several new media emerged during this era. One was the comic book,
which generated superheroes including drugs as well as Superman and
Superwoman (e.g., Hajdu 2008). Figure 9 is from *Penicillin* (*True Comics*
41, December 1944, 22, cited in Hansen 2004, 157). The potency of this
nonhuman actor is vivid as "the infection is checked" and "the little girl will
recover"; "what a story" indeed! Like the x-ray, this new category of medical technology—antibiotics—becomes mundane and remains transfor-

9 Penicillin comic, 1944. *True Comics*, no. 41 (December 1944).

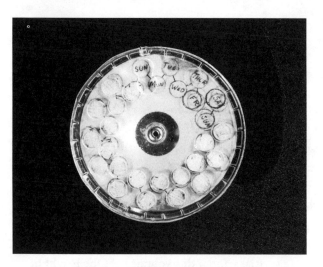

10 *Prototype Birth Control Pill Dispenser*, 1962. Courtesy of the
Smithsonian Institution, National Museum of American History.

mative. Because they make primary care more effective and surgery more
survivable, antibiotics were key actors in promoting medicalization after
World War II. "By the early 1950s, penicillin was the single most important prescription pharmaceutical; in the U.S., its sales alone accounted for
10% of the industry's total. . . . The antibiotic bandwagon began to roll. . . .
Pharmaceutical companies were the capitalist vanguard against microbes"
(Goodman 2003, 147; see also Tone and Watkins 2007).

Figure 10 features another major medical technological innovation still
echoing loudly and globally—the birth control pill. The image shows a

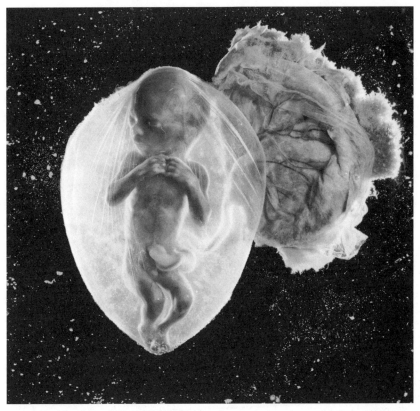

11 Lennart Nilsson, *Fetus at Eighteen Weeks*, 1965. Bonnier-Förlagen Stockholm, Agent for Lennart Nilsson.

prototype pill dispenser developed in 1962 by David P. Wagner, an early "compliance package" for prescription drugs "intended to help the patient comply with the doctor's orders" (Gossel 1999, 105).[22] This piece of material culture quickly became part of the everyday lives of millions upon millions of women as we pushed the pill out of its foil wrap and swallowed it. An equivalent male pill was developed but, remarkably, is still unavailable almost half a century later (Oudshoorn 2003).

The fourth image of the medicalization healthscape emerged as "the ideal of bodily transparency" followed Steichen's *The Family of Man* (1955) into the womb. Lennart Nilsson's famous and infamous fetal images appeared in both *Life* and *National Geographic* (Berlant 1997; Casper 1998; Cartwright 1998; van Dijck 2005). Figure 11 shows the image of a fetus that appeared on the cover of *Life* on April 30, 1965 (Newman 1996, 12). Jain

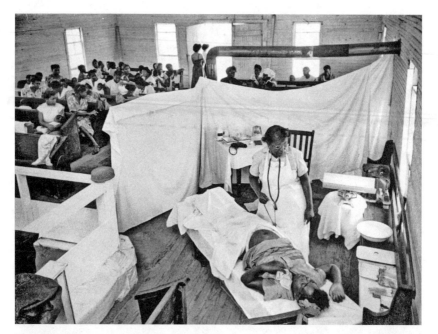

12   W. Eugene Smith, *Nurse Midwife Maude Callen Examining a Patient in a Church*, 1951.
From the photo-essay "Maude Callen: Nurse Midwife," by W. Eugene Smith. Time
and Life Pictures–Getty Images.

(1998, 373) captures it well: "Nilsson's images disrupt discursive boundaries
between science and popular media; they slide among genres of science
fiction, modernism, biological illustration, and surrealism. Each image is
in itself stunning, glistening with unabashed magnificence and visual excess." The film *2001: A Space Odyssey* (1968) used a similar visual in its final
scene, enhancing the iconic stature of the fetus as "the symbol of all mankind." By excluding both the womb and the woman-mother from sight, it
transcended the vivid visual differences that Steichen's *The Family of Man*
simultaneously documented and refuted.

After World War II, increases in the effectiveness of medicine began producing an essential tension regarding lack of access to healthcare for many
in the United States, especially poor people and those living in rural areas,
often racial and ethnic minorities. Their "plight" during and after the Great
Depression had been well documented by wPA photographers, whose
work appeared in major American newspapers and magazines.[23] Figure
12 is from W. Eugene Smith's photo-essay "Maude Callen: Nurse Midwife" in *Life* magazine on December 3, 1951 (Squiers 2005, 72–73). It shows

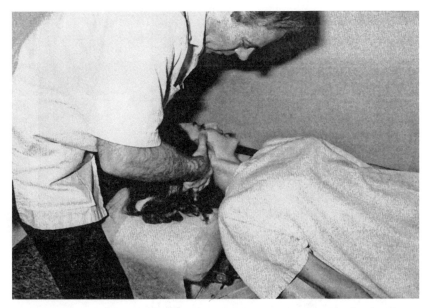

13 Chiropractic rotary atlas adjustment. Reprinted from Donald Gitelman and Bruce Fligg, "Diversified Technique," in *Principles and Practice of Chiropractic 2/E*, ed. Scott Haldeman (1992), 488. Courtesy the McGraw-Hill Companies.

Callen examining a pregnant woman in a corner of a church screened off by sheets. Others waiting to be seen are nearby. When this photo-essay appeared, advertising usually offered images of the country doctor, and Norman Rockwell was painting beloved family physicians. Yet many Americans had no physicians in their lives (Tomes 2001, 538). The poorest of the poor were fortunate in their midwives, whose actual scope of work was often far broader. The caption for this image in *Life* read: "She dreams of having a well-supplied clinic but has small hope of getting the $7,000 it might cost" (Squiers 2005, 84).

The "other" medicine on the rise during the medicalization era was chiropractic, an approach to healing by skeletal manipulation to correct dislocations and restore the nervous system to full functioning (e.g., Moore 1993). Initially developed in Davenport, Iowa, in the late nineteenth century by D. D. Palmer, the Palmer School of Chiropractic remains active.[24] Figure 13 shows a chiropractor adjusting the neck (Gitelman and Fligg 1992, 488). Today chiropractic treatment may be covered by insurance and is increasingly integrated into sports medicine (e.g., Theberge 2008).

Resistance to medical practices and to medicalization grew considerably

14 Cover of *Our Bodies, Ourselves*, 1973. Courtesy of the Boston Women's Health Book Collective.

across the medicalization era, coalescing into almost oppositional health-scapes (Schneirov and Geczik 2003; see also the discussion of counter-trends in the introduction). The feminist women's health movements formed the major such organizations, objecting to the overmedicalization of women's bodies. Figure 14 is deeply emblematic—the cover of the Boston Women's Health Book Collective's *Our Bodies, Ourselves* 1973 edition. First published on newsprint in 1970, the book broke new ground, providing women with all kinds of information about their bodies in health and disease. It has gone through five editions (plus an online version), sold over four million copies, and been translated and adapted into twenty-two languages.[25] The diagrams of female genital anatomy were and remain among the most radical and feminist ever published (Moore and Clarke 2001). At about the same time, the Black Panthers (Nelson, forthcoming) and the Grey Panthers (Kuhn 1991) were organizing around black and aging issues.

Both movements made access to healthcare and care that addressed their specific needs and goals important priorities. Both also provided food and health services at the community level.

One facet of the medicalization healthscape often overlooked is the development of the "self-help health and lifestyle" publishing sector. Of course, *Our Bodies, Ourselves* helped initiate this trend. Before the 1970s, aside from home remedies historically found at the back of cookbooks, most health and medical information dwelled solely in publicly inaccessible libraries of medical institutions. Today such books sell in numbers comparable to the romantic fiction market (Marchessault and Sawchuk 2000, 1). Further, the development of a wide-ranging "patient social movements movement" since the 1970s (Epstein 2007, 2008) and the vast proliferation of Internet health resources since the 1990s grew out of the legitimation of access to medical information for nonprofessionals (e.g., National Research Council 2000; Oudshoorn and Somers 2007).

The Biomedicalization Healthscape

> If things are 'speechless,' perhaps it is because they are drowned out by all the talk about them . . . showing how variously things knot together matter and meaning. — LORRAINE DASTON, "Speechless"

The biomedicalization healthscape emerged in the early 1980s, based in several shifts. First, computer and information sciences were becoming infrastructural for biomedicine — in epidemiology, molecular biology, and genomics, in diagnostic technologies, and in organizing delivery and payment for services via HMOs and third parties, both private and state. The corporatization (especially "managed care") and technoscientification (both infrastructural and clinical) that began to reorganize biomedicine from the inside out in the early 1980s led, according to some analysts, to the "end of the golden age of doctoring" (McKinlay and Marceau 2002).

The biomedicalization healthscape boldly and predictably features technoscience, its interventions, and its promises for the future.[26] The new millennium hailed the "century of biology," following the twentieth "century of physics" (nuclear bombs and energy). Various media hailed us with announcements of "breaking the genetic code" and finding the Holy Grail through the Human Genome Project. Special issues of *Time* and *Newsweek* were devoted to them, and docudramatizations such as David Suzuki's *The Secret of Life* also appeared (Marchessault 2000). Less-dramatic manifestations of biomedicalization such as "medicine in a bottle" also proliferated: more than $200 billion was spent in the United States in 2005 for prescription pills (Tone and Watkins 2007, 1).

On television during the biomedicalization healthscape: "Modern technology, featured prominently on realistic sets, allows viewers to identify elements from their own experiences with hospitals . . . [and] the near-miraculous abilities of TV doctors represent the current technomedical expectations of the American health system . . . *medical care to fit a public idealization*" (Cohen and Shafer 2004, 213; italics mine). *House, M.D.* (2004) focuses on using the scientific method in challenging medical diagnostic situations (parallel to the narrative series in the *New York Times Sunday Magazine*). "*House* is the ultimate expression of modernist medicine—where there is always a right answer to be found despite the fact that 'all patients lie'" (Strauman and Goodier 2008, 131). *Heartland* (2007) centers on extreme biomedicalization—organ transplant surgeries and the recruitment of donors (Bellafante 2007). In an era of organ shortages, the infomercial aspects of the show for the significant medical business sector of transplantation ramify.

In contrast, the popular *Grey's Anatomy* positions itself far from the bureaucratic and clinical, focusing instead on the personal and professional lives of surgical interns and their supervisors portrayed by a racially diverse cast (Strauman and Goodier 2008, 129). *Nip/Tuck* too is a long way from *Marcus Welby, M.D.* Set in a south Florida plastic surgery center, it "glorifies the profitability and corruption of plastic surgery" (Strauman and Goodier 2008, 130). These reinvented doctor/hospital programs are augmented by "reality medical TV," including documentaries and live coverage of various surgeries from "life or death" dramas to cosmetic transformations. The latest incarnation of "med TV" is the highly successful *The Doctors*, in which real doctors respond to patients' phoned-in or e-mailed questions. This is biomedicalization writ large, offered in presentational formats that work effectively as infomercials and consults. Medical educators have proposed using episodes of *House* in class for "teaching of 'soft skills' in medicine" such as ethics, interaction, and professionalism (Lim and Seet 2008, 193). Strauman and Goodier (2008, 131) conclude that "whether the shows offer a positive or negative image, or a mingling of both, they can and do offer lessons to an ever widening audience of consumers."

These medical dramas are "popular around the world: *ER, Dr. Quinn Medicine Woman, Chicago Hope, L.A. Doctors* (U.S.), *Urgence* (Quebec), *Casualty* (U.K.), and the surrealistic Danish series *The Kingdom*, to name but a few" (Marchessault and Sawchuk 2000, 1). The entertainment industry accounts for 6.5 percent of the U.S. GDP, and its products travel widely (Benson 2008, E2).

Going far beyond the traditional genre of police procedurals, a number of popular mystery novels (Patricia Cornwell's are perhaps most famous) and a plethora of television shows have focused on forensic science and medicine, dead bodies, and "secrets of the dead." Here the technoscience of identity is very much at the fore (see also Timmermans 2006). The *New York Times*, for example, has noted a "*CSI* effect" in the courtroom, when juries become "overimpressed" by forensic evidence; today prospective jurors may even be asked about their television-watching habits (Santos 2007, 7).

Another dramatic change in the biomedicalization healthscape involves patients representing themselves qua patients publicly — in art, photography, and nonfiction — taking multiple positions vis-à-vis the value of modern biomedicine, including contributing to oppositional healthscapes. For example, producing what Cartwright (2000a) calls "the public body," Jo Spence photographed her own trajectory of breast cancer diagnosis, treatment, and dying in the hospital (Bell 2002; Radley and Bell 2007; see also Willis 2005). A website offered women's experiences of cancer, often excluding all things medical to be "read as a defiance and transgression of the medicalization of women's bodies under the sign of cancer" (Alice 1996, cited in Radley 2002, 1). A bare-breasted, single-breasted woman on the cover of the *New York Times Magazine* in 1993 created a major stir, and many such women are portrayed in a collection titled *Art.Rage.Us* (Breast Cancer Fund of America 1998; Cartwright 1998, 126–27). One does not need to be ill to purchase personalized artwork based on one's own DNA.[27]

There is also a new media focus on the biomedical experiences of celebrity patients. Lerner (2006) argues that this phenomenon is actually changing how we look at medicine. One instance was Angelina Jolie's cesarean birth without medical indication with physicians and high-tech equipment flown in to operate on her in Africa. This example sits on the shifting boundary between the normal practice of medicine and enhancement or optimization of life through medicine (e.g., Block 2007). Celebrities are also paid not only for advertising particular medical products or even hospitals (like athletes for sports gear and designer items) but also for congressional lobbying efforts on behalf of particular diseases. Commoditization is everywhere.

In the biomedicalization era, the visual "ideal of transparency" extended clinically by the endoscope during the medicalization era is further extended by video, digitalization, and robotics. Telemedicine renders patients visible across long distances (e.g., Cartwright 2000b), and remote surgery is guided by living experts but executed by robots on site (Zetka 2003). Brain scans are increasingly interpreted as images of particular kinds of personhood (Beaulieu 2000; Burri and Dumit 2007; Dumit 2004; Joyce 2005a, this volume). Digital anatomies are produced for home and school consumption (often reinforcing highly traditional gender orders) (e.g., Moore and Clarke 2001). Children are also taken on biotourism trips in books like *The Magic School Bus inside the Human Body* (Cole 1990), some of which are made into television programs and rerun frequently.[28]

Digital cadavers allow virtual dissection by medical students and others (e.g., van Dijck 2005; Waldby 2000), while controversial privately owned exhibits of plastinated cadavers are traveling the world, riveting a wide variety of audiences (e.g., van Dijck 2005, 41–63). Here the bodies of (mostly?) Chinese people have been transformed through a process of plastinating different tissue and organ systems differently and positioning the cadavers to reveal different visual stories of embodiment. One such blockbuster show, "Body Worlds," has attracted over twenty million people worldwide since 1995 and has taken in over $200 million; the bodies for the show are produced in factories in Dailan, China (Barboza 2006, 2). Connor (2007, 852) sees this as "the reinvention of anatomical spectacle," asserting that "for both professional and lay audiences, now and in the past, bodies attract." While controversial in Europe, in the United States the show is "an homage to the wonders of the human body, a wonderful educational vehicle and an admonition to healthful living" (Schulte-Sasse 2006, 369). Such exhibits particularly appeal to nonphysician and alternative providers with little access to cadaver-based learning.

Websites are major sources of information and imagery—a new medium of popular culture of things medical (e.g., Oudshoorn and Somers 2007; National Research Council 2000). Reproductive technologies are featured with images of in vitro fertilization and infomercials on service providers.[29] During the biomedicalization era, new forms of museum and related public exhibits of (bio)medicine have also expanded, becoming much more a part of popular culture, rather than high culture or the hobby domain of older doctors. The San Francisco Exploratorium Special Exhibit "Revealing Bodies" presented early carved-ivory and other disassembleable bodies on view in European museums.[30] The Science Museum of London mounted a major multimedia exhibit called *Health Matters* (Bud, Finn, and Trishler 1999). Their display of a computer as a key post–World War II technology provoked some of our original analyses of biomedicalization. The U.S. National Library of Medicine has mounted a number of more accessible popular-culture-linked exhibits online as well as in Washington.[31]

Last, the cross-cutting theme of fear in popular and visual cultures of things medical is maintained, if not expanding, during this era. Altheide (2002) argues that the media produce a "discourse of fear"—a vigilant awareness that danger and risk lurk at every turn. The AIDS epidemic has been emblematic. Todd Haynes's film *Safe* (1995) portrayed horrifying symptoms of environmental illness (Reid 1998). The film *Gods and Monsters* (1998) revisited Frankenstein lore. The movie *Dirty Pretty Things* (2002) portrayed illegal kidney harvesting among the poor in London. And the influence of anxiety on health through media discourses of September 11 and bioterrorism is thematic (King 2003).

ICONOGRAPHY OF THE BIOMEDICALIZATION ERA

In ways and to an extent that surprised even me, who has argued for years both that a shift to biomedicalization occurred and that direct applications of technoscientific interventions are made earlier and faster, such changes are vivid in the biomedicalization healthscape. The central visual focus is neither physicians nor patients but the procedures and technologies per se, orchestrated by people in multiple positions in hospitals or other high-tech venues. The shift away from naming programs after particular doctors, so common earlier, appears linked to what McKinlay and Marceau (2002) describe as "the end of the golden age of doctoring"—the shift to corporate and technoscientific dominance in medicine that we call biomedicalization.

Distinctively American yet circulating globally, the television show *ER* established medicine per se as pop culture icon (figure 15). *ER*, created by Michael Crichton in the 1990s, had about twenty-four million viewers each week and featured super-high-tech interventions. Analyses of

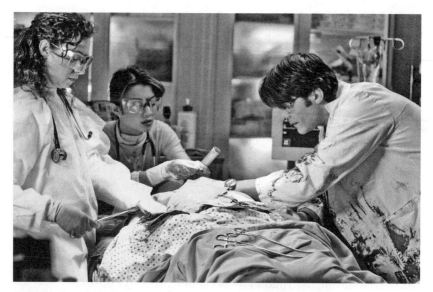

15 Still from the television program *ER* (episode 1401, "The War Comes Home").
ER © Warner Bros. Entertainment Inc. All rights reserved.

this and other programs revealed "unrealistically" high success rates for
CPR and other radical feats, as well as increasingly obscure illnesses (Lederer and Rogers 2003, 499–500; see also Timmermans 1999). One study
found that 15 percent of viewers surveyed had contacted a physician about
an ailment they had seen on the show (Brodie et al. 2001). A CDC study
in 2000 reported that more than half the Americans who regularly watch
prime-time TV said that they relied on medical dramas to present accurate
health information.[32] Faintly echoing the public-health films of the 1920s,
ER in the 1990s featured HIV testing, broadcasting this possibility across
the sixty-six countries where it aired (Mensah 2000, 139).

Direct-to-consumer (DTC) advertising of prescription drugs and procedures has produced the major visual cultural change vis-à-vis things medical in the biomedicalization healthscape, dramatically widening distribution of biomedical knowledge. Figure 16 is a DTC ad for Viagra.[33] Tomes
(2001, 546) notes that "the Food and Drug Administration justified its 1997
decision to loosen restraints on direct-to-consumer advertising of prescription drugs on the grounds that such advertising helps acquaint patients
with the best treatments available." DTC ads now appear widely. A recent
study of news coverage of DTC pharmaceutical ads found that the corpora-

tions selling those products were portrayed more prominently in the news than providers, consumers, or the state; not surprisingly, DTC critics in particular were minimally represented (Hartley and Coleman 2007, 107).

Viagra, a prescription pill for the treatment of male impotence, first appeared in 1998 and is now the most recognized brand name in the United States after Coca-Cola (Baglia 2005, 2). After the experiments with testes transplants noted earlier, promotion of male treatments had waned until Viagra appeared (Pfeffer 1985; Fishman 2004). While still widely used, it has also been noted that not all prescriptions are refilled (Berenson 2005), and fulfilling the male sex role through pharmaceutical assistance may also be dangerous to your health (Baglia 2005, 3–4; see also Mamo and Fishman 2001). The search for a drug for female sexual dysfunction has been more complicated and less successful (Fishman 2004). We will know instantly of its availability.

The second image that captures a key change in biomedical knowledge distribution is the material cultural object of a computer being used to research "my disease" (figure 17) or find a clinical trial on the Internet (e.g., http://clinicaltrials.gov). The Internet itself has become iconographic vis-à-vis accessing medical information. It frames knowledge profoundly and constructs patients in the process (e.g., National Research Council 2000; Oudshoorn and Somers 2007). Both images are very much part of the biomedicalization healthscape, making information that comes through such media appear extremely modern and up-to-date (Bolter and Grusin 1999), although it may well be "old wine in new bottles" (Forsythe 2001). The very mundanity of these images demonstrates the pervasiveness of biomedicalization.

A major intersection of consumption and clinical medical imaging is the use of ultrasound to produce the "visible fetus" (e.g., Blizzard 2007; Taylor 1992). Instead of waiting until birth to pass out cigars, today American dads-to-be are passing around ultrasound images as "the first baby pictures" (figure 18). Fetal ultrasounds follow from Nilsson's images from the 1950s, shifting the icon from the generalized fetus to the personalized family fetal snapshot. "Embodied quickening has been replaced with a kind of 'technological quickening,'" a family media event with fathers' active participation (Draper 2002, 781). Such "entertainment ultrasound," advertised in magazines for parents, is now discouraged by the American Institute of Ultrasound Medicine (Schmid 2006).[34] There is also a medical

16 Online Viagra advertisement. © Pfizer, Pfizer
website, October 1, 2008.

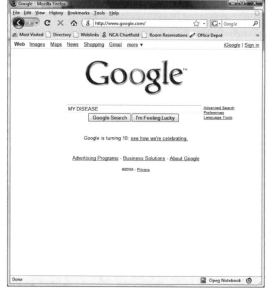

17 Googling "My Disease."
© 2008 Google, Google
website, October 1, 2008.

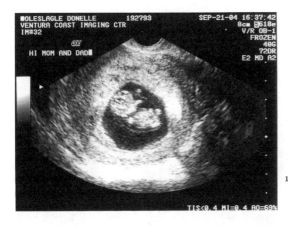

18   Fetal ultrasound image, 2007. Courtesy of the Woleslagle family.

critique of "boutique fetal imaging," arguing for exclusive control by physicians to eliminate lay access (Chervenak and McCullough 2005). The political implications of this fetal visual culture for debates around stem cells and abortion are vast (e.g., Draper 2002; van Dijck 2005).

The siting of things medical has also been changing. For a century, medical practices were traditionally sited separately and often within stand-alone entities—the doctor's office, the clinic, and the hospital. In contrast, within the biomedicalization healthscape, medicine moves to the mall, as well (Sloane and Sloane 2003). Bodily imaging technologies, outpatient surgery, medical and dental offices, and the like are increasingly visually and organizationally "just another" site of consumption, using savvy consumers' knowledge of how to maneuver in such places to facilitate access to medical care. Figure 19 shows the Westwood Surgery Center in Los Angeles in 1988, tucked in with a copy center and fast-food place (Sloane and Sloane 2003, 133). More recently, "docs in a box" are offering drop-in services inside mega-chain stores like Wal-Mart (Freudenheim 2006). The siting of things medical in the biomedicalization healthscape often clarifies their commodity status.

During this era, "other" forms of healing and medicines have also elaborated, grown tremendously, and to some degree are being integrated into/ subjecting themselves to the research and testing of biomedicine to facilitate that integration.[35] Known today generally as "complementary and alternative medicine(s) (CAMs) (Hess 2005), these range from chiropractic (now covered by many health insurance plans) to acupuncture (covered by fewer), to Ayurvedism and New Age approaches (not usually covered).

19 Westwood Surgery Center at the Mall, 1988. Reprinted from David Charles Sloane and
    Beverlie Conant Sloane, *Medicine Moves to the Mall* (Baltimore: Johns Hopkins University
    Press, 2003), 133. © 2002 David Charles Sloane and Beverlie Conant Sloane. Reprinted
    with permission of the Johns Hopkins University Press.

Figure 20 shows the cover of a recent book on clinical acupuncture (Stux
and Hammerschlag 2001) with contributions from Germans, Asians, and
Americans. Acupuncture has become such a part of American life that an
exhibition, "Acupuncture: A Transnational Tale of Medical History," was
mounted at the National Museum of American History in 2002.[36]

Last, organized resistance against particular forms of biomedical practice has also grown and elaborated. In this era in the United States, two
domains stand out: breast cancer activism (discussed earlier) and the AIDS
movement.[37] Figure 21 shows Keith Haring's famous AIDS poster from
1989.[38] Haring, along with other gay activists of the era, used a pink triangle
and the phrase "silence equals death" to invoke Nazi sentencing of homo-
sexuals along with Jews, gypsies, and others to concentration camps during
World War II. The grassroots organization ACT UP successfully advocated
speaking out, acting up, and intervening in NIH clinical trials guidelines
(Epstein 1996). This increasingly transnational patient movement has generally advocated more and better biomedicalization and expanded patient
roles in shaping both AIDS policies and their own care (Patton 2002). In

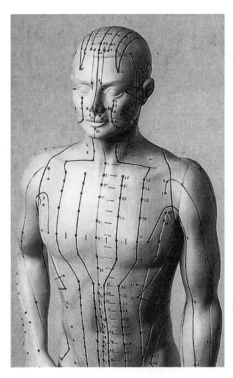

20 Front cover of *Clinical Acupuncture: Scientific Basis* (2001). Gabriel Stux and Richard Hammerschlag, editors. Reprinted with kind permission of Springer Science and Business Media.

sum, the biomedicalization healthscape is visually, politically, and biomedically intense, active, lively, involved, and very high-tech.

Conclusions

> Even if they do not whisper and shout, these things press their messages on attentive auditors—many messages, delicately adjusted to context, revelatory, and right on target.—LORRAINE DASTON, "Speechless"

Elaborating Appadurai's (1996) metaphor of "scapes" into "healthscapes" allows us to focus on all kinds of things medical as forming assemblages, infrastructures of assumptions as well as people, things, places, images. Synthesizing these into healthscapes of particular historical eras not only demonstrates their flows across time and space but also elaborates how they are generated by and through cumulative changes incorporated into each period. Their shape-shifting properties become more apparent, and we can see how they capture the era in which they appear.

21 Keith Haring, *Untitled*, 1989. © The Estate of Keith Haring.

Overall, the rise of medicine healthscape emphasized the authority of physicians, the practices of surgeries, and the institutions of hospitals—each as an increasingly successful and powerful site of medical care provision. During the medicalization healthscape, new social actors emerged as visible participants. Drugs, patients, and heterogeneous professional providers were incorporated into popular visual cultures of things medical as medicine became part of routine everyday life—not for emergencies only. Last, in the contemporary biomedicalization healthscape, the spotlight focuses on high-tech interventions performed by teams in complex biomedical settings. Patients as consumers face multiple networks of dispersed knowledges and new decision-making responsibilities. The cultural pervasiveness of things medical and the normalization of medicalization of both health and illnesses are palpable in new ways. These clearly fuel ongoing biomedicalization and the production of the innovative sciences and technologies that it requires.

In Jordanova's words (1998, 216): "These examples from popular culture are testimony to the success, not just over decades but over centuries, of sustained attempts to cultivate certain medical qualities, such as benevolence, to externalize them, so that they could assume an apparently independent life of their own, and then, once separate, be used to reinforce the original feeling. Here medical display gives shape to fantasies of power and authority, and actually helps mediate their acquisition." The three healthscapes vividly demonstrate how in the U.S. (bio)medicine has become a public "cultural good" like public education. The shifts of emphasis across the three healthscapes also underscore how healthcare and things medical more broadly are increasingly in our neoliberal times consumer "goods" (e.g., Briggs and Hallin 2007).

Biomedicalization is, however, not without its critics and sites of objection, rejection, and resistance. I have only been able to hint at what else was also always going on—from alternative approaches to healing and health promotion to explicit and organized challenges to (bio)medicalization. Across the three eras, these countertrends are growing in momentum into social-movement-like formations offering a mélange of alternative, anti-consumerist, and green healthscapes. Here the transformations of identity involved are not technoscientific but reflective, ecological, spiritual, and sometimes even fundamentalist religious (e.g., Schneirov and Geczik 2003), manifesting active alternative forms of "care of the self" (Foucault 1988). These sometimes wildly heterogeneous elements are already of considerable economic magnitude. But they are not necessarily organized into conventional oppositional institutional forms (such as directly confronting the NIH or HMOs). Instead many are rhizomatic—widespread and webbed loosely together in the invisible practices of daily life from food purchasing to exercise to recycling as healthy living that together constitute possible sustainabilities. They link issues of environmentalism, the public good, and "appropriate" (bio)medicalization in new ways and may—or may not— become more politically consequential.[39]

Yet as we (both as scholars and as patients/consumers) "mediate between, on the one hand, the distrust of high technology expressed by some cultural critics analyzing institutional fields such as medicine, and on the other hand the utopian embrace of new technologies as unproblematic tools of social transformation," we also need to keep in mind the "increasing disparities in access to them" (Cartwright 1998, 219). The startling unevenness of the rise of medicine, medicalization, and biomedicalization across the United States led us years ago to analyze "stratified (bio)medicalization" (Clarke et al. 2003). It is still understudied, still deserving of our attention.

The cultural, political, and other economies revealed through the healthscapes of the rise of medicine, medicalization, and biomedicalization in the United States also clearly have implications for the transnationalization or globalization of biomedicine. The institutionalized legitimacies of (bio)medicalization flow along digital and other infrastructures of popular cultural discourses already in place, easily capable of and already organizing things medical in faraway places. The framework offered here may be useful in conceptualizing empirical studies of healthscapes elsewhere, and the epilogue to this volume discusses these research possibilities. Clearly, as Cartwright has argued (1998, 220), we need to attend more carefully to "the centrality of visual media in medicine's cultural politics in the public sphere" wherever we are looking.

Notes

I thank a number of colleagues for valuable feedback: Vincanne Adams, Warwick Anderson, Monica Casper, Steve Epstein, Anne Fausto-Sterling, Carrie Friese, Laura Harkewicz, Jennifer Harrington, April Huff, Laura Mamo, Chandra Mukerji, Joao Arriscado Nunez, Robert Proctor, Vololona Rabeharisoa, Joao Ramalho Santos, Carlos Sonnenschein, Ana Soto, Keith Wailoo, Mary Zimmerman, and the amazing anonymous reviewers for Duke University Press. Many ideas that undergird this chapter emerged during my postdoc in sociology at Stanford with Richard Scott (e.g., Clarke 1988). Versions of this paper were presented at the following: the conference "What's Left of Life" (UC Berkeley); the Peter Hall Lecture, Carl Couch Symposium, Midwest Sociological Association; "Healthscapes and Body States" workshop, UC San Diego; East Asian Science, Technology, and Society Journal (EASTS) Conference, Taipei; Workshop on Life Sciences and Society: Biomedicine, University of Coimbra, Portugal; Seminar on Science, Technology, and Society, Stanford University; Foundation Lecture, School of Social Sciences, Cardiff University; and the Third U.S.-U.K. Medical Sociology Conference, Boston. Brandee Woleslagle provided invaluable assistance, preserving my sanity against all odds.

1 On why historians ignore images, see Clarke 2005, chap. 6; and Gilman 1995. See also Karpf 1988; Lederer and Rogers 2003; and Seale 2004.

2 Complex ongoing debates about how and why images "operate" on, in, and through us are far beyond the scope of this chapter (see, e.g., Daston 2004; Stafford 2007).

3 Other new "scapes" are also in circulation, e.g., Cusack 2007 on "riverscapes" and formations of national identity (the Seine, the Jordan). Sloane and Sloane's (2003, 5) project was provocative.

4 Starr's work has been ambitiously revisited in a special issue of the *Journal of Health Politics, Policy, and Law* 29, nos. 4–5 (2004).

5 I am indebted here to Mary Zimmerman's discussant's comments on my Peter Hall Lecture (2007). See also Jordanova 1998 and Hansen 2009.

6 There is an increasingly ambitious literature, mostly published since 1990: Bud,

Finn, and Trischler 1999; Carmichael and Ratzan 1991; Cartwright 1995; Cohen and Schafer 2004; Gilman 1995; Golden and Rosenberg 1991; Hansen 2009; Harrison 2002; Harrison and Aranda 1999; Jones 1981; Jordanova 1998; Marchessault and Sawchuk 2000; Porter 1985, 1999; Radley 2002; Reagan, Tomes, and Treichler 2007; Squiers 2005; Treichler, Cartwright, and Penley 1998; van Dijck 2005; and Waldby 2000.

7   See also the extensive chart in chapter 2 of this volume. Helpful discussion of our periodization occurred at the conference at UC San Diego with Keith Wailoo, Laura Harkewicz, April Huff, and Steve Epstein.

8   See, e.g., Gevitz 1988; Johnson 2004b; Risse et al. 1977; and Rosenberg 2007, chap. 7. On medical pluralism, see the epilogue in this volume.

9   See, e.g., Brown 1979; Risse 1999; Rosenberg 1995, 2007; Starr 1982, n. 5; and http://www.ama-assn.org/ama/pub/category/1923.html.

10 In contrast, Henderson and Petersen (2002) assert that commodification is very recent.

11 On advertising and consumption across these eras, especially but not only of things medical, see Apple 1996; Brown 2005; Hansen 2009; Helfand 2002; Laird 1998; and Williams 1980.

12 On visual imagery and the rise of medicine era in the United States, see especially Hansen 1997, 1999, 2004, 2009; Grey 2002; and Tomes 1998, 2001.

13 See also Emery and Emery 2003, 66–67; on elite private practice at the time (including visuals), see Crenner 2005.

14 See http://www.nlm.nih.gov/hmd/pdf/national.pdf (p. 20). Sloan was part of the Ashcan School of New York artists (Schiller et al. 2007).

15 On Bernays and related developments, see Tye 1998 and http://www.prmuseum.com/bernays/bernays_1915.html. The excellent documentary *The Century of the Self* also features Bernays.

16 Special thanks to Claude Zachary of the University Archives, University of Southern California.

17 Opened in 1933 and located at 1200 North State Street, Los Angeles, its Art Deco main building was the exterior of the hospital in the soap opera *General Hospital* and was used in the 1998 movie *City of Angels*. See http://www.usc.edu/about/visit/hsc. Thanks to Jon Weiner, USC Health Sciences Media Relations (Personal communication, February 14, 2008).

18 On hospital history see, e.g., Risse 1999; Rosenberg 1995; Sloane and Sloane 2003. Earlier hospitals tended to be small local religious or ethnic institutions. Their racial integration was required by the reimbursement regulations for receipt of Medicare and Medicaid coverage mounted in the 1960s (Reynolds 1997).

19 Forty-two countries currently have lower infant mortality rates than the United States. See https://www.cia.gov/library/publications/the-world-factbook/rankorder/2091rank.html.

20 On visual culture and medicalization, see especially Bud, Finn, and Trischler 1999; Duffin and Li 1995; Feldman and Douplitzky 2002; Friedman 2004; Karpf 1988;

Löwy 2001; Reagan, Tomes, and Treichler 2007; Stoeckle and White 1985; and Turow 1989.

21 See http://www.subcin.com/titicut.html.

22 See also http://www.pbs.org/wgbh/amex/pill/gallery/gal_pill_02.html.

23 See Grey 2002; Squiers 2005; Stoeckle and White 1985; and http://document ingamerica.org.

24 See http://www.palmer.edu.

25 See Davis 2007 and http://www.ourbodiesourselves.org. Special thanks to Judy Norsigian and Kiki Zeldes of the Boston Women's Health Book Collective for this image, and to Elana Hayasaka for information on editions and translations and adaptations.

26 See especially Sloane and Sloane 2003, and note 6.

27 See, e.g., http://www.dna11.com.

28 PBS's *Magic School Bus* initially aired between 1994 and 1997. See http://www .tv.com/the-magic-school-bus/show/1352/summary.html.

29 IVF was also the focus of the movie *Maybe Baby* (2000); see http://en.wikipedia .org/wiki/Maybe_Baby. Thanks to Jennifer Harrington.

30 See http://www.exploratorium.edu/bodies/index_about.html.

31 See, e.g., http://www.nlm.nih.gov/hmd/about/exhibition/talks.html.

32 See http://www.msnbc.msn.com/id/18233164.

33 See http://www.viagra.com. Pfizer provides extended warnings of side effects at www.PfizerHelpfulAnswers.com.

34 Ultrasound has expanded from two-dimensional to three- and four-dimensional versions. For permission to use this particular ultrasound image, I am indebted to Lance Woleslagle.

35 A PubMed search on acupuncture testing run on February 18, 2008, found 735 citations; chiropractic testing found 251; and homeopathic testing found 451. See, e.g., Adams 2002; Hess 2003; and Jackson and Scambler 2007.

36 Curated by Katherine Ott. See http://www.americanhistory.si.edu/about/staff .cfm?key=12&staffkey=223. See also Bivins 2002.

37 On breast cancer activism, see, e.g., Bell 2002; Ferguson and Kasper 2002; Klawiter 2008; Radley and Bell 2007. On HIV/AIDS see Epstein 1996; Preda 2005.

38 Found in Carmichael and Ratzan 1991, 358–59. On "silence equals death," see http://www.backspace.com/notes/2003/04/07/x.html. The AIDS activist group ACT UP had an effective graphics collective called "Gran Fury." See, e.g., http:// www.queerculturalcenter.org/Pages/GranFury/GFIntr.html and http://www .actupny.org/indexfolder/GranFury1.html. On Haring, see http://www.haring .com/about_haring/bio/index.html.

39 See, e.g., Brownlee 2007; Gevitz 1988; Hayden 2007; Johnson 2004a; Risse, Numbers, and Leavitt 1977; Rogers 1998; and Schneirov and Geczik 2003.

Elianne Riska

4 / Gender and Medicalization and Biomedicalization Theories

The concept of medicalization was introduced into social science research in the early 1970s. This seemingly modest term latched onto broad structural changes going on in medicine as a complex web of organizations, professions, and bodies of knowledge. Irving Zola, a sociology professor at Brandeis University, coined the term in an essay that addressed the notion of medicine as an institution of social control (Zola 1972). This aspect of his argument was not the reason for the success of his thesis. Instead the attraction lay in the argument that medicine had now been delegated tasks and authority that increasingly extended beyond its original mandate. This argument challenged the consensual view of medicine that reigned in mainstream contemporary functionalist medical sociology (e.g., Parsons 1951). Zola's argument, introduced in the phrase "the medicalizing of society," came to be known as the medicalization thesis or theory (both framings are used; see Conrad 2007, 5).

Significantly, Zola saw the medicalization of society not as an imperialistic act of the medical profession alone but rather, as he phrased it, as "an insidious and often undramatic phenomenon accomplished by 'medicalizing' much of daily living, by making medicine and the labels 'healthy' and 'ill' relevant to an ever increasing part of human existence" (Zola 1972, 487). The reason for the "medicaliz-ing of society process" was, according to Zola (496, 487), the increasingly complex technological and bureaucratic system that fed a reliance on the expert.

For Zola, the medicalization process was propelled more by a cultural climate that seeks technical solutions to essentially social problems rather than merely the desire of the medical profession to extend its professional domain (see also Koenig 1988). This view harbored a notion of *multiple* agents of medicalization. In Zola's analysis, medicalization is a tendency of society to reduce a social problem to an individual one and to find a technical — a biomedical — solution at the individual level. For Zola (1972, 500) this meant that "by locating the source and the treatment of problems in an individual, other levels of intervention are effectively closed." The medicalization thesis soon came to have more specific meanings than the broad process indicated by Zola.

In an effort to understand the consequences of such structural and cul-tural changes in twentieth- and twenty-first-century American medicine, this chapter examines subsequent sociological theorizing that has used "medicalization" as a conceptual tool. With the advent of the thirtieth anni-versary of the medicalization thesis, its tenets were reformulated as the biomedicalization thesis (Clarke et al. 2003), denoting the multiple ways in which technosciences and medicine are transforming disease, illness, health, and lifestyle.

I examine two issues in the sociological literature on medicalization and biomedicalization. The first issue is that the theoretical foundations under-girding the medicalization thesis are seldom spelled out. My argument is that the shift from medicalization to biomedicalization is not merely a matter of the phenomena studied but also demonstrates a shift in *theoreti-cal* perspective. While a social constructionist perspective seems to have been the underlying theoretical approach for both theses, the interaction-ist labeling theory that originally characterized much of the research on the medicalization thesis has been augmented by poststructuralist theories in research on the biomedicalization thesis. Hence theorizing the character of social control inherent in the two theses has changed as a reflection of the insurgence of new theoretical approaches in medical and health sociology.

The second issue I take up is the hidden gendered character of the re-search produced using these two theses. Both theses have drawn attention to the gendered body as the concrete "matter" of social control and have thereby challenged the notion of the abstract, disembodied, and unmarked subject in previous theorizing on deviance and mental and physical illness. The medicalization thesis has mainly been used to show how medical knowledge of the female body and the concomitant medicalization of women's health-related issues are part of a patriarchal control of women. In contrast, the biomedicalization thesis has highlighted the gendered character of health and illness and the way that the female body and the male body are constituted relationally. In this later phase of research on the medicalization processes, gender has been used as an analytical category to point to the multiple relationships between masculine and feminine (Jordanova 1993, 474). In addition, the biomedicalization thesis has pointed to the tendency of biomedicine and technoscience to decontextualize medical knowledge and medical technology and to construct a "natural" body. Research drawing on the biomedicalization thesis has deconstructed the knowledge making about discursively constituted bodies and health and illness in scientific discourse (Johnson 2005). The medicalization of men's health, especially their sexual health, has served as an example in illuminating this point in biomedicalization studies.

In short, the medicalization and biomedicalization theses have been contextualized and used to promote certain knowledge claims about the gendered body. My main argument is that the medicalization thesis was anchored in labeling theory and its explanation of social control, while poststructuralist theories informed the analysis of the governed body and the discursively inscribed body in biomedicalization studies. The nature of the gendering of the theses then proceeded from these theoretical positions.

Theoretical Origins of the Medicalization Thesis

While mainstream medical sociology of the 1960s gave an impression of being a monolithic enterprise represented by the functionalist approach to the social organization of medicine, there was diversity from the start. Known as the Chicago school of sociology, a conflict and symbolic interactionist perspective — a view of the negotiated order of social control — was represented in the works of Everett Hughes (1958), Howard Becker (1963), Eliot Freidson (1970), Anselm Strauss (Strauss et al. 1964), and Erving Goffman (1963) in their writings on deviance and occupations as agents of social control. The central notion was that certain occupations had the mandate and power to define what was normal, and normal for what and for whom. This was the labeling approach to deviance, which highlighted the social construction of deviant behavior. Deviance was a matter of definition, detection, and public response (Pfohl 1985; Conrad and Schneider 1980a/1992).

The symbolic interactionist legacy of the labeling approach pointed to the importance of typifications and to the social construction of categories in producing and reconfirming the social and moral hierarchies of groups. The integration of this perspective into the medicalization argument stems from the writings of Irving Zola and Peter Conrad, who have promoted social constructionist approaches to the understanding of the role of experts and expert knowledge in the organization and power of medicine.

It was Peter Conrad who in the late 1970s and early 1980s mapped the concrete pathways of medicalization in his empirical research on hyperkinesis (Conrad 1975) and in a broad historical review of the culture of medicalization (Conrad and Schneider 1980a/1992). Conrad's early work on the medicalization thesis related to two themes in the 1970s critique of the Parsonian model of medicine as an institution of social control. The first theme, often referred to as the professional dominance thesis (Dingwall 2006a), examines the special power that physicians have as medical experts. The second theme, labeling as a process, looks at how social categories are defined and acted on in society. During the past thirty years, Peter Conrad has developed the medicalization framework as an analytical tool to understand the changes in the power of the medical profession and patients in the contemporary healthcare system. Conrad has used the American healthcare system as a concrete case; he assumes that the American medical culture provides an opportunity to gain a theoretical understanding of the culture and power of medicine in Western societies.

In "The Discovery of Hyperkinesis: Notes on the Medicalization of Deviant Behavior," Conrad (1975, 12) defines the concept: "By medicalization we mean defining behavior as a medical problem or illness and mandating or licensing the medical profession to provide some type of treatment for it" (see also Conrad 1992, 211; 2000, 322; 2005, 3). His later book *Identifying Hyperactive Children: A Study in the Medicalization of Deviant Behavior* sets out to explore the question "How does deviant behavior become defined as a medical problem?"(Conrad 1976, xv). He defines his theoretical approach in the following terms: "I use a sociology of deviance perspective in focusing on the definition, identification, and attribution of hyperactivity to children who exhibit deviant behavior."

The crude version of the thesis gave all agency to the medical profession. It suggested that the status of physicians as moral entrepreneurs extended their medical authority by giving them the power to define more and more behaviors as falling within the purview of their expert knowledge and thereby changing the boundaries of the expertise considered legitimate to take care of social problems. Concurrently, the subjects of the social definition of behaviors falling within the domain of medicine were perceived as lacking agency and hence viewed as victims. Some have seen the process of medicalization in cultural and populist terms and posited a deskilling of consumers of healthcare because biomedicine and experts have created a new culture in the United States (Ehrenreich 1978; Illich 1976/1982; Rothman 1998). Others have tended to interpret this process in political-economy terms so that the profit motive of capitalist interests is presented as the major motor of medicine (e.g., Navarro 1976; Waitzkin 2000).

Over the past thirty years, Conrad's theoretical focus has shifted from deviance theory to a social constructionist position. His focus is now on definitional processes: how and by whom are new medical categories constructed that expand medical jurisdiction? Conrad argues that the "engines" behind medicalization have shifted from the medical profession to social movements and interest groups and more recently to market interests in the form of consumers and pharmaceutical companies. This three-way phasing of the major agents of medicalization is further developed in Conrad's latest book, *The Medicalization of Society: On the Transformation of Human Conditions into Treatable Disorders* (2007).

Conrad has defended his social constructionist position against post-structuralist theorizing represented by a number of Foucault-inspired writers (Lupton 1997; Clarke et al. 2003), who have annexed the term "medicalization" into their analysis. Conrad has argued that the biomedicalization concept is too broad and "loses sight of the definitional process, which has always been a key to medicalization studies" (Conrad 2005, 12; Conrad 2007, 14). For Conrad, the major task is to find the key actor who attributes a medical label to a behavior or syndrome because "it is the very processes of medical categorization that create medicalization" (Conrad 2007, 13). In this task, Conrad suggests that a new theoretical angle should today be added to the medicalization studies: "This means supplementing the social constructionist studies with political economic perspectives" (145).

Gender and the Medicalization Thesis

In terms of gender, research on the medicalization thesis can be divided into two phases (Riska 2003). In the first phase in the early 1970s, the thesis was used in a gender-neutral way to indicate the emergence of a new culture of medicine. It could be argued that during this phase the gender neutrality had a hidden gendered content, because the examples mentioned in the research literature were behaviors experienced more frequently by one gender. For example, alcoholism and hyperactivity among children are behaviors expressed more often by males than females.

In the second phase, between 1975 and 1985, feminists were the first to add a gender-sensitive component to the thesis. Feminists argued that the practices of scientific rationality made the female body an object of expert control (Ehrenreich and English 1978/2005; Riessman 1983; Clarke 1983) and hence the major site of medicalization. The focus was on female reproduction and how an ideology of biological determinism was used to oppress women because female biology was used as a "scientific fact" to construct sexual difference and sexual dualism (Sydie 1987). The medicalized phenomena were used as indicators of the way that women's biology and health were socially constructed.

In the mid-1980s, there was a shift in views of the capacity of consumers to resist physician-promoted medicalization. Research in the 1980s suggested that women were actively involved in defining their own health, and feminist research pointed to the importance of patients' own perspectives. As an influential reader in the field on women's health suggested, such a perspective "reveal[s] women not as passive creatures who can only be acted upon by others, but as active constructors of contexts and meanings which both reflect and affect their society" (Olesen and Lewin 1985, 10). Soon this capacity was presented within an empowerment model, which pointed to women's own capacity to "medicalize" symptoms previously not recognized as medical—for example, chronic fatigue syndrome, fibromyalgia—and gain control over their diagnosis and treatment. The empowerment version of the medicalization thesis points to the claims-making capacity of consumers, as opposed to repressive medicalization, which makes women the victims of medicine (Riessman 1983; Cussins 1996; Conrad 2005; Greenhalgh 2001).

The empowering aspects of medicalization have further been enhanced by access to a new technology: the Internet. The Internet is a source of knowledge about health, illness, and treatment, and it encourages an active and pluralistic approach to health (Hardey 1999; Pandey et al. 2003). The Internet offers consumers a way of finding and sorting out medical knowledge, a process that can be characterized as "navigating" the medical system (Pitts 2004, 44). "Cyber-agency" was optimistically first interpreted as giving rise to the "Internet-empowered patient" (Pitts 2004, 34; Seale 2005, 517). Early research on the use of the Internet in this capacity confirmed this notion. For example, a study of a breast cancer list showed that the online discussion group provided women "cyber-support" and a sense of empowerment (Sharf 1997). More recent research has been skeptical of the consumer's autonomy and has showed that websites have been appropriated by powerful commercial interests (Pitts 2004, 53; Seale 2005, 535). This trend supports the key arguments of the biomedicalization thesis (Clarke et al. 2003, 177–80).

The Theoretical Foundations and Arguments of the Biomedicalization Thesis

The formulation of the biomedicalization thesis can be attributed to Clarke, Shim, Mamo, Fosket, and Fishman's influential article in 2003 asserting the need to reformulate the key argument of the medicalization thesis to address the changing conditions of American medicine in the twenty-first century. The term "biomedicalization" is used to indicate "the increasingly complex, multisited, multidirectional processes of medicalization that today are being both extended and reconstituted through the emergent social forms and practices of a highly and increasingly technoscientific biomedicine" (Clarke et al. 2003, 162). The article suggests that the shift from medicalization processes to biomedicalization processes consolidated around the end of the 1980s. This shift, Clarke and her colleagues suggest, captured a broader cultural trend that they define in the following terms: "Medicalization was co-constitutive of modernity, while biomedicalization is also constitutive of postmodernity" (Clarke et al. 2003, 164).

The medicalization thesis focused on patients and their lack of agency in light of the new moral power of medicine. By contrast, the biomedicalization thesis privileges the material body as a major site of biomedical and public discourse. The focus shifted to knowledge-making practices, especially how the body is discursively constituted through biomedical knowledge and practices. The central argument is that the new technoscience and biomedical corporate enterprise influence not only how medicine is practiced but also how technoscientific discourses penetrate the public discourse. Medical and public conceptions reflect a conviction that the body's capacity is almost unlimited. Representatives of the biomedicalization thesis argue that biomedicalization has reconstructed the boundaries between the material body and social identity, so that medical interventions in the form of "technologies of the body" enable an enhancement of a certain type of revered notion of the self, the creation of "technoscientific identities" (Clarke et al. 2003, 184). Nevertheless the self is not a gender-neutral notion. Instead the construction of the enhanced body recaptures normative, heterosexual notions of femininity and masculinity and gendered expressions of sex and sexuality (Mamo and Fishman 2001; Loe 2001, 98; Marshall 2002).

While the medicalization thesis was founded on the legacy of the labeling theory, the biomedicalization thesis builds on this legacy and further traces its concepts and theoretical framework to poststructuralist theories, especially those of Foucault. In terms of gender, the medicalization thesis analyzed the gendered implications of the medicalization process, but it restricted its scope to women. This focus was exemplified by the research on gender and health in the 1970s and 1980s, when the theme of gender and health became synonymous with women's health (Kuhlmann and Babitsch 2002). Not only were the problematization and theoretization of men's health from a gender perspective curiously absent from most social science research in the 1970s, 1980s, and much of the 1990s (but see Pfeffer 1985), but moreover there was during those decades hardly any *sociological* research on the medicalized character of men's health (Rosenfeld and Faircloth 2006). Still men's health was normative in *clinical and epidemiological* research on diseases, such as heart disease and cancer. The "standard human" in biomedical research tended to be an adult white male (Epstein 2007, 47, 276).

Research on the process of biomedicalization reversed the trend. Most of the research on gender and health that has explored the biomedicalization process has focused on the male body, especially on men's sexual health. For poststructuralist theorists, men's health has been a means to elaborate the gendered character of knowledge making, to think of gender in relational terms, and to deconstruct binary notions of gender.

The biomedicalization thesis suggests an increased importance of the role of sciences, technologies, and experts in biomedicine per se and in enhancement of the body. Two theoretical perspectives undergird the arguments: the poststructuralist perspective and feminist theorizing about the body.

First, the main theoretical strand of particular import in the development of biomedicalization theory is poststructuralism, notably Michel Foucault's (1975, 1988) views on the role of science and experts in the construction of new regimens of social control. The argument here is that medical science and biotechnologies serve as hidden instruments of social control, so that a rational and self-disciplining governance permeates society at various levels. This is a form of biopower that is concerned with the management of life and governing of populations. The term "biopolitics" refers to knowledges, practices, and norms that have developed to regulate the quality of life of individuals and populations (Howson 2004, 126; Petersen and Bunton 1997; Rabinow and Rose 2003/2006). It extends the use of new technologies in the government of the self, new forms of self-regulation and self-disciplinary measures that are internalized so that society seems to be self-regulated and self-corrective. But in this kind of society, individuals are dependent on experts who provide the technologies for rational management of the self.

For Foucault, however, power is discursive and hence not monolithic. Power can be challenged, as Lupton (1997) suggests in her incisive analysis of the theme of medicalization in traditional and Foucault's writings, so that the discourse of biomedicine is just one of the discourses that are challenged by alternative discourses (Lupton 1997). The central theme of the Foucauldian approach is that where there is power, there is always also resistance, a theme that can be discerned in the literature on biomedicalization (see the discussion of countertrends in the introduction to this volume).

The methodological approach of the Foucauldian perspective is called genealogical: the task is to trace the origins of medical knowledge, especially the way that categories have been constructed and how they have constituted new instruments or lenses that have enabled new gazes or new ways of seeing the configurations of the "facts." Once phenomena have been decontextualized and universalized, they become normalized elements of scientific language and knowledge, and then the "true essence" and "nature" of the phenomena unfold. Hence the deconstruction of categories is the main feature of the archaeology of medicine, or what Foucault (1975) sees as a central task: to trace the genesis of medical knowledge.

Second, another important theoretical development relates to a sociological interest in the body in the mid-1980s and early 1990s. European sociologists identified the rise of a sociology of the body in the early 1990s (Howson 2004; Shilling 2005), but the connection of this field to theorizing on postmodernity resulted in gender neutrality in the writings in the new subfield. For example, Anne Witz (2000, 19) has noted that the first decade of European sociological writings on the body, largely written by men, failed to recognize and theorize the gendered character of the body. By contrast, American scholars on the body in the early 1990s focused on the female body and were inspired by feminist theorizing using social constructionist or poststructuralist perspectives (e.g., Martin 1987; Butler 1993; Grosz 1994).

Gender and the Biomedicalization Thesis

The medicalization of men's health is not a new phenomenon, although masculine medicalization largely escaped the sociological gaze of medical sociologists. The classic example of the medicalization of men's health in American medicine in the past was the construction of Type A and the Type A personality in the scientific discourses of biomedicine and psychology.

THE MEDICALIZATION AND BIOMEDICALIZATION OF THE MALE BODY

New modes of medical thinking emerged in the 1950s that defined traditional masculinity as a health risk for the development of coronary heart disease. This thinking medicalized traditional masculine behavior: it connected the emotional strains of executive work with the rising mortality of middle-class, middle-aged (white) men caused by the fatal outcome of heart disease. The prototype for this new thesis was the Type A behavioral pattern, later called Type A personality (Riska 2000, 2004). The Type A thesis predicted that a certain type of behavior, embodied in a man who was extremely competitive, inclined to work to deadlines, and never stopped seeking material gain doomed him to heart attack. The medical diagnosis of the behavioral pattern — Type A behavior pattern — was given by two American cardiologists, Meyer Friedman and Ray Rosenman (1959, 1960). Type A behavior pattern (Friedman and Rosenman 1974, 67) was operationalized as follows: "Type A Behavior Pattern is an action-emotion complex that can be observed in any person who is *aggressively* involved in a *chronic, incessant* struggle to achieve more and more in less and less time, and if required to do so, against the opposing efforts of other things or other persons."

The Type A thesis suggested that persons with this kind of behavioral pattern were prone to coronary heart disease, or as the inventors of the construct contend: "The Type A man is prone to heart disease; these characteristic behavioral habits identify the Type A man" (Friedman and Rosenman 1974, 70). The proneness to heart disease was assumed to reside in *men's* behaviors and attitudes.

More specifically, proneness to heart disease was assumed to reside in the negative emotions of Type A men. Hostility, anger, and aggression were defined as the gendered health risks of hard work required in entrepreneurial and executive positions in the American business world. It was not failure to conform to the norm of healthy middle-class manhood that was of concern to medicine, but rather overconformity to traditional masculine behavior. This was the first time that men's competitive and aggressive behavior was seen in pathological terms rather than as part of the normal cultural repertoire of masculine behavior. It was the toxicity of the hypermasculinity of Type A men that was constructed in medical discourse as a health risk that had implications for men's cardiac health.

The medicalization of masculinity was accomplished by pathologizing what had previously been considered "normal" behavior of white, middle-class men. The medical category Type A was a representation of men's personality and provided a conceptual framework for diagnosis and therapy. A social analysis was implicit in the construction of the Type A category: Type A was depicted as a man who responded to the norms and constraints of the specific historical and social context of post–World War II America.

But Type A man was seen as a victim of the major ideology constructing hegemonic masculinity (Connell 1995; Kimmel 2000; Connell and Messerschmidt 2005): the work ethic.

Type A is a case not only of physician-initiated medicalization but also of a physician-initiated demedicalization. When the Type A hypothesis began to be empirically tested with larger population samples than middle-class men — for example, women and other social classes of men — the constructed medical category could not be empirically verified. Hence the thesis was declared scientifically invalid and abandoned by academic medicine in the early 1980s (Riska 2004). The Type A thesis has continued to exist as a set of health beliefs or lay epidemiology in public discourse about "coronary candidacy" and its gendered character (Davison et al. 1991; Emslie et al. 2001).

Performance criteria of men's bodies and behavior undergirded the construction of the pathology of Type A man and claims for the intervention of medicine. A relocation of performance criteria has taken place. Between the 1950 and 1980, medicine and drugs were aimed at improving men's performance and function in the labor market, while current sexual-enhancement-producing drugs are related to men's performance in private life.

THE MALE BODY: ENHANCEMENT TECHNOLOGIES

The emergence of the "hormonal body" as a scientific and biological fact provides a case study of the Foucauldian approach to medical knowledge and the role of experts in defining the gendered body. Oudshoorn's (1990; 1994) documentation of the emergence of the "hormonal body" as a "natural" and biological fact provides a significant insight not only about the social construction of the body but also about the social construction of sexual dualism — femininity and masculinity.

Oudshoorn (1994) points to the emergence in the 1950s of a conceptualization that hormones were the "chemical messengers" of femininity and masculinity. Sex hormones constituted the chemical mechanism whereby sexual difference is expressed on the body and becomes a "natural fact." The chemical model confirmed the mechanical and reductionist notion of the body embraced by biomedicine. It suggested that the body was controlled by hormones, which as properly developed in the body were expressed as a sexual difference. The scientific discovery of the hormones implied that sex could be measured. The level of sex hormones in the body defined the proper and "natural" functioning of the body. This understanding of the workings of the body as "the hormonal body" itself medicalized the body. It introduced the notion of "hormonal deficiency," which became a core concept legitimating the regulation of the sexed body in the age of medicalization and biomedicalization (Fishman, this volume; Roberts 2007).

This way of pathologizing the body also suggests a biopharmaceutical remedy: hormonal deficiency can be corrected by a regimen of hormone replacement (e.g., Watkins 2007; Roberts 2007). Oudshoorn (1994) argues that the success of this perception of the body enhanced the position of obstetricians and gynecologists. These knowledge claims about female sex hormones thus fitted into existing and emerging institutional structures of medicine: professional groups, physicians, laboratories, and pharmaceutical companies provided a ready market for the application of the new knowledge about the female hormonal body.

Oudshoorn (1990, 19–20; 1994) also ponders about the "undermedicalization" of male menopause. She suggests that the same "facts" about male sex hormones lacked an equivalent institutional structure that could have used this knowledge as a legitimation for its practice and interventions. Despite the formal development of a male-focused specialty, andrology (Wöllmann 2004; Rankin 2005),[1] knowledge of the male "hormonal" body remained relatively dormant for decades. Today this situation has changed dramatically: new knowledge claims about male hormonal disorders have exploded since the late 1990s because of the convergence of scientific and commercial interests in medicalizing masculinity, especially in later life (Marshall 2007, 510). The discursive knowledge of the male hormonal body has become the focal point of a social and cultural phenomenon, here called the "viagracization" of men's health (Letiche 2002, 254; Riska 2003, 76). Representations of the hormonal bodies of men have set new standards of sexual health that call for the repair or enhancement of the "normal" and "natural" sexual health and performance of the male body. This trend is one manifestation of the gendered content of the research on the biomedicalization thesis.

The construction of the Viagra man (Loe 2006, 27) in current biomedicine promotes a life-enhancement approach (see Fishman, this volume). Compared to the work context of the Type A man, the Viagra man's sexual performance and life in the private sphere are the focus of the restoration project. The Viagra man appears as classless, raceless, and at times of any sexual preference, although the norms of the viagracized body have been viewed as enforcing hegemonic masculinity and prevailing social hierarchies. This is one gendered implication of what Clarke and her colleagues (2003, 170) have called "stratified biomedicalization."

THE "NATURAL" BODY AS THE MALE BODY

There is now a genre of studies that has examined the biomedicalization of men's bodies and sexual health (Potts 2000, 2004; Potts et al. 2004; Loe 2001, 2004a; Mamo and Fishman 2001; Fishman 2004, this volume; Marshall 2002, 2007). The arguments and findings of these studies tend to support each other. Studies on the Viagra man share three basic inquiries: what is normal, normal for what, and normal for whom.

First, biomedicalization has implied a reinvention of the natural body as the normal and universal body. The target of this reinvention is men's bodies, now located as residing *outside* the social realm. The focus on men's sexual health enables the reconfirmation of the naturalness of men's bodies. The assumption is that sex is natural and biological, and so man's sexually functioning body is constructed as a biological given and assumed to rest on the "human sexual response cycle" discovered in the studies of Masters and Johnson (Jackson and Scott 1997, 557). The *naturalization* of the male body is interpreted as a restoration project: the natural performance of the male body can now by means of biotechnological intervention be fully restored or even enhanced (Potts et al. 2004; Potts 2004; Jackson and Scott 1997). This thinking rests on a mechanistic model of the functioning of the body, a view that constructs male bodies as needing repair (Loe 2001, 113; Fishman and Mamo 2002; Fishman 2004, this volume). As Fishman notes in her chapter in this volume, the notion of impotence has come full circle: at the beginning of the twentieth century, impotence was viewed as an inevitable organic process, a view which shifted to understanding impotence as a psychological condition but then shifted again to reconceptualizing it as an organic process.

The medicalization of men's sexuality denotes a process in which men's sexual health comes to be defined and treated as a medical problem. But the new meaning of medicalization — the biomedicalization of men's sexual health — implies that the treatment of sexual dysfunction is by biopharmaceutical means: a potency drug, in common parlance called Viagra, although other brands have entered the market, such as Cialis and Levitra.

Viagra not only is a concrete drug intervention but has also become a metaphor for a new thinking about the male body. The Viagra-enhanced body is seen as more natural and real than the natural and real body (Mamo and Fishman 2001; Letiche 2002). In short, the Viagra man is a representation of a man whose natural sexual performance and thereby masculinity can be restored or even enhanced. As Marshall (2007, 522–23) suggests, the remedicalized climacteric in men reasserts a hormonal and biomedical basis to masculinity itself. The construction of the disorder of androgen deficiency not only confirms a discourse of biologically based sexual differences but also represents a tendency to locate gender in an asocial or presocial and naturalized male body.

Second, studies that have explored the Viagra phenomenon have grappled with the question: normal for whom? Some researchers (Adkins 2001, 48; Marshall 2002, 139) anticipated processes of biomedicalization as part of postmodernity that has created a culture of "responsibilization" and individualization of risks. The health promotion discourse defines subjects on the basis of the kind of responsibility and preventive measures they initiate for maintaining and promoting their physical and sexual health. But it is suggested that these endeavors feed an increased reliance on scientific expertise, which provides guidelines and means for achieving the revered status of healthiness. These researchers argue that health promotion constructs and reaffirms a hierarchy of sexual identities. Viagra is not only a pharmaceutical technology that exerts technoscientific control over men's bodies (Loe 2001, 98; Marshall 2002, 138; Potts 2004, 24); it is also a normative script that reifies traditional discourses of sexualities and masculinities (Mamo and Fishman 2001). The science of "sexual dysfunction," it is argued, renders invisible the assumptions about gender and sexuality it rests on (Potts 2000; Marshall 2002, 144).

Third, the question addressed in the research on men's sexual health in the age of biomedicalization is: normal for whom? Two interpretations have been put forward in the existing literature on the biomedicalized male body.

The first interpretation is Foucauldian and suggests that biomedicalization is a form of biopolitics (defined in the foregoing section). This is a rationalization of sexuality and reliance on experts who regulate and control sexuality through biopharmaceutical means. But the Foucauldian notion of power is not monolithic. Instead it suggests the existence of counterdiscourses and resistances to power. The emergence of resistances to a viagracization of men's bodies has been suggested in recent research. For example, Potts and her colleagues (2004) found diversity in the sexual experiences and interpretations of sexual dysfunction among male users of Viagra, who challenged the medical discourse on sexual dysfunction, for example, by holding to a normalization discourse. Furthermore, Loe's (2004b) study of women's responses to their male partners' use of Viagra challenged the functional and male-centered conceptualizations of the drug as a panacea for the heterosexual relationship. Women's responses ranged from views that men are incapable of satisfying women's special sexual needs to views that Viagra threatens older women's right to act in accordance with their lack of desire for sex.

The second interpretation suggests that biomedicalization promotes a hidden confirmation of men's superordination in the prevailing gender order and a way to reaffirm existing sexual and social hierarchies. Compared to the original medicalization thesis that assigned women the status of victim, the target of biomedicalization is not to victimize men but to keep them from any possibility of victimization, such as failing masculinity. Hence feminist researchers have seen Viagra as a euphemism for a pharmacotechnology that confirms hegemonic masculinity. For example, the medicalization of men's sexuality is argued to be concurrent with a view that sees men's restored sexual fitness as an imperative for the functional sexual relationship of the heterosexual couple. The functionalist interpretation of the underlying cause of the biomedicalization of men's sexuality supports a radical feminist interpretation that viagracization is a masculine medical conspiracy: sexual fitness is good for monogamy and patriarchy (Jackson and Scott 1997). However, Barbara Marshall challenges this radical feminist interpretation and offers a negotiation model. For her, the medicalization of men's sexuality is the result of a "constantly shifting coalition of actors — including scientists, doctors, patients, industries, media and consumers — operating within a cultural horizon of rationalization, medicalization, commodification and gendered heteronormativity" (Marshall 2002, 146).

The interpretation that sees the biomedicalization of men's sexual health as rooted in a consolidation of hegemonic masculinity is also put forward by Loe (2001, 98; 2004a), who suggests that Viagra is a biotechnology that "is being used to 'fix' or enhance heterosexual male confidence and power, and thus avert masculinity 'in crisis.'" This cultural interpretation has, however, limited explanatory value. The reference to a crisis of masculinity, or to an even broader construct "a crisis of patriarchy," is a rhetorical explanation that can always be called on but can seldom be demonstrated. But perhaps more important, such a broad explanation misses the essential point it is harboring: naturalizing the male body makes it unmarked by gender. The unmarkedness of men's gender serves, Robinson (2000) suggests, as a powerful discourse that constructs women as the other and confirms men's superordination. But, more important, the unmarkedness of the bodies of men serves a special purpose: what is invisible escapes surveillance and regulation (Robinson 2000, 1).

THE "NATURAL" BODY AND STRESS: THE RECONFIRMATION OF GENDERED SOCIAL HIERARCHIES

During the past decade, media and science have revitalized biological thinking about the body and emotions. Genetic and molecular research has further supported technoscientific theorizing about health and emotions. A number of scientific journals—for example, *Biological Psychiatry, Psychoneuroendocrinology, Neuropsycho-pharmacology, Brain Research*—have become major channels in constructing a scientific discourse that portrays the body, emotions, and behavior in essentialist terms. In these scientific forums, the body is conceptualized as a given, natural biological entity with fixed psychological and physical needs. A more recent version of this biosocial perspective presents a biochemical approach to emotions and moods. As Nikolas Rose (2007, 222) has suggested, this scientific discourse has constructed a neurochemical sense of ourselves.

This new conception of the neurochemical self is based on a representation of an essentialist universal selfhood. This conceptualization of the self obscures the hidden gendered narrative of this essential self. As I will show, underlying the scientific discourse of the biochemical or neurochemical conception of the essentialist self is a representation of a masculine self.

There is nostalgia for the distant, unproblematic days of men's superordination and efforts to recapture that era of manly existence in recent writings on psychosocial stress and health. Biomedical research on stress responses rests on a technoscientific discourse that points to the fixed capacities and limitations of the natural body and valorizes the existence of a generic body. In the age of biomedicalization, biological science has become the overarching scientific discourse that claims to explain both psychological and social phenomena. Hence neither the psychological nor the social science discourse is seen as a valid scientific framework for explaining emotions and moods. The biologicization and molecularization of emotions constitute a discourse that not only provides a biomedical definition of pathological emotions but also suggests a remedy: mood disorders can be corrected by means of psychopharmacology (Horwitz and Wakefield 2007; Rose 2007, 190). For example, in the case of depression, disorders can be addressed through a new generation of antidepressants called SSRIs, like Prozac, which correct neurochemical imbalances in the brain.

The discourse on the brain's role as a messenger in coding stress responses bears an interesting resemblance to the discursively inscribed "hormonal body." In the case of the hormonal body, sex hormones are chemical messengers that confirm male and female sexual difference (Oudshoorn 1994; Roberts 2007). In the case of the stressed body, heightened levels of stress hormones indicate pathology. In the stress discourse, the brain acts as the messenger, and a heightened level of those hormones serves as the marker of a threat to the autonomy and integrity of the body and the self.

According to the vocabulary of this technoscientific discourse, the neurotransmitters constitute the crucial mechanism that mediates messages between brain neurons and the body. Hence the translator of the body's "own" needs is the brain. According to the language of this discourse, the brain "interprets" the biochemical processes set in motion when the body "feels" threatened. A prominent proponent of this view, Bruce McEwen (2000, 173), defines stress in the following terms: "'Stress' will be used to describe an event or events that are interpreted as threatening to an individual and which elicit physiological and behavioral responses. The brain is the key organ involved in interpretation and responding to potential stressors." In this translation process, the brain "knows" what kind of physical and psychological burdens the body can endure: "The brain is the master controller of the interpretation of what is stressful and the behavioural and physiological responses that are produced" (172). In this scientific discourse, the body is portrayed as a system with a fixed biological capacity. This state is called "allostasis" (McEwen 2000, 174; 2003, 201): "Allostasis is the process that keeps the organism alive and functioning, i.e., maintaining homeostasis or 'maintaining stability through change' and promoting adaption and coping, at least in the short run." But the body is also portrayed as having certain limitations. The limitations are conceptualized in terms of an overload, which is defined as an "allostatic load" (McEwen 2000, 174; 2003, 201): "'Allostatic load' refers to the price the body pays for being forced to adapt to adverse psychosocial or physical situations, and it represents either the presence of too much allostasis or the inefficient operation of the allostasis response systems, which must be turned on and then turned off again after the stressful situation is over." Most of this research on psychosocial stress has been done on (male) rats and (male) monkeys. For example, evidence of the impact of sex is introduced in the form of experiments with male rats competing for food and for female rats. Results show that within two weeks, "it is not uncommon for one or two of the subordinate males to die, usually not so much from wounds as from autonomic collapse related to defeat" (McEwen 2000, 181, 185). The results of the research on psychosocial stress caused by a subordinated status of male rodents and vervet monkeys are believed to have implications for human behavior (McEwen 2000, 182): "The advent of magnetic resonance imaging (MRI) and positron emission tomography (PET) has opened the door for translating information based upon animal models of neurological and psychiatric disorder to directly studying patients with those disorders. In particular, changes in hippocampus in animal models provide insights into altered human hippocampal structure and function in depression and aging." A whole genre of brain imaging studies of the hippocampus relates alterations in cell number or morphology to mood disorders — in particular depression (Duman 2004). Hence representatives of this kind of neurobiological research on animals suggest that their findings have implications for humans.

The new neurochemical discourse on health and psychosocial stress suggests that there exists a primordial, generic body, unaffected by technological culture and by the fragmentation of society caused by modernity. This thinking promotes a degendered and unmarked body, which is located in nature and thereby inscribed as a natural kind. But this natural body is discursively constituted through biomedical knowledge and practices (see also Dumit 2004). As Danziger (1997, 190) reminds us, "scientific discourse is only able to represent objects as they have been constituted by the categories of that discourse." The generic body in stress literature is presented in universal and gender-neutral terms, although what is at issue is the male body. The brain as a messenger of stress hormones and pathology becomes in this scientific discourse a messenger of threatened masculinity as well.

The hidden character of gender in the bioscientific discourse on psycho-social stress appears in two ways. First, the biological discourse presents a generic body, unmarked by gender, race, or class. It is suggested that the generic body has the capacity to revolt against "unnatural" social environments, especially at work. Furthermore, work settings are portrayed as post-Fordist jungles, where the generic body's reactions to the signs of danger are those acquired in some original prehistoric jungle (e.g., Sapolsky 1998). As representatives of this biosocial thinking argue, stress is a vital sign of the "body's own wisdom," and disease is consequently seen "as a loss of bodily regulation" (Freund 1982, 24; Kugelmann 1992, 35; Sapolsky 1998). Still the assumptions of the body's needs are framed within a masculinist understanding of control, where it is assumed to be natural for men to be in control. Hence masculinist assumptions about an autonomous body and individualism are framed as a "standard human," which is the repository of the natural. It is argued that the brain senses when a man's body is exposed to a too-stressful situation and needs to return to normality—the condition of superordination. Rather than identifying hegemonic masculinity as the underlying social and cultural structure that defines men as the super-ordinated group in the current gender system, recent sociobiological thinking naturalizes individual men's reactions to a condition characterized by subordination.

There is also an implicit materialist view of masculinity in this biological construction of stress, which has the potential for deconstructing a notion of homogeneous masculinity. Viewed from a materialist perspective, men are being portrayed as victimized by their working conditions. This notion of masculinity locates men's status, sense of self, and bodies almost exclusively in economic conditions, an argument that could also be used to suggest that different economic locations produce different kinds of masculinities.

Second, men's subordination in the world of work is presented as unnatural, which, ipso facto, elevates their superordinated status as men in the gender system to a natural condition. This way of essentializing masculinity and showing the implications of a physically and emotionally threatened masculinity (even if the real cases have been rodents and monkeys) performs the cultural work of recentering dominant masculinity (see also Robinson 2002, 11). That women constitute the major subordinated group in the gender order is thus not only again made invisible but also renaturalized in the bioscientific discourse on stress. In sum, the bioscientific discourse on stress not only asserts a binary notion of gender but also legitimates existing hegemonic notions of masculinity.

The biomedicalization of stress is based on the knowledge claims about the fixed character of the natural body, which have been made transparent by means of new imaging techniques that enabled the mapping of emotions in the brain. These techniques have provided a lens for seeing configurations of scientific facts in new ways, and a new neuroscience of emotions and mood disorders has unfolded. As Dumit (2004, 158) notes about the appeal of brain images in today's medical and popular culture: "Facts have to find us, and we have to incorporate them as facts." Like the quotations cited earlier from scientific journals on brain research, the knowledge claims harbor "prescientific ideas" (Oudshoorn 1994, 11; Shim 2002a, 2002b; Jordanova 1993, 482; Epstein 2004, 192; Johnson 2005; Dumit 2004) about gender and gendered behavior that have become part of the representations of the new scientific facts about emotions and stress. The biomedicalization of stress contains certain gendered and normative notions of prevailing gender categories and gender hierarchy. They reinforce the current gender system by relocating and renaturalizing men's health in its traditional position of unmarked superordination. At the same time, this framework also intrinsically reproduces women's subordination as part of the gender system. In short, the current biomedicalization of stress tends to perpetuate and naturalize the traditional gender system.

Conclusion

This chapter has reviewed the central tenets of the medicalization and biomedicalization theses with a particular focus on their theoretical approaches and their salient gendered content and implications. My first argument is that the shift from emphasis on medicalization to biomedicalization processes not only represents a substantive shift but also reflects the paradigmatic shifts in sociological theorizing. When initially introduced, the medicalization thesis was used as a conceptual tool to redefine the traditional distinction in sociology between structure and agency as related to health, illness, and social control. Proponents of the medicalization thesis pointed to the social construction of new social categories of deviance as medical by means of expert knowledge. While Zola's (1972) original thesis asserted multiple agents of medicalization, initial work focused more narrowly on the medical profession as the agent of medicalization. Later consumers were (re)construed as agentic within an empowerment model of medical settings. Most recently, the biomedicalization thesis has restored Zola's original emphasis on multiple agents of medicalization.

In conclusion, the terms medicalization and biomedicalization have been analytical tools used to understand the dynamics of changes in American medicine. Both theories valorize social control aspects of medicine, broadly conceived, but emphasize different sociological perspectives. The medicalization thesis traced its theoretical heritage to symbolic interactionism and labeling theory. The biomedicalization thesis additionally relies on the conceptual tools and arguments of poststructuralist theorizing, especially the work of Foucault. From a sociology-of-knowledge perspective, one could therefore say that symbolic interactionism and labeling theory are constitutive of medicalization, while poststructuralism is constitutive of biomedicalization. Most recently, biomedicalization theory has also emphasized biocapital and bioeconomic processes (see the introduction to this volume).

Second, issues of agency and the targets of (bio)medicalization have been a common theme of both theses. Early work using medicalization theory often presented a victimization model, which was later changed to an empowerment model (see Conrad 2005; Conrad and Leiter 2004). During both of these two phases, women were implicitly or explicitly in focus—first as the major victims and then as the major empowered lay agents of their own and others' medicalization (Riska 2003). The biomedicalization thesis is more complex and could be said to represent a relativist model. The representatives of this thesis tend to propose that biomedicine and technoscience construct a hyperreality that obscures boundaries between agency and the target of this new kind of medicalization.

Research on the biomedicalization phenomenon has also included a major focus on gender and has deeply integrated men's health into the analysis. To contrast these, research on women's health vis-à-vis the medicalization thesis pointed unerringly to the social construction of women's bodies. Feminist scholars have argued that "scientific facts" about women's reproductive biological functions have resulted in the construction of women as the subordinated category of the universal (male) body. The incorporation of men's health in sociological research on biomedicalization theory has extensively examined how the male body has become the major site not only of naturalization but also of hypernaturalization. A deconstruction of the scientific facts, including the presentation of the male body as a natural category, has illuminated how knowledge and gender are discursively constituted in society (Jordanova 1993).

The medicalization and biomedicalization of men's health have been exemplified by two constructs: Type A man and Viagra man. Type A man was constructed in medical discourse by pathologizing hypermasculinity as a coronary health risk. I have used the term "viagracization" to understand the new biomedical and public thinking about the capacity and performance of the male body. The Viagra man represents a man whose sexual performance can, by means of new technologies of the body, be restored and enhanced to a more natural performance than the "natural" one. As the literature reviewed here suggests, the naturalization of men's sexual health locates masculinity outside the social and thereby confirms the unmarkedness of men's gender and body, although remedies like Viagra tend to confirm heterosexual norms of female and male sexuality. Several scholars have pointed out that recent discourses on the dysfunctions of the aging male body reassert a hormonal basis for masculinity itself (Marshall 2007, 522; Fishman, this volume). A reconfirmation of the heterosexual matrix seems to be part of a broader political climate and cross-cutting morality characterizing twenty-first-century America.

Similarly, the new neurochemical explanations of mood disorders locate emotions in the brain and in this way naturalize and degender emotions. Emotions were once the chief definer of women, who were constructed as emotionally knowledgeable but also therefore as different, in contradistinction to rational man (Sydie 1987; Jackson and Scott 1997, 568). The biomedicalization of stress is a scientific discourse suggesting that emotions are located in the biological system and hence outside the social—outside gender, racial, and class systems. The brain-governed body in biochemical stress discourse and the Viagra man in the sexual-dysfunction discourse are male-gendered bodies constituted in scientific discourse. Both scientific discourses point to the functioning of neurotransmitter pathways and altered hormone levels as having detrimental health consequences but also constituting a threat to traditional masculinity. The texts of the bioscientific discourse on sexual dysfunction and on stress tend to reproduce gendered notions of the naturalness of men's need for autonomy and control, normative views that tend to confirm men's superordination and women's subordination within the prevailing gender order. The implications of this kind of discursively constituted notions of the hierarchical relations of gender have been illuminated in the research characterized here as the biomedicalization thesis.

Last, the shift from medicalization to biomedicalization has also implied a new way of theorizing gender. Poststructuralist theories have redefined gender in relational terms and deconstructed the binary oppositions of man-woman, natural-cultural. Thus they have paved the way for a conceptualization of multiple categories of gender rather than merely two. The next stage of research on biomedicalization processes should therefore examine how other social categories used in the stratification of society— categories such as class, ethnicity, and race—are manifest in research on health and illness. Questions about how differences are constructed and where they are understood to be located are of fundamental importance (e.g., Shim 2002a, 2005; Epstein 2004, 191; 2007, 297). Particularly promising here, and building on the multiplicities of categories, is intersectionality theory (Schulz and Mullings 2006) as it meets biomedicalization.

Notes

I thank Adele Clarke, Laura Mamo, and Katherine McCracken for their comments on an earlier version of this chapter.

1 The term "andrology" was first proposed as the name of a medical specialty in the United States in 1891 but languished for decades (*JAMA* 1891). It gained momentum as a specialty in the early 1950s, sponsored by basic scientists in the fields of urology, endocrinology, and dermatology-venerology who had a common interest in the male reproductive system. Today andrology is a subspecialty area within urology and deals with problems of the male reproductive organs. The American Society of Andrology was founded in 1975, and its goal is to promote "knowledge of the male reproductive system" (http://andrologysociety.com; Schirren 1985, 1996; Rankin 2005).

▷▷▷PART II / CASE STUDIES: FOCUS ON DIFFERENCE

Laura Mamo

5 / Fertility, Inc.

CONSUMPTION AND SUBJECTIFICATION IN
U.S. LESBIAN REPRODUCTIVE PRACTICES

During the biotech bubble of the late 1990s, I was living in San Francisco and conducting research for my doctoral dissertation in medical sociology at UCSF. I was researching the lesbian baby boom that was under way. My analytic findings were straightforward enough: low-tech options including self-insemination were being joined with, if not displaced by, high-tech procedures. Lesbians, once marginalized as nonprocreative and doin' it (conception) themselves, were instead routinely labeled as "infertile" and undergoing all the medical technologies biomedicine had to offer. What was going on?

I too was being shaped by the shifts in late-twentieth-century biomedicine. Unlike today, it was the height of the hype, the beginning of the end of the dot.com era, and the beginning of the dawn of personalized medicine and other medical innovations resulting from the conjoining of computer informatics and biomedical sciences. I admit it: I was captured by it all. As I collected data by interviewing lesbians about their encounters with medicine, sperm bank personnel, and fertility specialists, I spent my evenings reading business news articles, surfing the latest biotech stock picks, and watching as UCSF planned a new biotech campus. Then one day I used all my savings ($400) to purchase stock in a start-up, an onco-mouse medical research company called Abgenix. Donna Haraway had warned us, but she also pointed us in the right direction. Consider this my conflict-of-interest statement. It is the only one and, I am fairly certain, the last one I will ever make as a scholar of science, technology, and medicine.

My retelling of buying this stock is also my feminist engagement in a politics of technoscience: We are all part of the New World Order, Inc. (Haraway 1997). The commodity I invested in—a genetically modified, hybrid mouse produced for research purposes only—was brought to us by Dupont. Ethical issues aside, I was (mis)taken in. This fetishized commodity made my heart race and my imagination run wild. It was no longer simply a paper transaction, it was more than a living entity or research material—it had become the future, the future of biotech, my future as a science and technology scholar, my future earning potential, and even my future identity as someone who had made the right pick. The oncomouse and I were going places . . . or were we?[1]

Making the Right Pick:
Technoscience and the Biomedical Imaginary

This chapter, part empirical case study and part rumination on biomedicalization, represents my attempt to theorize choice and subjectivity in the new world order of U.S. biomedical technoscience. It is a cautionary tale of buying in to the hype, to interpreting options as "choices," and to seeing everything as new and holding oh-so-much promise for the future. It is a rumination on imagining otherwise and being accountable. We were, are now, and will continue to be part of the (social science) machine constituted within the politico-economics of the U.S. New World Order, Inc.

Lesbian reproductive practices, as I argue in *Queering Reproduction* (Mamo 2007b), are pragmatic negotiations of biomedical institutions and practices punctuated by multiple constraints (economic, political, material, and cultural) as well as multiple enabling points. They are also exemplars of the uneven, stratified power of U.S. free-market technoscience. In the United States and increasingly elsewhere, commodified sperm offers recipients the chance to "pick a winner" (Schmidt and Moore 1998) and "choose" the right sperm (donor) to imagine and realize future subjectivities as mothers, fathers, parents, or the kinship and social relationships I term *affinity-ties*.[2]

My analysis of this case complements central processes constitutive of biomedicalization theory: political-economic conditions of privatization, the rationalization and devolution of services, and the broader notion of a consumer society (Clarke et al. 2003) produce current biomedical social forms, subjectivities, and practices.[3] My case, however, emphasizes the ways consumer processes call forward not only consumers' desires and pleasures but also their will to imagine the future: there is no choice but to exercise choice. Further, in medical discourse and practice there is no longer a prerequisite to pathologize the body to maintain medical authority and jurisdiction; instead biomedicalization extends its reach to include any and all issues concerning life itself, culminating in a moral imperative to be healthy. While cosmetic and other lifestyle issues are already part of U.S. consumer discourse, their place as objects of U.S. biomedicine is intensifying.

This case study illustrates the ways that consumer discourse — the promises and hopes embodied in the choice among fetishized commodities, and the imperative to exercise choice — saturates the infertility marketplace with its buying and selling of sperm, the practices of conception and getting-to-pregnant, the affective relations produced, and the subjectivities of the users of these services. Discipline through a medical diagnosis dissipates; subjects are made and regulated through processes of maximizing health, minimizing risk, and producing oneself anew. Subjectification processes continue to be produced through biomedicine, yet these no longer (only) rely on medical classification; instead I argue that these are cultural scripts inscribing bodies, identities, and ways of being.[4] Subjectification processes are, at least partially, driven by an expanded jurisdiction of medicine from treating patients to meeting so-called market demands of consumers.

In what follows, I argue that lesbian reproductive practices are constitutive of U.S. neoliberalism and its discourses of self-reliance, self-enhancement, and individual responsibility in the context of privatization, free-market regulation, and the devolution of public state services. Using the empirical case of the interactions of lesbian recipients of sperm bank services, I illustrate the ways in which consumer subjects are shaped by neoliberal discourse yet also push back, reflecting yet variously taking on such discourse. Theorizing neoliberalism along with subjectification allows a theoretical extension of what I see as an underdeveloped area in the sociology of health and illness—the relationship between biomedicalization, consumption, and subjectivities. In the United States, biomedicine, within a consumer-healthcare delivery system, brings forward consumers and offers putative choices. These are not without political consequences, nor are they driven by a neutral set of economic and social forces.

U.S. CONSUMER SOCIETY AND NEOLIBERAL DISCOURSE

Lisa Duggan, in her groundbreaking book *The Twilight of Equality* (2003), defines neoliberalism as a morphing of the original vision of free markets and limited government purported by classical liberalisms upheld in the United States in the form of New Deal social welfare programs and politics of the 1930s to the 1960s.[5] Neoliberalism, in contrast to liberalism, gradually emerged out of "conservative" activism from the 1950s to the 1970s and emphasized a pro-business ethic over welfare state politics. Its effects include the social distribution of wealth upward with its associated downsizing of democracy, attacks on redistributive social movements, a pro-business ethic based on global competition, domestically focused culture wars that rhetorically ensure the false separation of culture from economic issues, and an emergent U.S. multicultural equality politics that effaces economic inequality. Neoliberalism as a discourse and ideology emphasizes ideals of ownership, competitiveness, investment, and individual responsibility.

Accompanying neoliberalist discourse and policies is a "social surge of individualization" (Beck 1992); citizens are self-responsible for the quality, shape, and direction of their own lives, and the role of welfare states and social communities diminishes. Do-it-yourself projects emerge that often require and rely on cultural, social, and economic capital to achieve. Citizens must be self-enterprising and engage in constant self-monitoring and self-projects. The neoliberal mantra, to use the words of Anthony Giddens (1991), becomes "We are, not what we are, but what we make of ourselves." Consumption serves as the means to "make ourselves," to achieve a status, an appearance, an identity—a self.

Fertility biomedicine is constitutive of neoliberal discourse and shares ideological space with its technologies and services: pharmaceuticals and other medical technologies and procedures that, like consumer products more generally, promise subjective ideals. Ideals of ownership and individualism punctuate reproductive practices and services as reproduction becomes another do-it-yourself project enabling us to transform our selves, identities, and social lives through consumption. Fertility biomedicine produces new subjectivities: lesbian mothers, gay fathers, and new family arrangements brought into being through consumption. For many lesbians, buying sperm, and all that sperm embodies, becomes a route not only to achieving parenthood but also to realizing their imagined sense of self, their hope for the future, and a way to communicate their shared affect with others.

CONSUMPTION AND SUBJECTIFICATION

As a neoliberal consumer practice, lesbian reproduction follows similar principles of choice, desire, and what Marx long ago described as the fetishization of commodities—in this case, the fetishization of the biomaterial sperm.[6] Fertility biomedicine commodifies sperm, transforming it into an object that holds monetary and symbolic value. As Scheper-Hughes (2005) so eloquently and horrifically illustrates with the case of the traffic in kidneys, with global capitalism comes the emergence of fetishized biomaterials, and with them the complicity of medical professions and professionals in processes of commodifying bodies and our biomaterials.

This resonates with reproductive medicine. The meanings of commodities become matter-of-fact things in themselves. As Baudrillard (1996) argues, the individual consumer takes on a group identity with each purchase, merely fulfilling the needs of the productive system under the illusion that he is servicing his private wants.[7] Baudrillard's critique resonates with the long-standing sociological question of whether we, our agency, or something external to us—social structure—dominates the shaping of our social lives.[8] In technoscience studies this has been converted into a question of whether users matter. Do free markets generate any agencies in their valuing of individual choice, autonomy, and economic exchange? While we may agree that commodities are often fetishes of human desire, they are processes of meaning beyond their material realities as things. While this evokes the science and technology studies (STS) point that technologies have embedded scripts and meanings built in, it also raises the issue of whether people, as users, consumers, patients, and so on, are able to modify or subvert a technology's intended meaning or ideal use (Moore 1997), thus subverting the expectations and scripts of the developers and marketers of the technologies (Cockburn and Furst-Dilic 1994).

Fertility, Inc.: Commodity Culture and Assisted Reproduction

Fertility, Inc., is a $4 billion dollar a year business in the United States. It comprises free-standing fertility clinics, mostly private sperm banks, a medical specialty, and a growing population of patients seeking services. Given the big business of fertility, strong competition exists for patients among doctors, clinics, and sperm banks. Yet the consumer base of sperm banks, while slightly varied and increasing, has not been elastic. To increase profits, these services need to offer additional services, increase unit costs, or establish new markets. This is especially the case as heterosexual couples increasingly shift from recipients of donor sperm to users of advanced reproductive services such as in vitro fertilization (IVF) with intracytoplasmic sperm injection (ICSI) to maximize their chances of biological relatedness (Becker 2000a).[9] Services are expanded with offerings of in-depth donor profiles, computerized matching, donor photographs with handwritten responses to interview questions, and video-recorded interviews. Fees often include additional unit costs for sperm washing, storage, freezing, and shipping, as well as container rentals and other services. The consumer base has expanded with niche marketing to single women and lesbians as well as international markets through the reach of the World Wide Web.

The histories of this industry make this clear (see Daniels and Golden 2004). The sperm bank industry in the United States shifted from a physician-dominated model (from the 1920s to the 1960s) to a consumer-dominated model (from the 1960s to today). Donors once selected by physicians are today marketed and sold directly to the consumer. By the early 1970s, sperm banking in the United States was primarily a commercial enterprise emphasizing economic logics over social ones, and in the 1980s, with neoliberalism, a biotechnology boom with its concomitant global commerce in eggs and sperm began. The Scandinavian Cryobank, for example, operated as a multinational corporation with offices on Wall Street and in Scandinavia. Its main offering included "Viking" and "Nordic" semen marketed to women in the United States and elsewhere.

With innovations in computer technologies and the Internet, donor sperm purchasing and selection in the United States has largely become an online trade as recipients surf the Web to choose the "right" donor for the job. Assisted reproduction and sperm banks' selection practices are expanding economies of exchange steeped in a rhetoric of neoliberalism, market competition, and individual choice (see Waldby and Mitchell 2006).[10]

In contrast, outside the United States assisted reproduction is largely based on a "gift" model, with assumed voluntary donation and distribution based on need. The selling of human tissues, including sperm, is prohibited in Great Britain, China, Israel, Denmark, Canada, and some countries that follow the Shi'a branch of Islam. The countries that oppose the commodification of gametes do so for a variety of reasons and efforts are underway in the U.S. to regulate payment to egg donors.[11] Significant, and worth further research, are the ways the forms of state regulation proposed or enacted shape how sexuality intersects with gender, nation, race, and ethnicity. For example, in the Danish case, a policy forbids doctor-assisted artificial insemination for either lesbians or single women. Family and reproductive policies are constitutive of sexuality norms and regulations, even in nations proud of their same-sex policies.[12]

In the United States, where my fieldwork took place, lesbian reproduction (and all reproduction) is driven not by a diagnosis of pathology but by a desire to have a child. Health (or everyday wellness, in the language of normalization), not pathology, is the object of intervention. Health is the commodity and process through which to achieve (new) technoscientific identities. The emphasis is less on places of consumption (the clinic is diffuse) than on shifting imperatives, from curing and treating illness to a moral imperative to improve one's health, one's life, one's future. While medicine has a long history of pathologizing childlessness and constituting the childless woman (and her male partner) as the object of disciplining through the application of technoscientific tools designed to "cure" this presumably unwanted state, this work is no longer necessary. In a culture of biomedicalization, with its emphasis on commodification and *health*, biomedicine as institutional practices, culture, and knowledge becomes a key means for transforming oneself from child-free to with-child. No biophysiological "problem" need exist.

Driving forces guiding consumer choice are success rates, range of technologies rendered, and the variety and availability of sperm and egg donors offered. Such biomedical choices are part of consumer culture — the expansion, and at times displacement, of commodity production in the last century and the subsequent shift to commodity consumption. This shift, encompassing advertising and the mass media, has taken on major symbolic and cultural centrality. Direct-to-consumer (DTC) advertising of prescription drugs and other medical products stands as a fundamental case in point of this shift. Advertisements for pharmaceutical drugs are everywhere: magazines, television and radio, billboards, Internet pop-ups, and e-mail spam. DTC advertisements have been described as a "wonder drug" for the drug industry itself because of their ability to influence patient demand — and in turn to influence doctors' behaviors (Jenkins 1999). These advertisements ask consumers to exercise individual choice, express desires, imagine futures, and become that which we are not now. In doing so, drug companies not only bring information directly to the consumer and generate demand for their particular therapies, but also produce consumers and their imaginary futures.

Processes of subjectification, the making of particular forms of being, knowing, and identifying, are constituted within and through assisted reproduction. But we are not yet at the place where advertisers declare, "Ask your doctor about lesbian insemination," solicit you to "find out which semen donor is right for you," and provide a link to http://www .spermbanksofamerica.com. Or are we?

As another do-it-yourself project, assisted reproduction requires self-enterprising individuals with cultural, social, and economic capital to become "what we make of ourselves" (Giddens 1991). Consumption is a means to transform ourselves, to achieve an appearance, an identity, a self, and a degree of physical control — an efficient performance of the self. Fertility biomedicine calls lesbians forward as consumers and appeals to their free will: with economic resources and insurance coverage in hand, choices are offered. What kind of child, future, subjectivity, are you seeking?

Such imaginings have been depicted vividly in popular and academic texts showcasing assisted reproduction. Lee Silver (1997), a molecular biologist at Princeton University, presented a futuristic (circa 2009 and beyond) account of a world divided into two social classes, the genetically enhanced, gene-enriched "GenRich," and the "Naturals" who are unable to financially afford genetic enhancement. Not only does Silver's account assume that all who can, will use genetic technologies, but he overtly argues that many lesbians will come to occupy a subset of the GenRich class.

Such predictions of a highly technologized first world populated by white, tall, rich people continually enhancing their progeny must be read against what we know about the demography of same-sex, cohabitating couples and their children. The 2000 U.S. Census, for example, found that children raised by same-sex couples in the United States reflect a greater racial and ethnic diversity than the population as a whole.[13] Compared to different-sex couples with children, same-sex couples with children have fewer economic resources to care for their children: they have lower household incomes ($12,000 lower per year), a 15 percent lower home ownership rate, and lower levels of education than different-sex parents, indicating that the economic benefits provided by legal marriage would be helpful to these families. Finally, lesbian and gay parents are raising a higher percentage of adopted children than same-sex couples (Sears, Gates, and Rubenstein 2005). This picture is not only far from the popular misconception that gay people are predominantly male, affluent, urban, white, and childless; it is also far from the misconception among some queer communities that gay and lesbian families are mere replications of heteronormativity, instances of "homonormativity" (Duggan 2003).[14] Yet what we know is also a product of power.

While the desire to procreate may or may not be accompanied by a desire to genetically or otherwise enhance the next generation, consumer rhetoric and an increasingly narrow (though at once seemingly vast) array of options shape lesbian reproduction. Perhaps fulfilling Silver's prediction, in *Buying Dad: One Woman's Search for the Perfect Sperm Donor*, Harlyn Aizley (2003) describes her and her partner's adventure in alternative family planning and sperm donor selection. If the title isn't revealing enough, an early chapter is titled "Shop, Shop, Shop 'Til You Drop," in which she likens buying sperm to cruising down the aisles of Sam's Club or Costco with a "genetic shopping list" for height, weight, and engineering and math skills. But what are Aizley and her partner Faith really buying? Are they buying a dad for the baby-to-be? Or are they buying a means of becoming a new identity—a mother, an imagined husband, a genetic profile, cultural legitimacy, or something else? *Buying Dad* reveals a curious imagining of the future, the future human, the future family, and a future of relatedness brought to us through free-market commodity culture with its emphasis on choice and seeking a perfect life. *Buying Dad* also highlights the ways the body, in this case the material-semiotic body part, semen, figures as a central actant in processes of meaning making.

Questions remain, however, about what kinds of subjects are self-made and being made when lesbians enter Fertility, Inc. In what ways is Fertility, Inc., changing them and what they believe in ways that they see and don't see? How is it changing their notions of self, family, body, future? How are processes of subjectification occurring? With what stratifications and consequences? And with what geneticized meanings?

Managing Risk, Maintaining Security: The Self-Project(s) of Assisted Reproduction

Lesbian reproductive practices reflect a heightened concern with security and the rhetoric of risk management that has saturated much of public discourse. These practices are inseparable from neoliberal emphases on ownership, individual responsibility and choice, and consumption as means to fulfill one's desires, identities, and life goals. A goal of biomedical technoscience, its knowledges, and its technical application lies no longer in delaying, curing, or treating one's illness but in calculating and minimizing risk; achieving, maintaining, and enhancing health; and driving consumers to buy into the services and dreams offered.

In today's biopolitical times, power is no longer organized around death but is "carefully supplanted by the administration of bodies and the calculated management of life" (Foucault 1980, 139). Individuals must exercise freedoms and responsibilities through self-governance and risk management (Novas and Rose 2000). Subjects are made; identification and disidentification are produced through strategies of self-governance. Such processes take place in and through assisted reproduction. Here individuals are embarking on forms of risk management. They are configuring not only their self-identifications but also their (potential) children's risk, health, and the meaning and substance of future relationships. Such a future project is produced by and through the technological choices offered up in the industry that Gina Kolata (2002) termed "Fertility, Inc." Lesbian reproductive practices, and all assisted-reproductive practices, demonstrate the production of consumer subjectivities as Fertility, Inc., calls its consumers forward, appeals to their free will, and asks them to engage in body projects including risk management. We can surf the Internet for sperm donors, egg donors, and surrogates who will carry out the gestational arrangements; we can direct search engines to sort large databases of donors to match our preferences for eye color, hair color, education, health history, risk profile, race and ethnicity, occupation, and other desired attributes of those supplying such services and biomaterials. We can bank biomaterials, buy in bulk, and own the rights to future sales. The choice is (y)ours: what kind of child, future, subjectivity, are you seeking? The technological practices of insemination, in vitro fertilization, and hormonal therapies do much of the same: What practices do you choose to enact en route to your future self and social connections? What responsibilities do you assume for your health and the future health of your offspring?

While sperm banks offer a range of electronic and print donor catalogs for consumer ease in choosing a sperm donor, it is the ability to surf the Internet that has come to dominate selection processes. Search and sort components allow inputs to become outputs with speed and accuracy. As consumers list desired characteristics and combinations (i.e., of a donor's race or ethnicity, height, weight, eyes, hair, body build, complexion, health history, hobbies, interests, and educational attainment), matches are generated via calculation. These allow and encourage recipients not only to compare across donors but to imagine and produce desired combinations. A lack of sperm transforms from a problem to an opportunity. From a reproductive politics perspective, however, the question remains: an opportunity for what?

As Cynthia Daniels (2006) found in her research, sperm donors are usually taller, leaner, and healthier than "average" men and usually possess interests and hobbies associated with masculinity such as golf, baseball, and other sports. Thus the winner usually represents the "ideal," not the average, man in the United States. Such ideals not only shape the donor output but extend to ideals about consumption, social relations, and family forms. Consumer choice is shaped by hegemonic notions of beauty (Bordo 1993), and race is increasingly a matter of style, moving from a biological category to an aesthetic one (Lury 1996). Picking a winner has sociopolitical consequences.

Becoming a sperm recipient is a process of becoming a consumer of sperm and all that sperm embodies. As one of my respondents, Dana, stated, "It's odd to go shopping for a donor." Another respondent, Esther, said, "It feels really weird to walk in off the street and pick up a bottle of sperm and pay for it. . . . It is a whole market." While both women expressed hesitancy about the consumer-driven market in sperm, they continued to describe how they made their picks: Dana by looking for donors who were similar to her partner, and Esther by looking for Jewish donors with interests and hobbies similar to her own.

For both women, and the others I interviewed, making meaning of and managing risk also emerged as central to their choices. Managing and minimizing risk emerged through women's perceptions of the promise of technoscientific advances in sperm testing, washing, and storage and through selecting sperm with the right genetics. Carla said: "Even before we look through the list of donors, we know they have screened out anyone who isn't extremely fit, smart, and healthy." Yet even with such pre screening techniques, when choosing donor sperm from a sperm bank, consumers are encouraged to enhance the next generation by reducing not only the risk of potential illnesses but also the risk of perceived cultural liabilities such as shortness, acne, lack of academic, athletic, or musical abilities, or having a certain skin color, eye color, or ancestral background. In other words, these users are presumably averting not only disease risk but also the risks of social mediocrity. As a result, minimizing physical risks is part of maximizing social fitness. In becoming users of Fertility, Inc., lesbians who can financially afford such choices are shaping themselves and their futures through their choices.

Sperm banks construct discourses of reproductive risk, capitalizing on consumer concerns about hazards to one's health, risks of birth defects, and increased time to conception. "Technosemen" is used as a means to overcome sperm limitations such as "uncleanliness," poor mobility, low velocity, and unpredictability (Schmidt and Moore 1998). To promote technosemen, sperm banks highlight screening procedures both to ensure trust in the integrity of the sperm and to attest to the selectivity of men invited to participate in reproduction. Specifically, such procedures include testing for sperm count and motility; infectious disease screening (i.e., HIV-1 and 2, HTLV-1 and 2, syphilis, hepatitis B and C, CMV, chlamydia, gonorrhea, etc.); a six-month semen quarantine for retesting; semen "washing" (a procedure to remove seminal plasma and nonsperm cellular material, thereby reducing the risk of uterine cramping and infection); and semen analysis to measure parameters such as liquefaction, volume, viscosity, pH, motile sperm concentration, total sperm concentration, percent motility, progression, percent abnormal morphology, and white blood cell concentration. Together such manipulation builds trust in a supernatural commodity and promotes a real (and perceived) sense of selectivity.

In selecting donor characteristics, nature is not only "enterprised-up" (Strathern 1992, 30) through selecting positive characteristics but also enabled by minimizing potential disease risk through a careful selection of health histories. Such processes highlight the ways in which biomedicalization includes new regimens of risk classification and practices designed to minimize, manage, and, in the case of breast cancer prevention, treat risk (Fosket 2004). Risk management entered as women constructed probability assessments for future health and illness of potential children based on their own and the donor's health histories.

Embedded in practices of sperm selection are late-twentieth-century discourses of geneticization that are central to the ways women imagine future connections by rematerializing donors as objects through which genetics flow. Part of today's heightened scrutiny of genes as presumed causal factors in one's health, sperm banks rely deeply on genetic discourse and knowledge in their services. Most sperm banks provide health histories of donors going back at least one, and often up to five, generations. In selecting donors, women engage in processes of understanding their own health and family health history in relation to the donors' family narrative. Their decisions become a way of reducing risk for the potential child: if breast cancer is present in a bio-mom's family, a donor would be chosen with no cancer in his family.

Dominant cultural discourses about genetics, heredity, and health were often transparent to the women I interviewed. Joyce said, "My family and I were calling it genetic engineering." All respondents thought they knew that some aspects of health are inheritable and identified certain illnesses as proof. Tina said, "It did feel like genetic engineering, though. How tall would we like him? Do we prefer a graduate student or an athlete? What about physique, intelligence, and health? Is there a history of cancer?" Raquel, Paula, and Esther indicated knowledge about what is and is not inheritable. While Raquel emphasized cancer and schizophrenia as genetically determined risks, Paula raised her concern about alcoholism as a genetic risk, and Esther perceived good eyesight as genetically inheritable.

RAQUEL: We wanted to pick someone who, even though we trusted that [the sperm bank] probably wouldn't have in their catalog someone who had a huge amount of schizophrenia in the close relatives or something like that. We wanted to stay clear of people with even an appearance of some kind of cancer.

PAULA: A really big issue for me was people who had alcoholics in their family. I don't know that there's hard conclusive evidence, but there's a lot of [studies] that have shown that there is an inheritance of alcoholism. And I guess because I've seen situations of families who've adopted kids whose parents were alcoholics and what's happened to their child. . . . And there isn't any alcoholism in my family.

ESTHER · My really big issues are health and then eyesight. I've really gotten stuck on eyesight lately. I want to give him a chance and figure if the donor has 20/20 vision, then I figure they've got a shot, you know.

In all, these comments exemplify self-management strategies used to maximize current and future physical and social well-being. Genetics have emerged as a key means through which life is understood and by which disease will be cured. As the mapping of the human genome has uncovered several gene markers for disease joining BRCA1 and 2 breast cancer genes, and as the media purports the discovery of genes for Down syndrome, Alzheimer's disease, prostate cancer, and others, cultural understandings of health and illness are increasingly geneticized (Lippman 1992).

Sperm, as commercial sperm banks tell us, can be sick or healthy. Socially and physically dominant donors were selected assuming that their sperm would help create socially dominant offspring. At times donors were also selected to enhance familial qualities (i.e., choosing a tall donor if one is short). In addition, social health (such as education, hobbies, and interests) also emerged as an important quality in donors. In both cases, selection decisions mirrored dominant U.S. cultural understandings of physical and social power. Key indicators of ideal physical and social power include health, as well as height, weight, body build, sports, occupation, grade point average, and years of college. As Schmidt and Moore (1998) argued, these are social indicators of one's ability to be physically and socially dominant. June described the importance of maximizing health and her belief that health is genetically inherited:

We were looking for someone with remarkable health that extended out to maternal, paternal grandparents and aunts and uncles too; someone healthy without a cancer history or Alzheimer's or other things that seemed genetic. My mother said, "You know more about this donor than I ever knew about your father when we started having kids." But we figured, as long as we have a choice, we might as well try to go for the most remarkable health stuff.

Similarly, Judith linked the commodification of sperm with the ability to "buy" health:

> What really affects me is how much I have to pay to get healthy sperm that can survive freezing. Sperm that has been quarantined and tested for all the diseases. I know a lot about this person in terms of health. I know what he's not carrying. If I met him at a bar or he was my best friend who didn't tell me a few things, I wouldn't know. I think it's important to know these things. . . . It's a matter of health. It's a matter of viruses. It's a matter of self.

Women are making these choices "because they can," understanding the commodification of sperm, health, and genetic inheritance. This raises the questions: What are screened sperm worth? What are good genes worth? What opportunities for the future are offered? Sara describes the variety of qualities that came into play as she made her selections:

> Health was first and then music. Intelligence wasn't really an issue because the people they recruit, at least among the Asian donors, are very intelligent already. They all had over 1,400 on their SAT scores and were in graduate school or they were undergraduates and had amazing grade point averages. They all seemed really intelligent already. If you can pick, we figured we definitely wanted someone that can do what they'd like in life, whatever they'd like to make them happy. Handwriting and creativity were also important issues. I guess it extended to include general creativity and what they seemed like, their personality. Part of this was what they thought of the women in their lives. They write opinions of their relatives and siblings, and what they said gave us an idea of their personality. What is their mother like. We asked ourselves, "Do they feel really positively towards the women in their life?"

Asking themselves what they wanted, what they preferred was a matter of considering the promises and pitfalls held by the choices offered. Cultural ideas concerning genetics and heredity were mobilized as inheritable. Qualities such as a health condition were knowable, selected, and mapped through blood ties. Sara said: "We know how biological all this is," and "I think my primary concern was genetics." Throughout, she emphasized what she believes we "know" and "don't know" about heredity. Somewhat discounting environmental explanations, she eliminated donors with any history of mental illness and substance use in their ancestral lineage. Health histories — almost regardless of the weight of the evidence as to the heritability of different disease conditions — were used to enhance potential offspring and reduce risk. Further, social characteristics such as musical abilities, intelligence, not being a misogynist, being creative, flexible, and possessing strong coping skills emerge as possible hereditary attributes although they are more commonly construed as acquired rather than hereditary characteristics.

Finally, risk was reduced and security maintained by navigating legal policies. Maintaining security is evidenced through women's access to medical-based and legal-based services. Each is enabled by access to, and ability to pay for, the legal advice and services necessary to protect one's parental rights. Under the law in California, where my research took place, a known sperm donor is legally considered the father of the child except "where sperm is provided to a licensed physician and surgeon for purposes of insemination in a woman not his wife."[15] Risk is not only a technoscientific construction and identification; it is also a perceived threat to one's future family. In this research, one way to minimize the risk of custody battles, and therefore losing a child to a legally defined father, was to use sperm bank services protected under the medical umbrella with its associated definitions and regulations. Although the language "licensed physician and surgeon" eliminates from protection most family practitioners and internists who are not surgeons, as well as all nurse practitioners, sperm banks as medical clinics under the supervision of an MD are included in such protections. Recipients who used sperm banks were highly aware of this added security in their insemination practices.

If a sperm bank was not used, however, as was the case for Diana and her partner, then the significance of the physician took on added importance. Diane stated: "The donor and his partner came over, they produced. And my other friend, she's obviously a doctor, came over and handed it to me. Then we [she and her partner] went and put it in using a needleless syringe. The four of us went out to dinner. You know, it was really nice." In this insemination story, the pass-off of semen to a physician to give to the user ensures that legal risks are "managed." Thus Diane not only understands California law (i.e., a known semen donor is considered the father of the child except "where sperm is provided to a licensed physician and surgeon for purposes of insemination in a woman not his wife") but has actively negotiated its parameters to maximize future security. Thus, while consumers are being agential, self-enterprising consumers, they are doing so within the structural design of not only how the technologies were intended to be used but also by whom and for the production of what kinds of families.

Final Ruminations: From Reproductive Choice to Justice

This is my past-present-future rumination on the power of technoscience. It's been almost a decade since I left the San Francisco Bay Area and since the dot.com bubble burst, and three years since the publication of my book. Things look very different from over here. What seemed like normativity in San Francisco appears as a far more nuanced queering in Washington. Yet this queering also seems to be a rallying cry for inclusion. From the point of view of my previous situation in a largely nontraditional urban landscape, having families seemed to be the most traditional thing one could do. My current situatedness in a more tradition-bound area allows me to see the diversity of lesbian and gay family forms and the self-enterprising means required to achieve these family forms as queering the boundaries of normalcy. As a result, my analysis is more subtle today than it was then and there. Yet at the same time, I hear the even louder calls for inclusion as marriage and family become central to mainstream LGBT politics.

The thrust of the story remains the same: Lesbian reproduction is quickly moving from a do-it-yourself alternative practice to complex engagements with, and consumption of, a panoply of biomedical services. The importance of economic and cultural capital plays a larger role in the story. Accessing these biomedical services is produced by health insurance status and the ability to pay. While lesbians seeking pregnancies can look for sperm donors among friends, pick sperm from a catalog, or have the fertilized egg of one partner implanted in the other, each pathway not only entails compromises about cost, safety, control, and legal relationships but varies and is driven by cost and ability to pay. Economics stratify reproduction. Fertility, Inc., shapes lesbian subjects as flexible citizens ready and willing to participate in the self-projects offered by consumer culture (see Davis-Floyd 1992; Strathern 1992; Edwards et al. 1993/1999). Yet the more these "free choices" are expected, the less choice remains. With a vast array of choices (IVF, egg donor, sperm donor, home insemination, IUI, etc.), the choice is the same: biological reproduction.

In sum, lesbian practices speak to a consumer society marked by ideals of ownership, presumed individual choice, and consumption as means to fulfill one's desires, identities, and life goals. Consumption infuses all aspects of our lives, including our reproductive lives. Healthcare has extended in ways that emphasize consumer processes of pleasure and transformation; there is no choice but to exercise choice. Healthcare's object encompasses any and all issues concerning health and lifestyle with a goal of meeting assumed market demands. The respondents describe interactions that are pragmatic negotiations taken as they reach toward an end goal or desire: to become pregnant and attain motherhood. For many, buying sperm—and all that sperm embodies—becomes a route not only to achieving parenthood but also to realizing their imagined sense of self, their hope for their own future, and the future health of their children.[16] While it may not be enhancement, lesbians are managing cultural norms as they make their selections: what height, color, eye shape, and hobbies will we choose? Such selections are fetishized as each choice becomes an opportunity for the future.[17]

The contours of who is constrained from such choices are effaced, yet everywhere all too apparent. In my interviews, for example, those who stopped trying to become pregnant faced economic barriers they either could not or were unwilling to overcome. Some, however, negotiated this constraint by using alternative insemination methods (for example, Diane and her partner engaged the [free] services of a physician friend).

As I debate the degrees of queering, it cannot be argued that at the center of such subjectification is the production of one of the most durable identities in the United States and elsewhere: mother. The practices I analyzed were in the service of achieving a long-range self as a parent. Further, it is through technological means that parents are made. While old stratifications continue in the form of discrimination against lesbians, access to reproductive technologies, legal regulations defining family, and an embedded heterosexual script in the technologies and services themselves, these actors largely were able to secure their own future. Absent here are those who choose not to reproduce and those who do not align with liberal ideals of family embraced by U.S. mainstream LGBT politics. As a result, a more leftist analysis of inclusion and choice is missing from this paper as well.

I have argued that for many lesbians, buying sperm, and all that sperm embodies, becomes a route not only to achieving parenthood but also to realizing their imagined sense of self and their hope for the future. It is also a means of communicating their shared affect with others and thus, a legitimacy in a social order that legally and socially privileges heteronormativity. Finally, buying sperm is shaped by and through stratified biomedicalization and cultures of consumption, risk, and a will to health. The meeting of bodies with technological and scientific practices is part of culture and power; they do not exist outside culture and power. Bodies—and subjectivities—are conceived, technologized, and debated within politically and socially meaningful contexts by people who face different and multiple situations of power. Judith Butler, for example, argues that it is politically crucial that we lay claim to intelligibility and recognizability while simultaneously maintaining a critical and transformative relation to the norms that govern what will and will not count as an intelligible and recognizable alliance and kinship (Butler 2002, 28). My concern with whether or not a queering is taking place is reworked as a question about what price producing recognizable and legitimate subjectivities brings.

The implications of neoliberalism and its emphasis on free-market regulation have not been fully theorized in the context of assisted reproduction and consumption, nor has the question of whether or not fertility services should be regulated at all. There are many good reasons for this. Most notable is the history of forced sterilization; equally important is the agenda item of some feminist health organizations to include reproduction as a human right: that all women should have control over not only if and when they reproduce (through the availability of contraceptives and abortion services) but also their choice to reproduce (through assisted reproductive services).

The emphasis of neoliberalism on the rights of individuals gave lesbians, gay men, and single people increased access to reproductive services under the name of privacy rights. Privacy rights have largely generated rights such as contraception information (the 1965 *Griswold v. Connecticut* decision to strike down a state law prohibiting the use of contraceptives by married couples), the right to choose to have an abortion (the 1973 *Roe v. Wade* decision), and more recently the right to engage in same-sex practices (the 2003 *Lawrence v. Texas* Supreme Court decision to appeal state laws criminalizing consensual sex between two adults of the same sex). While these have collectively expanded reproductive citizenship, they have also magnified the dividing practices between those who can afford such services and those who are unable to pay and are not covered by the state or private insurance. These rights are largely regulated by the free market. Charis

Thompson (2005) has argued that cost is the main driver of biomedical citizenship in the reproductive arena. I agree. By becoming a major site for the unfolding of the practices of Fertility, Inc., lesbian and gay reproduction is constitutive of the production of consumer citizenship. Queers of all kinds, as consumers with identities and lifestyles expressed through purchase and acquisition, are being incorporated into consumer culture (Freitas 1998; Richardson 2001).

There is much to embrace here. Through consumption of biomedical knowledges, technologies, and services, lesbian mothers, mothers-to-be, relationships, and family forms are no longer abject; instead these actors are recognized as members of culture (even if not always legitimate ones). The social institutions of biomedical services (fertility specialists, sperm banks, assisted-reproductive technologies and practices) and the law (via marriage, domestic partnership, and adoption rights) are central avenues through which parenthood and ultimately equality, normality, and citizenship are secured. This turn to biomedicine and the law is indicative of a sexual citizenship and is, in part, constituted within a culture of consumption.

Several have critiqued these advances, including those who argue that these produce a potential reinforcement of heteronormativity or a new "homonormativity" (see Duggan 2002; see also the 2005 special edition of *Social Text* edited by David Eng, Jose Munoz, and Judith Halberstam). Of course, the critique is not about any individual action or choice but about the social and political forces shaping social lives. I argue that lesbians as consumer citizens are produced within and through the shift from lesbian, gay, and queer struggles for sexual liberations to battles to reproduce, marry, parent, and form familial relations in the same way as heterosexuals (Dominus 2004). While one may interpret the growing access to consumer-based reproduction, parenthood, and even legal marriage among lesbians (and gay men and others) as inclusion, or a broadening of citizenship, such transformations pose fundamental problems concerning the production of difference and normalization within the social.

In thinking about free-market regulation, I close by returning to a statement we raised at the end of our ASR paper. In that conclusion, we call for the use of biomedicalization as a critical lens, as it allows for the mapping of momentary spaces of negotiation and the possibilities of democratic interventions (Clarke et al. 2003, 185). We advocated for justice and a feminist politics. The future is here. Forty-five million people in the United States have no medical insurance, and a billion people globally have no basic healthcare. These numbers are on the rise. At the same time, healthcare is an increasingly commoditized, for-profit market, based in and driven by consumerism. Many biotech-based treatments as well as basic healthcare provisions are extremely expensive, thereby increasing health and other disparities.

The commercialization of the human exists as human beings are rendered perfectible through the market. Examples are numerous, including the rise of sex selection, the sale of organs from the poor to the rich, the boom in enhancement technologies such as cosmetic surgery, and gene doping for athletes. Furthermore, social issues are increasingly being defined as strictly genetic or biomedical problems, not social or environmental phenomena. These include disability, obesity, sexual orientation, gender variance, poverty, violence, breast cancer, osteoporosis, and rickets. In all, existing social divisions are exacerbated. As Lee Silver predicted almost ten years ago, this may create a biological basis for difference leading to inequalities based on such difference and a possible emergence of genetic castes.

Finally, I end this chapter by calling attention to an activist-academic alliance that produced a statement and emergent social movement titled Beyond Marriage (see http://www.beyondmarriage.org), a campaign that not only joins but is organized through Queers for Economic Justice (QEJ, online at http://q4ej.org). My politics of technoscience are accountable to the coexisting, mutually shaping social forms variously addressing and producing social stratifications. I am accountable both to an organization such as QEJ, which asks us to imagine otherwise, to develop an agenda for queer rights that moves beyond (yet includes) reproductive and marriage rights, and to those lesbian subjects seeking some inclusion in normativity. Yet in researching lesbian reproduction, the QEJ broad agenda of coalition building with groups advocating for economic, racial, gender, immigration, and other justice concerns has remained in the shadow of queer politics. I yearn for my own and our collective politics of technoscience to hold all these forms of politics as possible futures. The issue items on the QEJ agenda, for example, include homeless shelter organizing, immigrant rights, welfare organizing, and others are each important to family values. Their platform for LGBT communities seeks to move beyond marriage and reproduction as single-issue concerns and instead to maintain these issues and build on them. I'll let you surf your way to these and other possible futures.

Notes

1 Although my $400 did magically become $10,000 during the boom, alas, I forgot to sell. In 2006 Abgenix was purchased by Amgen for a cash payout of $22.50 a share. My $400 investment was redeposited in my account as $2,250. While this was an incredible rate of return, the maximum return remained elusive.

2 In the United States, sperm is incorporated into the Tissue Guidelines as a biomaterial available for sale. It follows the same model and policies as the commodification of other therapeutic tissues. In contrast, the selling of human tissues (including sperm) is prohibited in most other developed countries, including Canada and the Netherlands (countries who pride themselves on liberal same-sex policies). The United States is unique in its treatment of the commercialization of this biomaterial.

3 I intentionally use the term "consumer" throughout the chapter to highlight the emphasis on commodification and the business mentality of U.S. Fertility, Inc.

4 Jennifer Fishman and I showed how the medical category of erectile dysfunction (ED) was replaced with everyday erection difficulties—part of being human (see Mamo and Fishman 2001). In *Queering Reproduction* (2007b) I argue that lesbian reproduction (and all reproduction) is driven not by a diagnosis of pathology but by a desire to have a child: biomedicalization, with its emphasis on commodification and *health*, is a key site for transforming oneself from child-free to with-child. In a current project, Jennifer Fosket and I argue with the case of Seasonalle and other emergent birth control pills that pregnancy prevention has been replaced with being and living free from menstruation (Mamo and Fosket 2009). In all, the discursive shifts from sexual disorder to difficulties, from childless to achieving parenthood, and from pregnancy prevention to living menstruation free, produce new subjectivities.

5 The global economic collapse that began in 2007 with the housing bubble followed by the banking crises has produced a return to state programs. This, however, remains to be studied. Early signs indicate bailouts for industry, not social programs for publics. Nonetheless, free-market economics are in the process of their own demystification.

6 Ivan Illich (1992), drawing on Marx, argued that as an object of manipulation, the ultimate fetish is life itself.

7 In *The System of Objects*, Baudrillard (1996) offered a cultural critique of the commodity in consumer society, arguing that there is a psychological imperative to consumption. It is meaning, not use, that gets transferred through consumer objects. The body enters into this discourse as text, as a "material-corporeal" unit of exchange, and as a technological project. Biomaterials become semimagical and symbolic representations, heavy with social meanings and significations (Appadurai 1986; Scheper-Hughes 2005).

8 Biomedicalization processes, as meso-level institutional practices and new social forms, are ripe for analyzing questions concerning the degrees of agency and

structure. I examine lesbian practices of achieving pregnancy as forged with many agencies; my argument is that despite the most idealistic vision, structural constraints are imposed and negotiated. I attend to the messy workings of power and culture in shaping these practices.

9 ICSI involves injecting a single sperm into an egg in the IVF lab. It is used in cases where the man has weak sperm or an extremely low sperm count. If the ICSI procedure is successful, the fertilized egg is transferred to the woman's uterus using normal IVF procedures. Recent statistics show that ICSI is used in about 47 percent of IVF procedures in the United States (Centers for Disease Control 2000).

10 In *Tissue Economies* (2006), Catherine Waldby and Robert Mitchell examine what they see as the rapidly expanding economies of exchange in human tissues. Their argument is marshaled using the cases of blood banks, stem cells, umbilical cord blood, the market in human organs, and others. In doing so, they argue that an emergent commercial market of human bodies, bodily processes, and biomaterials is constitutive of late capitalism. Waldby and Mitchell conclude that a blurring has taken place from what was once considered "donation" and a gift exchange to the more recent incursion of free-market values resulting in "tissue economies."

11 The U.S. market model that largely views gametes as commodities is not universally replicated in other countries. In China, in 2006, the Ministry of Health tightened regulations over sperm banks and banned the commercial donation and supply of human eggs (*China Daily*, April 13, 2006). In Great Britain in 2004, the Human Fertilization and Embryology Authority (HFEA) discouraged sperm banks and fertility clinics from using paid donors, thereby promoting an ideal of altruistic donation. Israel, Denmark, and Canada similarly enacted policies to discourage the use of paid gamete donors. In Canada, the Assisted Human Reproduction Act of 2004 put into law a prohibition on the purchase of sperm or eggs. This move has fueled a transnational trafficking of eggs and sperm to Canada, largely from the U.S. (Daniels et al. 2006; Reid, Ram, and Brown 2007). In Israel, third-party sperm donation is legal as a result of the conferral of Jewishness through the maternal line. It is also evident that the American consumer model of free-market reproductive medicine has not (yet) taken hold in Israel (Kahn 2000) or in the Middle East and Sunni-majority Muslim countries (e.g., Egypt), where artificial insemination with a husband's sperm is allowed, while sperm donation is not. This distinction upholds the belief that marital functions such as reproduction should not be interfered with by a third party. In Iran the boundaries around a heterosexual couple are not strongly upheld. The Shi'a Islam issued a fatwa permitting gamete donation. This fatwa is followed by many Shi'a, including leaders in Lebanon. Any child born by sperm donation will follow the name of the father, not the sperm donor. Yet regulations persist and any couple requiring and requesting assisted-reproductive services must come before Shi'ite religious courts who ultimately make decisions case by case.

12 While policies on donor reimbursement cross-nationally reflect the corresponding value of gift versus commodity, the gift-commodity dichotomy, as Waldby and

Mitchell (2006, 9) illustrate, is an "inadequate way to conceptualize the political economy of tissues in the modern world of global biotechnology." Debate over donation versus gift of biomaterials is not new; however, intervention by state governments into medical practices is.

13 The census data shed no light on how many single gay men and lesbians are raising children, given that sexual orientation and sexual behaviors are not asked on the census itself. Further, it is estimated that the number of gays and lesbians declaring their partnership status is an undercount given either an unwillingness to disclose their sexuality or a motivated decision not to declare until the "marriage" box is allowed by law.

14 Another way to read these practices is as a queering of family demography. Whether or not reproductive technologies and practices in general represent a queering is driven by one's definition of queering itself. If the benchmark is contesting dominant heteronormative assumptions and institutions, then lesbian- and gay-parent families may adhere to Lisa Duggan's (2003) analysis that these family forms are creating a "homonormativity." If queering is defined as subverting a norm, as I think it is, then these families can be read as producing social change.

15 California Family Code 7613 reads: "(a) If, under the supervision of a licensed physician and surgeon and with the consent of her husband, a wife is inseminated artificially with semen donated by a man not her husband, the husband is treated in law as if he were the natural father of a child thereby conceived. The husband's consent must be in writing and signed by him and his wife. The physician and surgeon shall certify their signatures and the date of the insemination, and retain the husband's consent as part of the medical record, where it shall be kept confidential and in a sealed file. However, the physician and surgeon's failure to do so does not affect the father and child relationship. All papers and records pertaining to the insemination, whether part of the permanent record of a court or of a file held by the supervising physician and surgeon or elsewhere, are subject to inspection only upon an order of the court for good cause shown."

16 The phrase "achieving parenthood" is borrowed from Ellen Lewin (1993) and Faye Ginsburg (1989).

17 Yet many of their practices reflect anxieties about social connection and belonging. Their actions take place in a culture in which healthcare services facilitate their reproduction while state policies simultaneously legally and politically contest their right to exist. In this context, lesbians are seeking to redefine and create family in ways that render them legible as full participants in the United States and in ways that might be used to gain legitimacy in a largely heterosexist culture. In doing so, the bounds between family and community are blurred as the field of procreators extends beyond traditional mothers and fathers.

Kelly Joyce

6 / The Body as Image

AN EXAMINATION OF THE ECONOMIC AND
POLITICAL DYNAMICS OF MAGNETIC RESONANCE
IMAGING AND THE CONSTRUCTION
OF DIFFERENCE

Since 1985 the organization and practice of medicine have shifted from medicalization to biomedicalization (Clarke et al. 2003)—scientifically dense and dynamic processes. This transformation in biomedicine, while uneven and incomplete, calls attention to the increasing technoscientization, corporatization, and privatization of healthcare; the expansion of risk categories and surveillance practices; the heterogeneous production and distribution of knowledge; and the production of new technoscientific individual and collective identities (Clarke et al. 2003, 161, 163). Woven and webbed together, these processes co-produce the contemporary medical field and mark the turn toward biomedicalization.

In this chapter, I argue that analysis of magnetic resonance imaging (MRI) illuminates a particular material form of the increasingly technoscientific nature of biomedical innovations—a tendency toward visualization. Occurring concurrently with the transformation to biomedicalization processes, MRI technology was introduced to clinical practice in the early 1980s and was initially used to visualize the brain, the spine, and the ligaments, muscles, and tendons around joints. MRI today is used to create anatomical pictures of blood vessels, tendons, ligaments, muscles, organs, and soft tissues throughout the body.

New research aims to further extend MRI use into areas typically controlled by psychiatrists and psychologists. Demonstrating how technoscience is crucial to the production of individual and collective identities, research projects now explore if MRI use can help legitimize identities such as schizophrenia, obsessive compulsive disorder (OCD), and attention-deficit (hyperactivity) disorder (ADHD). Research also uses MRI and other imaging technologies to examine differences between men's and women's brain activity and between disease presentation across various ethnic and racial categories (e.g., Kakigi and Shibasaki 1991; Kolata 1995; Liao et al. 1997). Part of a broader cultural move to visualize identity, MRI illustrates just how technoscientization takes place in contemporary healthcare.

Examination of MRI also demonstrates the continuing importance of the state (in its local and federal forms) in the technology's expanding use in the United States. The use of visualization apparatuses like MRI machines in research and clinical practice is fundamentally rooted in contemporary political and economic relations, or what Clarke et al. (2003, 166–67) refer to as "the Biomedical TechnoService Complex Inc." MRI examinations exist as commodities that simultaneously establish cultural identities such as man, woman, cancer patient, and schizophrenic, as well as produce profit and income for manufacturers and imaging centers. This chapter shows how U.S. federal programs and institutions such as the Centers for Medicare and Medicaid Services and the National Institutes of Health participate in and help produce the expanding political economy of MRI technology in the United States. Government agencies in conjunction with private actors such as manufacturing companies and imaging centers create a lucrative imaging market in which the buying and selling of visual culture takes place at an intensifying speed. State agencies, as key actors within the Biomedical TechnoService Complex Inc., aid the growth of MRI industries in the United States and should not be ignored in analysis of biomedical practices.

Computer-Based Visualization Technologies

Clarke et al. (2003, 173) argue that biomedical practices and innovations depend increasingly on hybrid configurations of science and technology. Innovations in computers, data banking, genomics, pharmaceuticals, and medical technologies are integral to the growing technoscientization of biomedicalization, and they create new objects and subjects in biomedicine. The material forms and practices of new technoscientific transformations and innovations, however, require investigation (176). The growth of medical imaging technologies demands opening up the biomedicalization theory to another trend in medicine, one that has continued for centuries and in recent times become inseparable from the increasing technoscientific nature of biomedical practices: the ever-expanding visualization of biomedicine.

Visualization and the desire to see the inner body are crucial to knowledge production in contemporary biomedicine. Scholars of science, technology, and medicine provide critical analyses of medicine's visual culture and examine the deployment of cameras and other visualizing devices in healthcare and research (e.g., Beaulieu 2000, 2002; Cartwright 1995; Cohn 2004; Dumit 2004; Hartouni 1997; Joyce 2008; Treichler, Cartwright, and Penley 1998; Waldby 2000). The move toward visualization has a long history in medicine—one that is tied to an increasing emphasis on surveillance and control. Michel Foucault locates the emergence of the insistence on visuality, or what he calls the clinical gaze, in eighteenth-century Europe. This period is one in which "illness, counter-nature, death, in short, the whole dark underside of disease came to light, at the same time illuminating and eliminating itself like night, in the deep, visible, solid, enclosed, but accessible space of the human body. What was fundamentally invisible is suddenly offered to the brightness of the gaze, in a movement of appearance so simple, so immediate that it seems to be the natural consequence of a more highly developed experience" (Foucault 1973, 195).

From this era onward, medical knowledge and practice insist on rendering inner spaces and processes seemingly transparent. Privileging a way of knowing that centers on "a world of constant visibility," the clinical gaze insists on making the interior body exterior (Foucault 1973, x). The gaze normalizes the manipulation of the inner body, allowing the customization of bodies at the cellular and genetic level (Clarke 1995; Clarke et al. 2003;

Squier 2004). What matters now is the reconfiguring of bodies through chemical, surgical, and lifestyle interventions and, more recently, genomics.

Computer-based visualization technologies are crucial to the legitimization and deployment of surgical, chemical, and lifestyle interventions and are part of the larger trend of biomedicalization that, as Clarke and her colleagues (2003, 181) rightly suggest, works "from the inside out." After World War II, researchers—as part of a broader social investment in science and technology—explored ways to extend and transform existing scientific techniques such as sonar, x-ray, radioactive tracers, and nuclear magnetic resonance to produce knowledge about the internal body and health (Blume 1992; Kevles 1997; Wobarst 1999). This work culminated in imaging technologies such as ultrasound, computer-assisted tomography (CT), and magnetic resonance imaging (MRI), which were introduced to hospitals and clinical practice throughout the 1970s and 1980s.[1] Imaging techniques are hybrid technologies, as each relies on innovations in computer processing and information technologies to produce and display pictures in a timely fashion. Without such innovations, the development and integration of MRI, CT, and ultrasound into medical care would have been impossible.

The development and expansion of visualization technologies in biomedicine occurred in a broader cultural context that emphasizes the visual. Throughout the 1900s, the use of pictorial images proliferated in private and public spaces in the United States (Clarke 2005; Mirzoeff 1998; Sturken and Cartwright 2001). X-rays, photographs, and film cameras all provided new ways to visualize life. The transformation of daily life into visual form, while gaining momentum throughout the last century, intensified in the 1980s and 1990s as video cameras and recorders, photocopying machines, cable television, and computer visualization technologies became more available. From mirrors to cameras to picture-producing cell phones, we are now expected to understand ourselves and our relations to others through technologically produced images. The use of images in all areas of social life provides the technological support and cultural recognition needed to produce medical imaging as a familiar practice. The invention of new medical imaging technologies, in turn, aids visualization by producing more visual artifacts for consumption in scientific practice and popular culture.

Computer-based medical imaging technologies are now integral to contemporary epistemologies of the body. Located within a cultural desire to visualize (and thus transform and reconfigure) the interiority of the body, these techniques represent one of the material forms of biomedicalization and are, as Clarke et al. point out, "jewels in the clinical crown of biomedicine" (2003, 162). The imaging armamentarium, as medical professionals sometimes call it, includes x-ray, ultrasound, CT, MRI, positron emission tomography (PET), and single positron emission computed tomography (SPECT). Continuing efforts in research laboratories offer a high probability that new imaging apparatuses will join the existing panoply in future years.

Technoscientization thus takes particular forms in biomedicalization processes, and one way it is enacted, visualization, taps into broader cultural conventions of imagining identity in pictorial form. Narrowing the focus to MRI, one of the most expensive and high-status imaging techniques, fosters a deeper understanding of the links between visualization and the political economy of biomedicalization processes. The knowledge produced by MRI is not innocent but instead is used to help produce technoscientific identities such as ADHD or cancer patient as well as create income for hospitals, manufacturers, and the owners of imaging centers.

The Political and Economic Dimensions of MRI

MRI is highly regarded by policymakers, medical practitioners, and the broader scientific community. The importance of the technology was recently recognized by the Nobel Prize committees. In 2003 two developers of MRI technology—Paul Lauterbur and Peter Mansfield—were awarded the Nobel Prize in Physiology or Medicine for their role in creating the modern MRI machine. Prospective patients are taught the machine's value through such awards as well as through news stories that celebrate the purchasing of new machines or the creation of new applications (e.g., Johnson 2004; Kerber 2005; Poling 2004). Stories that link sports stars' diagnosis and recovery to the use of MRIs also help legitimize MRI examinations in the public eye (e.g., Associated Press 2002; Battista 2002; Olney 2002).[2]

Such news stories act like advertisements, as the celebratory stance they often adopt blurs the line between information and marketing. Assisting in the *selling* of MRI, the news stories uncritically promote the technique and

signal how MRI examinations, in addition to being a valuable medical procedure, are simultaneously commodities that circulate within robust systems
of economic exchange. The growth and continued use of MRI technology
serve as an entry into empirical analysis of the political and financial dimensions of biomedicalization or what Clarke et al. (2003) call the Biomedical
TechnoService Complex Inc., linking the sociotechnical trend toward visualization with the economic and political dimensions of healthcare.

The idea of the Biomedical TechnoService Complex Inc., through its
use of the terms "incorporated" and "complex," calls attention to the significance of healthcare-related industries to the national economy. Representing "13 percent of the $10 trillion annual U.S. economy," healthcare is
perceived by some researchers and policymakers as "the main engine" of
economic growth (Clarke et al. 2003, 167). MRI-related industries make up
an important component of the Biomedical TechnoService Complex Inc.
These industries have expanded and grown as the use of MRI has steadily
increased since its introduction to clinical practice in the 1980s.

Data about the increase in the number of MRI examinations conducted
each year in the United States are not readily available because of another
key trend in biomedicalization — the privatization of knowledge. State and
national government organizations do not collect statistics about the total
annual number of MRI examinations. Instead for-profit and nonprofit organizations, such as Information Means Value (IMV) and European Magnetic Resonance Forum (EMRF), collect, distribute, and often sell these
data. IMV, for example, sells its "2006 MRI Census" market summary report
for $7,750; access to the entire 2006 MRI database is available for $65,000
(IMV 2007). Here the use of the word "census" links their reports to the
U.S. Census, a freely available public database that tracks information
about people. However, IMV's census creates a new kind of knowledge —
one that substitutes images for people and is for sale.

There are different ways to track MRI use. The federal government col-
lects data on the technology's use through Medicaid and Medicare reimbursement statistics, although such information is available only in raw
counts that then need to be calculated. According to these data, the number of examinations submitted to Medicare steadily increased throughout
the 1990s and 2000s, reaching over seven million in 2005 (figure 1). These
procedures generated a constant stream of revenue for imaging centers, as
payments cost Medicare close to $1.75 billion in the same year (figure 2).

Like the Medicare statistics, calculations of total annual MRI use show

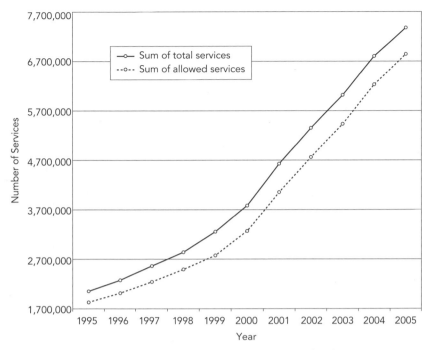

1   Total and allowed services by year. Centers for Medicare and Medicaid Services.

a steady increase over time. Since 1983, MRI usage rates (expressed as the number of examinations per 1,000 of population) have increased every year except for 1993 (figure 3). The healthcare debate in the first Clinton administration is thought to have slowed MRI use in 1993 as physicians and healthcare professionals became more aware of possible scrutiny (Bell 2004).

MRI represents a lucrative financial market and is an engine of economic productivity for hospitals, imaging centers, manufacturers, and medical professions such as radiology. The business of MRI—worth billions of dollars in total—includes two levels of economic exchange: the production and sale of machines and the production and interpretation of examinations.

LEVEL ONE: THE PRODUCTION OF MACHINES, COILS, COOLANT, AND CONTRAST

In 2006 the United States had between 8,000 and 11,000 MRI systems (EMRF 2007; IMV 2007). The United States operates more MRI machines per capita than any other country in the world with the exception of Japan

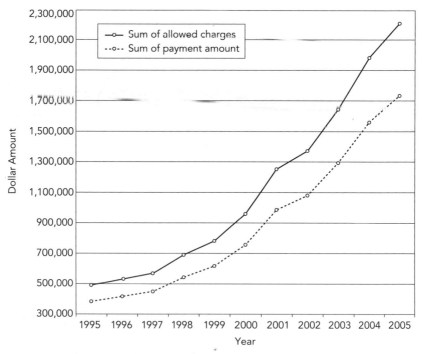

2   Allowed charges and payment amount by year. Centers for Medicare
and Medicaid Services.

(figure 4).[3] In contrast to the United States, many countries limit the
purchase and allocation of costly medical procedures such as MRI (Oh,
Imanaka, and Evans 2005). Using formal mechanisms (e.g., review processes, price controls) and informal mechanisms (e.g., limited healthcare
resources, healthcare reimbursement systems), governments intervene in
the use and distribution of MRI systems (Altenstetter and Busse 2005; Korogi and Takahashi 1997; Lavayssière and Cabée 2000).[4]

Transnational U.S., Japanese, Dutch, and German companies, such as
General Electric, Hitachi, Philips, Siemens, and Toshiba, control most
of the medical imaging manufacturing market, as well as the broader
computer-based visual technologies sector. These highly successful corporations produce, market, and distribute MRI machines and other components of visual technology, including televisions, DVD players and VCRs,
and video cameras. Challenging cultural distinctions between home media,
entertainment, and the private sphere, and medicine, science, and the public sphere, corporations such as GE, Hitachi, Philips, Siemens, and Toshiba
manufacture and profit from the visualization of contemporary life.[5]

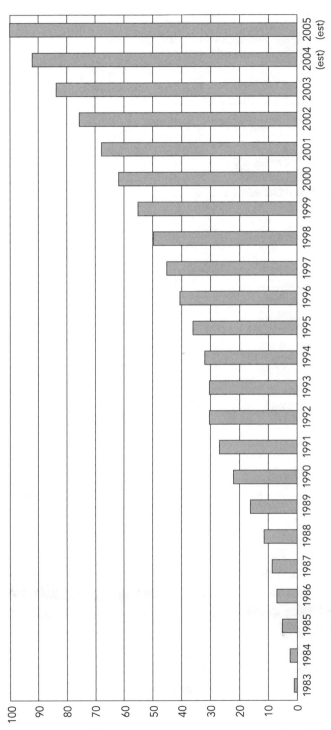

3 MRI utilization rates (expressed as the number of examinations per 1,000 of population) by year in the United States. Robert Bell, "MRI Utilization Mystery," in *Decisions in Imaging Economics: The Journal of Imaging Technology Management*, May–June 2001.

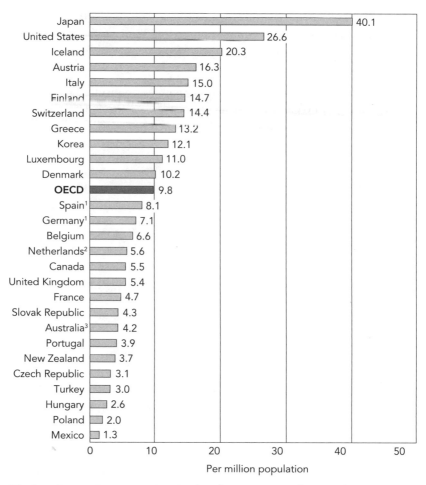

4   Number of magnetic resonance imaging (MRI) scanners per million population
    in selected OECD countries. Canadian Institute for Health Information,
    Medical Imaging in Canada, 2004 report.

Biomedicine and the components of the imaging technologies used in
it generate huge revenues. MRI imaging centers in the United States can
pay manufacturers anywhere between $1 and $3 million to purchase and
install an MRI unit. For example, 1.5 Tesla MRI systems currently cost between $1.5 and $2 million to purchase and install. Most clinical systems in
the United States use machines that operate at 1.0 or 1.5 Tesla (EMRF 2004;
IMV 2007). However, a variety of MRI machines are available, and prices
vary. Open MRI machines, for example, generally use .3 or .5 Tesla magnets
and cost less to purchase. MRI machines with stronger field strengths of 3.0

Tesla represent a more significant financial investment, with initial costs approaching approximately $2 million to $2.5 million.

In addition to the basic unit, imaging centers must also purchase coils, which are small pieces of equipment that add to the content of the image produced, and other accessories that are required to produce images. The price of new coils ranges from $18,000 to $90,000 each. Coolant is required to maintain the appropriate temperature needed by the magnet to function properly. The coils, coolant, and the actual machine represent the bulk of the MRI equipment.

Another lucrative market has also developed around the sale and maintenance of MRI units: the manufacture and sale of contrast agents. Contrast is a dye that is injected into the patient; it illuminates the anatomical area being scanned, producing, as the name suggests, greater contrast between body parts. Initially, most MRI exams were conducted without contrast agents. Their use, however, rapidly increased throughout the last decade. Approximately 45 percent of MRI procedures used contrast agents in 2006 (IMV 2007). Many companies (including machine manufacturers such as General Electric) develop and sell contrast agents, and even more want a piece of this market (Wolski 2006). Spending on contrast media was approximately $342 million in 2003, and the market sector is expected to grow (DeJohn 2005).

LEVEL TWO: THE PRODUCTION AND INTERPRETATION OF MRI EXAMINATIONS

The production and interpretation of MRI examinations make up the second level of economic exchange. This level of exchange highlights the service component of the Biomedical TechnoService Complex Inc. and moves the discussion from the manufacturing industry to the economics of imaging centers and their production of anatomical pictures.

Millions of MRI examinations are conducted each year in the United States, and the number continues to increase with no immediate end in sight. The IMV, a private research corporation, estimates that 18 million MRI procedures were performed in 2001, 21.9 million in 2002, and 26.6 million in 2006 (IMV 2005, 2007). The EMRF, a nonprofit organization dedicated to the exchange of MRI-related information and research, estimates that over 24.9 million examinations were performed in the United States in 2003 (EMRF 2004).

These examinations represent a significant flow of capital and knowledge, as they generate millions of dollars in revenue and are used to create diagnoses of health and illness for Americans who have access to MRI technology. In the United States, imaging centers and units charge per examination. A health insurance company may pay anywhere between $400 and $700 for a typical MRI exam; this amount is based on each company's unique negotiation of a contract with a site (Dell 2004; Poling 2004). Hospitals often run MRI machines twenty-four hours a day, and imaging units may do one to three exams per hour. The fee-for-service structure of imaging payment creates a financial incentive to produce more exams. As the health economist Robert Bell (2001, 1) notes, "To paraphrase the real estate dictum, the key to MRI financial viability is now volume, volume, volume."

Investment in MRI technology is costly and requires considerable capital. Reimbursements, however, not only pay off the cost of equipment in time but subsequently generate considerable income for the owners of an imaging facility. The income goes directly into the pockets of owners, or in the case of hospitals, it is used to offset other healthcare costs. It should be noted that the cost of contrast agents is now routinely added to the reimbursement fee for MRI examinations, shifting the cost of using contrast agents in exams from the MRI centers to the insurance companies and patients.

Physicians benefit in this level of exchange. In addition to reimbursement fees for the examination and contrast agent, there is also a "reading fee." This amount goes to the physician, usually a radiologist, who produces the written interpretation of an MRI examination. Radiologists, central figures in this sector of the Biomedical TechnoService Complex Inc., are physicians who specialize in the production of written translations of anatomical pictures that, along with the images themselves, are forwarded to the referring physician and patient. The radiologist's report attempts to fix the meaning of the image, making its meaning to other doctors and their clients clear.

Demand for MRI exams is therefore also demand for radiologists' services. Radiologists are among the best-paid physicians in biomedicine. Radiologists' salaries range from $193,000 to $386,000 (AAMC 2005). The Radiological Society of North America recently announced that radiologists "are the highest paid players on the medical team," earning a similar median income if not more than other highly paid medical specialties such as cardiac surgeons (RSNA 2003).

Millions of dollars circulate between MRI-related industries, thereby demonstrating the weightiness of the financial dimension of healthcare in biomedicalization. Capital flows from imaging centers to manufacturers and from patients, the government, and health insurance companies to imaging centers and radiologists. It is a lucrative healthcare venture that, at least in the current fee-for-service climate, guarantees a healthy stream of revenue.[6] MRI, like other diagnostic practices in biomedicine, is big business, and for those who run imaging units, this understanding is obvious. One manager made this point apparent in the following exchange:

KJ: What are the first words that come to mind when you think of MRI?
TECHNOLOGIST WHO OPERATES THE MACHINE: Noise.
MANAGER OF THE CLINIC: You know what I think of?
KJ: What?
MANAGER: Money.

The Biomedical Technoservice Complex Inc. and the State

The current expansion of MRI industries and services in the United States embodies a main tendency of biomedicalization—the corporatization and privatization of healthcare. The expanding economy of MRI enrolls an array of private companies, ranging from visualization technologies manufacturers to for-profit and nonprofit imaging centers and hospitals, and directly benefits medical professions such as radiologists. Moreover, the economic surge of the late 1990s provided both "the will to consume" and the capital needed by individuals and hospitals to invest in MRI technology. However, the expansion of the MRI sector of the Biomedical TechnoService Complex Inc. also required the support of the state. While Clarke and her colleagues rightly suggest that the role of the state changed after 1985 as it allowed "private corporate entities to appropriate increasing areas of the health-care sector under private management and/or ownership" (Clarke et al. 2003, 167), the state still remains a crucial player in biomedicalization processes.

The biomedicalization theoretical framework acknowledges that the state matters in healthcare. Working in tandem with corporate interests, the state primarily functions to transfer healthcare activities and monies to the private sector. For example, Clarke et al. discuss how the state invited "corporations to provide services to federally insured beneficiaries" in programs such as Medicare and Medicaid and how the state is "increasingly socializing the costs of medical research by underwriting start-up expenses of research and development yet allowing commodifiable products and processes that emerge to be privatized—that is, patented" (Clarke et al. 2003, 167). Discussion of the state and its relationship to the private sector, however, requires more in-depth analysis. While the biomedicaliza tion thesis importantly highlights the increasing role of private corporations in biomedicine, in doing so it could potentially move critical analysis away from the role of government (at local and federal levels). Here I want to examine more closely government's contributions to the political economy of MRI specifically and to biomedicalization processes more generally.

Individual states directly helped to enable the MRI expansion project by relaxing policies that required government approval before machines could be purchased. Initially put into place in the late 1970s, Certificate of Need (CON) and Certificate of Public Need (COPN) policies aimed to control spiraling healthcare costs thought to be related to big-ticket items, such as CT machines, by regulating the purchase of imaging technology (Blume 1992, 186). By the late 1990s, however, state priorities changed. Protecting state and patient expenditures was no longer a primary concern, and the legislative focus shifted toward increasing availability and accessibility of healthcare procedures. Rhode Island and Massachusetts, for example, loosened state restrictions, which in turn permitted hospitals and imaging centers to buy more scanners. Not surprisingly, the number of MRI machines increased after this policy change (Kowalczyk 2002). The state therefore partially relinquished its role as the arbiter of healthcare policy and purchases, allowing profit motives to more directly drive the adoption of MRI technology.

In addition to deregulation, the federal government also actively promoted imaging practices through Medicare and Medicaid policies. During the 1990s, for example, Medicare, a federal program, approved reimbursements for a range of procedures, including MRI examinations of coronary arteries, MRI as a first choice of image for disc disease, and use of specialty coils to enhance anatomic detail (Centers for Medicare and Medicaid Services 2006). During approximately the same period, Medicaid, jointly administered by the federal and state governments, covered established procedures such as MRI examinations of the head and spine as well as innovative techniques such as magnetic resonance angiography of the head, neck, and abdomen (e.g., Novello 2002; Sullivan 2000). Requiring reimbursements for such services helped legitimize the technology's use by sending a message to insurance companies, medical professionals, and patients that access to them is a feature of standard care.

Today, as new uses for MRI technology and other imaging technologies are created, new reimbursement requirements for these procedures are developed by federal programs. For example, the Centers for Medicare and Medicaid Services expanded their reimbursement policy to pay for the use of PET scans to diagnose and follow all types of cancers if the provider either participates in a PET clinical trial or submits data to the PET Registry. This decision, as Lisa Baird, the spokeswoman for CTI Molecular Imaging, a company that makes PET machines, notes, "is something we have been hoping for forever. It's a big deal for us, and it's a first step toward broad coverage reimbursement" (Brass 2005, C1). Medicare and Medicaid policies and fees are an important "first step" and often transform other health insurance companies' practices, leading to the reimbursement of new procedures and setting standards for fees.

The State and the Construction of Technoscientific Identities

The state, via the National Institutes of Health, also funded research projects that may promote new uses of MRI and a further expansion of the sector in the future. In the 1990s, the NIH called for and financially supported MRI research that focused on the construction of psychological identities through analysis of brain images. By this period, MRI was fully established as a technology of knowledge in relation to cancer and muscle impairments, and the NIH-funded research offered the possibility to extend the ever-expanding medical imaging field to new markets.

Throughout the 1990s, most mass-media attention focused on the Human Genome Project. However, there was another large scientific initiative that existed alongside the Human Genome Project and received little press. In 1993 the National Institute of Mental Health and four other government organizations launched the $14 million Human Brain Project, which aimed to construct a digital archive of brain images for research and possible future clinical practice (Kahn 2001). This project, in conjunction with the new focus on brain-imaging technologies, led President George H. W. Bush to proclaim the 1990s "the Decade of the Brain."

One of the main sites for the Human Brain Project is the Laboratory of Neuro Imaging (LONI) at the University of California, Los Angeles, School of Medicine. The LONI archive includes over seven thousand brain images that are organized by race, ethnicity, age, and gender. Initially UCLA press releases celebrated the ability of the archive to distinguish brain images and function by race, gender, and ethnicity, as well as other categories of difference. Examples such as the claim that researchers "can see how men and women use their brains differently" were used to illustrate the archive's potential (Hotz 2001). The rhetoric used by the LONI researchers has now shifted to emphasize their ability to demarcate the ADHD and schizophrenic brain, although categories of race, ethnicity, and gender are still used to label and organize images (e.g., Adler 2003; Huget 2003; Talan 2002).

The Decade of the Brain also included government research grants to evaluate whether MRI technology could be used to diagnose disorders such as schizophrenia, ADHD, alcoholism, and obsessive-compulsive disorder in the brain. During the 1990s, the National Institute of Mental Health, the National Institute of Child Health and Human Development, and the National Institute of Neurological Disorders and Stroke funded numerous studies that sought to investigate whether supposed "mental health" illnesses could be clearly visible in MRI scans (CRISP 2005). The studies found that in all these situations, MRI did not produce the certainty hoped for, since it could not reliably show correlations between certain brain patterns and the presentation of particular illness in an individual. For example, while MRI is believed to show associations between certain brain patterns and schizophrenia in groups, there is too much variation for it to be used as an individual diagnostic technique. Certain individuals have the brain patterns associated with schizophrenia but do not present any psychotic symptoms. Moreover, certain individuals who are diagnosed as schizophrenic do not exhibit the visual anatomy associated with the disease. Therefore, although research attempted to use MRI to produce these identities, it has yet to be translated into routine clinical practice.[7]

While resistance to MRI use as a diagnostic technique exists among contemporary psychiatrists and psychologists, it remains unclear how this resistance will change. MRI scans of the schizophrenic and ADHD brain are now included in psychology and medical textbooks and lectures (e.g., Bernstein and Nash 2001; Myers 2003; Plotnick 2001; Weiten 2003). Researchers continue to search for patterns between MRI scans and patients

diagnosed with certain mental disorders. This, coupled with a new generation of scientists and medical professionals trained to interpret the body
via MRI scans as definite knowledge, means that the group patterns associated with certain disorders could be reified as fact for individuals.

Government-funded research on mental illness—like the LONI
project—also used race, gender, and ethnicity categories to organize image
data and in doing so fixed these identities as biologically produced and
located.[8] The brain, in other words, became gendered and raced, though
the primary goal of the research was to explore other topics like schizophrenia. The use of these categories comes in part from the National Institutes of Health Revitalization Act, which required NIH-sponsored clinical
trials to include enough women and minorities "to make it possible to tell
if a proposed medical treatment worked differently for men compared to
women and for members of different racial groups" (Fausto-Sterling 2004,
3). This mandate, in addition to what Janet Shim (2002a, 2002b) calls the
"black boxing" of race and gender in clinical trials, promotes an often un-
reflective use of these categories in medical research.

The government has therefore funded new research that pushes MRI
into questionable areas that doctors themselves may not even want, such
as its use in psychiatric diagnostic work. Moreover, the government has—
inadvertently in the case of general research, or purposely as in the case of
financially supporting the LONI project—increasingly led to the gendering and racing of the brain through MRI imaging. Research throughout the
Decade of the Brain may produce new commercial uses for MRI scans in
coming decades; it is too soon to predict the outcome of these investments.
If it does, the state will have directly enabled the expansion of MRI into new
markets.

The move into new markets simultaneously illuminates another key biomedicalization process—the transformation of bodies and the production
of new individual and collective technoscientific identities (Clarke et al.
2003). As Clarke and her colleagues explain, "technoscientific identities
are produced through the application of sciences and technologies to our
bodies directly and/or to our histories or bodily products," and this practice occurs "whether we like them or not" (182). MRI plays a role in the construction of technoscientific identities. State-sponsored research provides
the material needed to access and perform established identities such as
man and woman or African American and white, as well as new identities

such as ADHD. The knowledge produced is political, as it both creates nar-
rower definitions of health by extending and shoring up disease categories
and reifies biological definitions of race, ethnicity, and gender. Part of a
larger project of reconfiguring the body from the inside out, MRI helps create identities and thus legitimize the deployment of pharmaceutical, surgical, and lifestyle interventions.

The twin processes of market expansion and identity construction continue to spiral as the participation of the state in the promotion of MRI
technological use persists. On December 29, 2000, the National Institutes
of Health created the National Institute of Biomedical Imaging and Bioengineering (NIBIB). The NIBIB's stated mission is to "improve health by
promoting fundamental discoveries, design and development, and translation and assessment of technological capabilities," and it is primarily dedicated to imaging innovation.[9] In addition, in 2004 the National Institute on
Aging (NIA) launched a $60 million, five-year venture—the Alzheimer's
Disease Neuroimaging Initiative—to explore the use of imaging to assess
the progression of Alzheimer's disease (Cahan and Dollemore 2004). This
project, which began in April 2005, may enroll a new group of users, people
with Alzheimer's, and may further create corresponding technoscientific
identities for them. This project may also expand the routine use of imaging
technologies in biomedical practice. The use of MRI technology, a profitable and important component of biomedicalization processes, continues
to move into new medical and psychological terrain.

Into the Shopping Mall?

Beyond the ever-expanding clinical uses, a new possible market for MRI
and other imaging techniques is direct-to-consumer sites. While the high
cost and special safety considerations associated with MRI prevent it from
moving easily into the shopping mall at the present time, MRI technology
has moved into storefronts at high-end hotels in Japan and may soon expand to the marketplace in the United States. Other less expensive medical technologies such as ultrasound and CT have already moved into sites
where physician referrals are unnecessary. Prenatal ultrasound facilities,
such as Peek a Baby, Prenatal Peek, and Fetal Fotos, are in malls and stores
across the country (Garza 2005, A1). Offering keepsake ultrasound photos,
DVDs, and videos with accompanying musical soundtracks, fetal sonogram

businesses are profitable and booming (Brand 2004, B1). Enterprising physicians and businesspeople have also brought CT scanners into shopping malls. Companies now offer body scans to consumers willing to pay out of pocket for them, and direct advertising to potential patients is now occurring (Barnard 2000, A1). While there are individuals who resist the increasing use of visual technologies, consumer demand for imaging techniques is steadily growing.

The role of the state in the expansion of imaging to direct-to-consumer markets is also important to analyze. The Food and Drug Administration (FDA) indirectly enables the move of imaging into the shopping mall by not demanding or having the power to control such establishments (Lewis 2001). The FDA primarily combats direct-to-consumer imaging establishments through information available on its website. For example, the FDA puts out a pamphlet called "Full Body CT Scans: What You Need to Know," which states that body scans provide "no proven benefits for healthy people" (FDA 2003). Other state officials like the attorney general may intervene, but this type of action is rare. Limited resources combined with lack of licensure programs make it hard for government bodies to regulate or control medical images in the marketplace (Garza 2005, A1). As the blurring of medicine and consumption continues, the relation between government and medical imaging will be crucial to examine. The presence of medical images in the mall suggests that they will be crucial to the performance of a range of technoscientific identities. That is, medical images will enable access to, and performance of, parent, mother, healthy, ill, and the like.

Conclusion

Biomedicalization provides a useful analytical lens for understanding the political economy of MRI. Its attention to both the state and for-profit businesses allows for sustained analysis of the intricate relations that co-produce the Biomedical TechnoService Complex Inc. This analysis shows that U.S. government agencies and policies legitimized and expanded MRI use through deregulation, Medicaid and Medicare reimbursement practices, and research funding. In the move of imaging from clinical practice to the shopping mall, the state continues to enable direct-to-consumer marketing. The FDA, despite efforts to critique boutique imaging practices, does not have the resources to intervene in a substantive way and consequently allows the medical imaging marketplace to expand. While privatization and corporatization mark biomedicalization processes, the state still matters.

This analysis also shows how MRI technology is part of a broader sociotechnical trend toward technoscientization in biomedicalization processes and how visualization is one of its key material forms. The twentieth and twenty-first centuries are noted for the increasing presence of visualization technologies in all parts of life—medicine, science, popular culture, and intimate relations (Clarke et al. 2003; Mirzoeff 1998; Mitchell 1995; Sturken and Cartwright 2001). From picture-producing cell phones to DVD players and recorders, visualizing machines are central to the construction and performance of identities and relationships. While access to such techniques is stratified and uneven, the presence of the visual in contemporary life provides the technological support and cultural familiarity needed to legitimize MRI as a producer of knowledge and thus technoscientific identities.

Finally, the biomedicalization theory attends to the *interactive* nature of biomedicalization processes and offers the opportunity to understand how technoscientization, the production of technoscientific identities, and the Biomedical TechnoService Complex Inc. are reciprocal and reinforcing. In the case of MRI, the material body is produced in digital, visual form, thereby creating a culturally familiar knowledge that helps race and gender the body as well as produce identities such as Alzheimer's or cancer patient. The state, a key actor in the Biomedical TechnoService Complex Inc., promotes the proliferation of MRI in the United States through deregulation, reimbursement practices, and research funding, which supports an increasing use of the technology by companies and organizations. The actions of U.S. government agencies, combined with both manufacturing companies' and imaging centers' need to produce income and a cultural and technical milieu that uses the visual to access and perform identity, make MRI a sought-after commodity in an ever-expanding biomedical marketplace.

Analysis of MRI shows the dynamic reciprocity between technoscience, identity, and economics always at work in biomedicine. While this chapter focuses on the United States, future research should untangle how visualization technologies, economic institutions, and government agencies uniquely interact in other countries. Investigating imaging use and government interventions in more national contexts will offer insight into the different roles government agencies can play in biomedicalization processes.

Notes

1   For social and cultural histories of particular technologies, see Golan 1998 and Pasveer 1989 for a discussion of the emergence of x-ray, Yoxen 1987 for an analysis of the development of ultrasound, and Dumit 2004 for an examination of PET.

2   In-depth social-scientific analyses of patients' reactions to MRI technology have yet to be published. In my fieldwork at three imaging centers, patients had mixed reactions to the procedure. MRI examinations are loud and, if using a closed machine, require the person to lie still in an enclosed space for twenty to forty minutes. While most patients finished the examination, some were unable to stay in the machine for the duration of the procedure. Research is needed to understand the complex ways in which people make sense of MRI technology.

3   Japan's high number of machines does not necessarily equal high use rates. For example, the number of exams conducted each year in Japan was almost half that of the United States in the 1990s (Korogi and Takahashi 1997; Hisashige 1994). For a comparison of perceptions and use of MRI in Japan and the United States, see Joyce 2005b.

4   Cultural beliefs (e.g., ones that suggest fewer interventions are better, caution is a sensible tactic) also contribute to a country's (lack of) use of MRI (Payer 1988/1996). In the United States, popular beliefs such as "doing something is better than doing nothing" and "technology use equals good healthcare" contribute to the use of medical procedures.

5   For a discussion of the emergence and critical analysis of contemporary visual culture, see Clarke et al. 2003, chap. 6; Mitchell 1995; and Sturken and Cartwright 2001.

6   Centers could see a decrease in annual revenues if the demand for MRI exams does not keep up with imaging-center growth. A glut of imaging centers could saturate the market, ending the possibility of unlimited demand and twenty-four-hour service schedules.

7   I interviewed practicing psychiatrists, psychiatrists associated with university hospitals, and medical students to better understand the contemporary psychiatric field.

8   According to Computer Retrieval of Information on Scientific Projects (CRISP), federal institutes funded 145 projects on sex differences and MRI and 84 projects on race and MRI between 1990 and 2000. These projects mainly explored MRI use in relation to schizophrenia, although projects examining Alzheimer's disease, alcoholism, and other conditions were also funded (CRISP 2005).

9   The NIBIB's Internet address is http://www.nibib.nih.gov. For a full listing of the National Institutes of Health, see http://www.nih.gov/icd.

Janet K. Shim

7 / The Stratified Biomedicalization of Heart Disease

EXPERT AND LAY PERSPECTIVES ON
RACIAL AND CLASS INEQUALITY

Juanita lives on a quiet residential street in a predominantly African American community in the
San Francisco Bay Area. Her home is on the bottom
floor of an older two-story home now split into smaller
apartment units, and it takes her a long time to come
to the door. A tall black woman, she moves slowly and
gingerly, breathing laboriously because of her congestive
heart failure, and wincing in pain from her arthritis and
an infection on her leg—a complication of diabetes that
almost required amputation.

My interview with Juanita turns out to be the longest
one I conduct. Her medical history of arthritis, diabetes,
hypertension, multiple heart attacks, and congestive
heart failure is lengthy and complicated. But in addition,
her telling of this history involved extended forays into
her social biography: her drive to surpass the diminished
expectations of family, teachers, and employers; her sub-
jection to racial and gender discrimination on the job
and at home; her experiences of domestic violence and
single motherhood; and her lifelong engagement in
community activism. Juanita does not mince words as
she attributes her bad health to a lifetime of strains and
stresses. In her view, "who has the most checkers, that's
who has the better chance of surviving." Her implication

is that she has not been given many checkers in life and has had to fight for those she does have. At the end of our interview, I comment on how informed she seems to be about her medical conditions, and how much I learned from her. She replies, "I've been the landlady of this body for over fifty years. I *know* what makes me sick."

Thinking about Juanita's words while heading home, I was reminded of my encounter with an epidemiologist at a scientific conference some five months earlier. She was exhibiting a poster of her research on the disparities in cardiovascular disease (CVD) risk factors between black and white women, and we began to discuss her findings. My field notes describe our interaction:

> One of the findings of her study is that black women have more traditional CVD risk factors. When I ask her why she thought this was, she responds that socioeconomic status, including occupation, family income, educational level, and insurance status "might have a lot to do with it." She also hypothesizes that diet, sedentary lifestyle, culture, education, and perhaps access to health care may play a role. The researcher then adds that she was involved in another study . . . exploring what prompts people experiencing cardiovascular symptoms to see a medical provider for evaluation. She notes this to me, she said, because black women have been shown to have high rates of delaying care and of symptom misrecognition, which to her sounded a bit like "blaming the victim." . . . Her thinking then was that delaying care had more to do with lack of time, child care, transportation, access to care, and so on. She then relates that two nursing coordinators at the center, who were black, had told her that that was not it, that "black women could come in but don't." They argued that there were also poor white women who had as many barriers yet still sought treatment for symptoms. The principal investigator subsequently speculates to me that maybe "there is a black culture, welfare moms, or whatever, with low education who maybe missed out on the public health messages of the past ten years."

In this encounter, the epidemiologist briefly considers conventional dimensions of socioeconomic status, but then attributes the disproportionately high cardiovascular risk of black women to a presumed propensity to delay care and to a lack of skill in symptom recognition. These traits are, she suggests, rooted in a pathological "black culture" combined with insufficient education. In advancing these arguments, she neglects the powerful influence of race, gender, and class in stratifying opportunity and material conditions and their subsequent effects on well-being, the very kinds of forces that Juanita had so emphatically implicated.

Why such divergent views? How do Juanita and this epidemiologist come to such different conclusions about why heart disease befalls one individual or group and not others? In this chapter, I analyze how epidemiological experts on the one hand, and people of color with heart disease on the other, tend to conceptualize and understand race, class, and their consequences for cardiovascular risk in different ways. I argue that biomedicalization theory is a useful way to understand this science-lay divide as a story about how epidemiology defines and manages bodily differences in certain ways while simultaneously marginalizing alternative conceptions of how illness is produced. Cardiovascular epidemiology is one site where we can witness the boundary work involved in biomedicalization—the selective ways in which it works, and how it ultimately reinforces and sustains social inequalities. At the same time, cardiovascular epidemiology is also a site where the instability and partiality of biomedicalization are evident, in the uneven uptake of epidemiological conceptions of risk by laypeople and in their own constructions of causal accounts of heart disease.

This case study exemplifies how stratified biomedicalization (Clarke et al. 2003) works in at least three interrelated ways. First, epidemiological classifications of race, socioeconomic status, and sex have reemerged as legitimate biomedical concerns, rearticulated as markers of risk and potential pathology, and as targets for needed intervention. Second, such representations contribute to sociocultural views regarding "different" kinds of bodies and the ways in which individuals so marked think about themselves and their health. Thus another dimension of stratified biomedicalization is revealed in the distinct consequences of biomedicine's gaze for subjectivity and identity. Finally, lay knowledges about the social etiology of heart disease reveal how biomedicalization is continually contested and negotiated. At the same time, this case also troubles biomedicalization theory's conception of how knowledge is produced and transformed in this new era. More specifically, I wish to raise some questions regarding expert-lay engagements in knowledge production and their consequences for scientists, clinicians, and people with illness.

These findings emerge from a sociological analysis of how race, social class, and sex and gender are understood by scientists and laypeople of color as meaningful (or not) for the incidence and progression of heart disease (Shim 2002a, 2002b, 2005). The data gathered in this multi-sited ethnography came from several sources. First, I conducted several hundred hours of participant observation at scientific meetings and health education events. Second, I conducted in-depth, semi-structured interviews with a purposive sample of ten scientists who had been or at the time were involved in epidemiological research on disparities in cardiovascular disease. Third, I conducted in-depth interviews with eighteen African Americans, Latinos, and Asian Pacific Americans who had been diagnosed with hypertension and coronary heart disease. Finally, I conducted a content analysis of published articles and commentaries on theoretical and methodological debates in epidemiology, the status of the discipline, and its role in public affairs. All materials were coded and analyzed using the principles of grounded theory (Strauss and Corbin 1998).

The Stratified Impact of Heart Disease

Epidemiology by most accounts is considered the basic science of public health, a field that, historically and contemporarily, emerges from and sustains social processes central to the construction of difference: the classification of populations, struggles over definitions of people and bodies as "problems," and interventions that are simultaneously scientific, clinical, and political in nature. The history of cardiovascular epidemiology is marked by a series of significant investigations that placed large samples of individuals and entire communities under the epidemiological gaze. These studies revolutionized the wider discipline of epidemiology as well as associated fields of public health, clinical medicine, and social science. They were made possible only through new computer and data management technologies and the complex social arrangements that accompany them, and are thus emblematic of the extensive technoscientific transformations in medicine and clinical research that biomedicalization theory highlights.

What emerged from the large statistical studies conducted in the latter half of the twentieth century was a picture of heart disease as an epic public health dilemma. Although mortality rates from CVD have been dropping since 1970, it remains the leading cause of death in the United States (American Heart Association 2001; NCHS 2001).[1] Heart disease has also come to be known as a disease of poverty and minority status. Consider the following:[2]

▷ Hypertension is disproportionately more prevalent among African American women and men, Mexican American women, and American Indian women and men. In fact, African Americans' levels of hypertension are among the highest in the world.

▷ Heart disease mortality is higher for black men between twenty-five and sixty-four years of age than for white men regardless of income. Mexican Americans also suffer from higher rates of heart disease mortality than non-Hispanic whites (Hunt 2002).

▷ Among men aged twenty-five to sixty-four, heart disease mortality for those with incomes less than $10,000 was 2.5 times greater than for those with incomes of $25,000 or higher.

▷ The poorest women between twenty-five and sixty-four years of age were 3.4 times more likely to die from heart disease than those women with the highest incomes.

Clearly, morbidity and mortality from heart disease have not been equally distributed across the racial and socioeconomic spectrum. Rather, their negative impacts have systematically and disproportionately fallen on people of color and those with fewer economic resources.

In its attempts to make sense of this statistical picture, cardiovascular epidemiology has explored a range of potential explanations. These have focused predominantly on the differential distribution of risk factors for heart disease, inequalities in access to health services, and biological differences. There also exists a much smaller body of research on the effects of institutionalized discrimination and other experiences of marginalized social status. In so doing, the discipline has identified and refined measures of various population differences — in particular race, sex, and socioeconomic status — and legitimated the scientific practice of differentiating individuals on the basis of race, social class, and sex (Shim 2002a, 2002b). In the process, epidemiology has become a key site of stratified biomedicalization.

Epidemiology and the Rearticulation of Risk

In this section, I summarize my analysis of the meanings of race and class and the ways in which these meanings are constructed and sustained. Laypeople tend to speak of race and class in structural terms in their accounts of disease causation. In contrast, epidemiologists tend to conceptualize differences of race and class as individualized, demographic variables. This science-lay divide, I argue, contributes to the biomedicalization of social differences of race and class, their transformation into markers of behavioral and biological risk and targets for public health and clinical surveillance, and the exclusion of consideration of social inequality itself as a cause of cardiovascular disease.

CONSTRUCTIONS OF RACE: STRUCTURAL EFFECTS OF RACISM VERSUS THE CULTURAL PRISM

While their constructions of racial differences and their effects on cardiovascular risk are often highly variable, expansive, and even contradictory and paradoxical, people of color with heart disease tend to articulate race in structural terms. Many directly attribute either their own or others' illness to the multiple effects of racial discrimination in education, housing, employment, and everyday life. These effects range from the emotional and physical experiences of stress and suppression, the damage to one's sense of self, to the drive to work extremely hard to overcome stereotypes and succeed in white- and male-dominated domains.

How might this connection between the constructions of race and their consequences and the health of implicated groups and individuals work? People of color spoke often of the deep conflicts between their desire to fight back and resist racist treatment and the need to survive. Geline,[3] an anti-Marcos political activist who had emigrated to the United States from the Philippines because of threats to her safety, describes some of the embodied repercussions of making such compromises:

> I remember . . . being conscious of trying to say the right things the right way . . . so that you won't . . . be ridiculed or lose an opportunity. . . . Just being very different, everything about you is looked at and either questioned, praised in a condescending way or criticized in an unfair way. . . . That's stressful. . . . I think it made me quiet, which was kind of weird because, here's this flaming young activist . . . and then [she] tries to disappear in, let's say, work meetings. . . . You force yourself to be what you are not. . . . You keep quiet . . . simmering . . . instead of speaking up because you don't feel safe to do so. . . . You calculate things — what are the tradeoffs? What are the risks?

The theme of self-preservation that Geline evokes is echoed by many others. Active resistance to racist treatment — aside from being professionally and socially damaging — is often seen not as psychically or physically liberating but as stress inducing. At the same time, the tension of adopting a new persona for safety's sake and the need to curb emotions simmering under the surface are seen as equally corrosive to one's health.

Participants also describe how racism shapes entire communities and their resources and opportunities. David, an African American with heart disease, offers a picture of some of these effects on health:

> No matter what your station in life, no matter what your economic, social [position]—you could look at every aspect, [race] had an effect on you. . . . You carry a scar for life. . . . And I don't think it's speculative. I think it's actually a fact. . . . If you're a product of an oppressed environment, quite naturally you're going to have some health problems that another group of people would not necessarily have in an environment that was entirely different. When I say the environment, that takes it in as a whole. You have to take in the school system . . . the availability of health facilities. . . . You have to take all of these and put them into that pot. You just can't extract one. . . . You would see that if that environment is oppressed, the people in that environment are going to come out with some problems, not only emotionally, mentally, social problems, but they're going to come out with a multiplicity of health problems.

Here, David makes an emphatic claim to authority, using his experiences as the basis for his expertise in the health effects of racism. He also articulates an understanding of the interlocking effects of race: it deeply affects multiple aspects of one's lived experience, and any one dynamic cannot be separated from the others. Racialized discourses and practices have constructed social, economic, political, and cultural infrastructures that synergistically sustain and reproduce one another.

Finally, Joe articulates a particularly compelling picture of how structural racism can act as a significant risk for CVD:

> [Not to have gotten all the chances you deserve,] that's what irritates so much. That's what makes you angry, mean, nasty, and unforgiving. . . . That's all part of the oppression thing, to be denied that chance. . . . It's got to weigh big with some people. . . . There are some people who just cannot shake that, you know, and it sticks in their craw, and it destroys them. . . . Because that's all they've got is the opportunity. . . . And you take away that, man, and there's nothing else. . . . You can go through life saying, "If I had just got this shot or these things happened," two or

three times, but these things continually happen. . . . How many times can you go through this? How many times can you turn your other cheek? How many times can you say, "Oh, it'll be all right. I'll get it the next time." After a while it hammers on you, it lays on you, it beats you down. . . . And eventually maybe destroy[s] you. . . . Is it genocide? I don't know. . . . It's the most worse thing there is, is to know that . . . your parents didn't have a chance, you didn't have a chance, and your children's not. . . . It's a monster thing! It's unforgivable. . . . You take that away, you destroy the people. You've destroyed yourself.

Thus participants with heart disease attest that their experiences of discrimination and the differential distribution of educational and economic opportunities are deeply and powerfully embodied.

Some epidemiologists do at times invoke structural dynamics as possible sources for racial inequalities in CVD. But most often these scientists interpret the meanings of racial differences through a cultural prism in which racial disparities are attributed to cultural differences. Researchers repeatedly refer to differences of a "cultural" or "ethnic" nature, ones they perceive to be related to the customary beliefs and practices of a racially or ethnically defined minority group: "their thinking process, how they make decisions," "cultural habits of how they eat, whether they exercise, those kinds of things."

The culturalist construction of race holds significant currency for cardiovascular epidemiology as an organizing concept and a collective approach.[4] It also reveals some of the precise ways in which stratified biomedicalization works: through inclusion, exclusion, embeddedness, and in its effects. First, stratified biomedicalization occurs through the explicit targeting of racialized populations and essentializing notions of their behavior, as in epidemiology's focus on "ethnic" groups and their "cultural" behaviors. That is, biomedicalization is stratified because of the *specific inclusion* of racial difference (from the normative "white" category) as an object requiring biomedical attention (see Epstein 2007).

However, this inclusion and attention are of a particular kind: that racial difference is understood (or at least enacted) by epidemiologists to be culturally mediated indicates that structural conceptions of racial dynamics are *strategically excluded* from consideration. Historically, culturalist constructions of race were aimed at opposing biological discourses. However, in this contemporary iteration, the cultural prism instead tends to reinstitutionalize individualized, astructural notions of racial difference (see Omi and Winant 1994). The selective embrace of cultural and other seemingly apolitical parameters of difference tends to uphold extant social hierarchies. An outcome of the biomedicalization of race by cardiovascular epidemiology is thus a relative lack of attention to structural sources of health inequality, and a continued focus on causes of, and consequences to, health on an individual level rather than a group or collective one.

Third, the outwardly observable and thus seemingly factual nature of behavioral differences makes it exceedingly difficult—and gives the appearance of it being conceptually unnecessary—to problematize and further explore the associations between racial groups and behavioral differences. Thus the capacity of biomedicalization processes to effect a particular definition of the situation or framing of the "problem"—in this case, that racial disparities in cardiovascular risk are attributable simply to racial differences in behaviors—means that notions of stratification become *embedded* into our "biomedical common sense" and, increasingly, into popular logics about disease causes and solutions.

Finally, the affinity between the cultural prism and prevention discourses shows how biomedicalization can be stratified *in its effects*: even when health promotion imperatives are not racialized on their face, their consequences are systematically uneven for different racial groups. Because risk factors often carry connotations of culture, ethnicity, and other dynamics commonly perceived to be related to race, CVD risk and prevention discourses often invariably, almost tautologically, become racialized.

This is *not* to say that some behaviors and practices commonly viewed as culturally rooted do not play a role in cardiovascular risk, because undoubtedly they do. However, what I critique here are the routine attribution of racial difference to cultural difference, the pathologization of such behaviors, and the lack of attention to the social structures and forces that give rise to and sustain them. Epidemiologists, in these respects, tend to treat culture as static, separable from and stripped of its historical contexts, and thus curiously intrinsic to racialized individuals rather than the fluid product of social, economic, and political life.

In this light, the relative silence on structural causes and the popularity of culturalist constructions of racial inequalities on the part of epidemiologists appear very problematic indeed, standing in sharp contrast to the experiences of those who live with CVD. At the least, the ritualized inclusion of race as a taken-for-granted and unexamined variable referring to cultural difference, and the continued study of health behaviors in CVD risk, neglect the role of race in organizing social relations of power and defer work on the effects of structural racism on health. In so doing, racial inequality structures inclusionary and exclusionary processes of biomedicalization (Clarke et al. 2003; Ehrenreich and Ehrenreich 1978), becomes embedded into the fabric of biomedical knowledge and practice, and subsequently manifests with systematically distinct consequences for different racialized groups. Much like de jure and de facto racism, then, biomedicalization is stratified both in name and in effect—in itself and in its consequences.

CONSTRUCTIONS OF CLASS: INTERSECTIONALITY AND THE STRUCTURAL DISTRIBUTION OF RESOURCES AND RISKS

Unlike race, the appropriate conceptualization of class is seen as indisputably and manifestly about differences that are social and structural in nature. Both researchers and those living with CVD argue that social class influences cardiovascular risk primarily via *structural* mechanisms. With varying frequency, they speak of the effects of class in at least two respects. First, class is seen as stratifying access to heterogeneous resources important to health, such as access to quality health services and information and nutritious foods. Second, class stratifies exposure to multiple kinds of risks, such as the stress of economic insecurity, constrained occupational and educational opportunities, and unhealthy lived environments.

Laypeople, for example, describe how class distributes access to myriad kinds of social capital and informational resources that shape what one knows and can do to promote their own cardiovascular health. For example, Diane, a middle-class African American woman with hypertension,[5] lists the benefits that her class status affords her:

> We have a computer at home, so I can always go online and get the latest news on what's happening. The ability to buy books and read on things. The ability to go to any doctor that we want to go to, and, if we're pissed off with that one, go to another one. . . . That's a major benefit of having money, that you can actually go where you need to go and where you want to go, and you're not told to go somewhere. And you have more of a voice in your health.

Many epidemiologists concur that access to informational resources and the ability to navigate and evaluate health-related material are crucial factors in shaping the socioeconomic gradient in cardiovascular health.

Both epidemiologists and people with CVD also cite the chronic mental and emotional toll of economic insecurity and the constant struggle for economic survival as significant bodily stressors and important contributors to class inequalities in CVD. As one epidemiologist argued:

> I think the biggest issue is that when you have so many other prevailing issues, like where will the next meal come from for the children, will we have a roof over our head—what happens to you in 20 or 30 years from high blood pressure or cigarette smoking or high cholesterol seems almost irrelevant. . . . It's almost like it's a privilege of the middle class and the wealthy to be able to worry about health and the future.

People of color with CVD further invoke *intersecting* processes of educational stratification, and labor market segmentation by class and gender, to explain the constraints they face by virtue of their economic, educational, and occupational positions. They recount long and complicated histories to explain the cyclical consequences of tenuous economic existence on their education and other aspects of their life that constitute risks to their cardiovascular health. For example, Mabel, a Latina with heart disease, had a tumultuous childhood and was able to attend school only sporadically. She speculates that had she been able to further her education, "My life would have been much easier. . . . I wouldn't have to go to laboring jobs and work like a fool for minimum wages. . . . It's hard to . . . think about . . . how are you going to eat, how you're going to take care of your kids." Thus the long-term consequences of childhood poverty linger as limited educational attainment in turn constrains participants' abilities to earn a livable wage. This then creates a lifetime of chronic stress that exacts a considerable physical, mental, and emotional toll. From the perspectives of laypeople, the conditions of life structured and imposed by their economic position, social status, and opportunities constitute significant causal pathways to bodily stress and ultimately to their cardiovascular health.

Many of these educational and occupational forces are the manifestations of complex interactions of class with race and gender (see, e.g., Dill, McLaughlin, and Nieves 2007; McCall 2001). For example, when women participants did enter the workforce, they confronted a racially and sex-segregated labor force and market that often further constricted their employment opportunities to low-paying jobs, with little potential for advancement, minimal job stability, and little power over the conditions of their hours, pace of work, or the nature of the work process. For instance,

Mercedes worked as a seamstress and janitor; Carmen was an entry-level office assistant; and Mabel had been a housekeeper, home health aide, cannery worker, and in food service. Mabel describes how discriminatory hiring practices produced an occupational hierarchy that reflected racial hierarchy: "Oh, my days, people were very prejudiced! I mean [the whites] got the best jobs. . . . It was all underneath the table. . . . I know what my place was there. . . . We got the dirty jobs while the others got the clean jobs. It's always been that way." Racism, classism, and sexism intersect to define and structure the terms under which people of different classes, races, and sexes are made to do different kinds of work. People of color construct such experiences of intersecting inequalities as exacting material consequences on their everyday lives and health (see, e.g., Krieger et al. 1993; Schulz and Mullings 2006).

The gendering of class inequality is also invoked as a significant structural contributor to cardiovascular distress in the unequal division of both paid and unpaid reproductive labor. Feminist scholars (e.g., Glenn 1992/1996) have brought attention to the racialized, classed, and gendered burden of social reproduction—the array of activities involved in the cultural, social, and physical maintenance of people, families, and households. The occupations of several of the women I interviewed exemplify the commodification of such reproductive labor. Women of color are disproportionately employed to carry out such lower-level "public" reproductive labor. Alternative occupational possibilities for women have further diminished with the shift from a manufacturing economy with relatively better-paying jobs to a service economy with increasingly bifurcated occupational and income hierarchies. Such dynamics contribute to what Piven and Cloward (1993, 397) call "the rise of a predominantly female and minority service sector proletariat."

But in sharp contrast, epidemiologists are far less inclined to conceptualize the interconnections of class, race, and gender in these deeply intersectional ways. Researchers almost invariably express concern over statistical confounding between class and race; they frequently observe, for example, that adjusting for socioeconomic status "tends to wash out" much (but not all) of the disparities by race. Many researchers thus appear to understand the role of class in CVD risk as separable and independent from that of race, such that their respective effects should be uncoupled through methodological techniques, despite their nonfungible nature in social life.

People of color in this study, by comparison, when asked how they would describe their class background, often interjected experiences related to *race*; or when asked about their racial identity, they would recount aspects of their *class* circumstances. The repeated use of racial indicators to give shape to their descriptions of their class backgrounds, and vice versa, conveys the intimate linkages between the meanings of one dimension of difference to those of another. Thus many researchers' framing of the "problem" in terms of how to isolate the distinct effects of class from race may be less relevant to the biographical experiences and embodied identities of people with CVD than how race and class intersect and interact with each other.

Despite epidemiologists' views of class inequality as a structurally mediated contributor to heart disease, in practice social class is by and large operationalized as one or more individual-level variables of income, occupation, and educational attainment. These measures constitute components of socioeconomic status, or SES, which, according to all the researchers I spoke with, is a key standard ingredient in almost any epidemiological investigation.[6] The ritualized use of SES variables seems to encourage and perpetuate a kind of conceptual devolution about the effects of class difference on cardiovascular risk.

As a case in point, scientists simultaneously intermingle structural accounts of CVD causation with contradictory explanations of individual fault and responsibility and invoke reductionist dynamics to account for what are perceived as individual-level pathologies of attitudes, knowledge, and behavior. For example, they discuss the impact of social class on individuals' awareness of health issues, their knowledge of when and how to engage with the biomedical system, and an agential attitude toward the consumption of health information and healthcare services (Shim 2010).

Epidemiologists also understand class differences as stemming from variations in the prevalence of risk factors—such as smoking and unhealthy dietary habits—that they theorize have something to do with lack of health education, problematic "role modeling," and an aberrant cultural environment in which such risky behaviors are tolerated and even promoted. Scientists fuse structural and more individualistic explanations for class differences in a contradictory manner, reflecting the pervasiveness of ideologies that in effect "blame the victim" and echo "culture of poverty" arguments.[7]

How epidemiological considerations of social class stratify the biomedicalization of heart disease therefore parallels many of the same dynamics as those related to race. Processes of strategic inclusion and exclusion occur as epidemiologists routinely use specified, narrow understandings of class, such as individual-level differences in income or occupation, in their investigations and in their own conceptualizations of etiology. At the same time, they perceive more structural notions of class relations to be outside their arena of concern and practice — beyond the appropriate jurisdiction of epidemiology.

In addition, culturalist constructions of class have been remarkably conducive to being taken up by epidemiological projects that tend to focus on proximate risk factors and health behaviors. They thus become embedded within health promotion rhetoric and risk-oriented, individual-level "common sense" about heart disease causation that argues that class inequalities in CVD are largely due to class differences in risk behaviors. As such, cardiovascular epidemiology and the risk reduction advice it generates disproportionately target people of lower social class, rather than being evenly applicable to the population at large. Clarke and her colleagues (2003) argued that biomedicalization promotes health maintenance and biomedical responsibilities for all, sustained by an ideological infrastructure that asserts that incorporating such mandates into our everyday lives is a matter of individual will, rather than the structural outcome of the differential distribution of material and other resources.

In sum, the process of producing epidemiological knowledge about heart disease biomedicalizes differences of race and social class in specific and strategic ways. By constructing race and class in culturalist, behavioral, and individualistic terms, epidemiological knowledge production devolves social, political, and structural dynamics and consequences into risk factors and other disease mechanisms over which biomedicine has claimed jurisdiction. Discourses on health and risk as individually determined veil the stratified nature of the biomedicalization of cardiovascular disease.

The Production of Heterogeneous Knowledges and the Problem of Engagement

The conceptual divide between laypeople and epidemiologists on the meanings and significance of social differences for cardiovascular health provokes two reconsiderations of biomedicalization theory's perspectives on transformations in the production of knowledge (Clarke et al. 2003, 177–80). First, I consider the potential of the heterogeneity of knowledge sources to disrupt the division of "expert" and "lay" knowledges. Second, I reflect on the extent to which the tentative attempts by some epidemiologists to deal with inequalities in heart disease are examples of the co-optation of competing knowledge about etiology (Clarke et al. 2003, 179) or are serious efforts to revamp the practices of epidemiology in ways that expand our notions of the connections between social inequality and disease causation.

Numerous scholars have grappled with the complex issues of knowledge and expert-lay engagement in various ways.[8] Arksey (1994), for example, argues that laypeople with repetitive strain injury have the potential, through their practical experience and "insider" knowledge, to exert influence in medical fact building, though their claims are vulnerable to co-optation by credentialed medical experts. Brown (1987, 1992, 1997) coined the term "popular epidemiology" to describe the collection, analysis, and use of data by community activists and families affected by toxic waste. He outlines how lay and professional "ways of knowing" vary in their positions regarding problem definition, study design, interpretation of findings, value neutrality, standards of proof, and uncertain and emerging diseases and conditions. Significantly, popular epidemiology, in contrast to mainstream epidemiology, emphasizes social structural factors as centrally implicated in disease causation. Finally, Epstein (1995, 1996) uses the term "lay experts" to refer to AIDS activists who acquired the vocabulary and frameworks of biomedicine and parlayed other significant social and cultural resources to become genuine participants in the design and execution of clinical trials. These studies thus examine lay-science interactions in cases where "auto-didacts" (Epstein's term) within social movements absorb enough scientific jargon and concepts to forge and force connections with the credentialed establishment and position themselves as credible interlocutors. That is, the grounds on which laypeople can be understood to possess expertise, in the extant literature, are based on their scientization.

This study raises similar possibilities that relationships between lay and scientific communities can change. But, in the case explicated here, there is currently no explicit, organized engagement or relationship between lay and expert forms of authority. Despite its status as the most common cause of mortality, cardiovascular disease does not have associated with it a visible hybrid activist-expert community of the kind that scholars have found in other medical and scientific arenas. There is thus no active arena where expert and lay social worlds can come together.

This lack of engagement extends to the clinical encounter: people with heart disease widely perceive that the structural forces rooted in racial and class hierarchy that they view as crucial contributors to their cardiovascular risk are of little interest and practical utility to their health care providers. When asked, almost none of the participants I interviewed said that they had mentioned their ideas about causation to their clinicians. On the one hand, their reluctance stems from their belief that social inequality is an intractable problem; in the absence of quick fixes, then, a social etiology of their heart disease is perceived to be of little real utility. But this self-censoring is also a conditioned response to their being disciplined as patients. Carmen, for example, describes her interactions with her providers:

> The doctors that I've spoken to . . . all pretty much tell me the same thing—they tend to take your condition less seriously if you're overweight. . . . [It's] almost like a recurring theme, that if you lose weight, your blood pressure will probably stabilize. . . . I don't know if that's necessarily true. . . . Those stresses from my job, from my personal life and everything else, those are still going to be there. . . . It bothers me that they emphasize the fact that losing weight is going to make you all better. . . . To me it's almost like a vicious circle.

Although Carmen expresses skepticism in the validity of her doctors' assessments, the experience of surveillance and judgment weighs heavily on her. Yet her clinician may well feel he or she has acted responsibly in this situation: having an individual-level recommendation to offer is wholly appropriate in clinical care. At the same time, however, it leaves Carmen, who sees her cardiovascular risks as rooted in her social and economic conditions of life and shaped by her social structural position, with little recourse that she believes will be effective.

While biomedicalization theory discusses health social movements and patient groups (see also Epstein 2007; Morgen 2006) and the link between technoscientific identities and biosociality, it does not address the "null" case. That is, the problem of non-engagement between expert and lay knowledges is a complication not anticipated by biomedicalization theory. But it is one that has significant consequences for whether and how expert authority will be disrupted or sustained in sites of epidemiological knowledge production. In the absence of engagement, then, the challenges of conceptualizing, much less achieving, a science that considers race, gender, and class as intersecting, social dimensions of difference are especially acute in the case of heart disease — despite its being a health concern that touches so many. However, a juxtaposition of lay and epidemiological knowledges seems a logical first step. Given the heterogeneous perspectives, even among epidemiologists, on how racial and class differences produce cardiovascular health or disease, the historical persistence of racial and class inequalities in health and their relative imperviousness to policies aimed at their reduction, and our ongoing confusion about the sources of and solutions to racial inequalities in health, I argue that official expertise on CVD causation — as currently embodied in the scientific practice of epidemiology — might have much to gain from the diverse knowledges constructed by laypeople with CVD. As Popay, Williams et al. (1996, 1998) contend, laypeople, through their experience of life events and their conditions of existence, accumulate an alternative body of knowledge that is authoritative in its own ways. Analysis of its content may therefore contribute to our understanding of the relationship between racial and class differences and individual and community health.

Thus the case of heart disease and its stratified biomedicalization explicitly raises the question of how to understand and even talk about the legitimacy and validity of knowledges in ways that do not themselves replicate experts' take on expertise but instead suggest alternative modes of engagement. Reserving the label of "experts" for those who possess specialized training, skills, language, and credentials, and for the few laypeople who acquire them, highlights that relations of expertise are relations of power. At the same time, however, defining "experts" and "expertise" in these ways reifies scientists as producers of "expert," official knowledge, while laypeople, in contrast, construct "subjective," "anecdotal" interpretations to create personal meaning from their experiences. In the case of racial and class inequalities in cardiovascular disease, the relative absence of an arena in which lay and scientific causal claims come into explicit interaction and contestation means that we largely take for granted the very bases of expertise and how it privileges certain constructions of racial and class differences over others.

Mainstream epidemiology has therefore been relatively free to define the problem of racial and socioeconomic disparities in consonance with their construction of social differences through culturalist and behavioral lenses. Like the epidemiologist quoted at the beginning of the chapter, the problem has been framed as being rooted in laypeople's insufficient understanding and uptake of epidemiological information. But Irwin and Wynne (1996), among others, argue that a "public ignorance" framing of the problem impugns the general public rather than the institutions, practices, assumptions, and conduct of science. By displacing concerns about the dynamics and distribution of power and resources to maintain health with an individualized emphasis on people having and understanding the "right" information, discourses surrounding the production, distribution, and application of epidemiological knowledge sidestep the question of "who, under what circumstances and with what resources, are able to influence their own health and that of others" (Radley, Lupton, and Ritter 1997, 12). Instead the problem is constructed as one of the public "misunderstanding" and "misapplication" of epidemiological science, and of the need for individuals to more fully accept their responsibility to acquire biomedical knowledge and use it to promote their own health. Such claims serve to fortify the demarcation between "expert" scientists and the "lay" public. Such boundary work and reinforcement and reinterpretation of its jurisdiction are a foundational move of stratified biomedicalization (Clarke et al. 2003, this volume).

Another implication of the lack of direct engagement between expert and lay knowledges is that any change in the near to medium term must come about from within epidemiology itself. This may well happen. Social epidemiology—a subfield within epidemiology that is also referred to as ecological or ecosocial epidemiology—is calling for transformations in the ways in which disease causation is conceptualized, the very definitions of what constitute "causes" and "risks," and the procedures through which disease risks are identified and studied. Social epidemiologists argue that social relations may affect individual health outcomes independently from individual factors (e.g., Berkman and Kawachi 2000b; Diez-Roux 1998, 2007; Susser 1998). Disease is therefore seen as *socially* produced, and the relative positions of socially designated *groups* and the structural processes and institutions that maintain such positions are highly consequential for their health. A more comprehensive understanding of disease etiology must therefore include the investigation of not only biological causes but also social, cultural, and political forces as causes of illness. These organizing concerns were in part born out of critiques of the continuing failure of traditional epidemiology to fully explain individual and group-level variances in disease incidence or to generate recommendations of practical relevance to the public other than lifestyle changes and behavior modification, which have been largely unsuccessful in reducing racial and socioeconomic disparities.

The growing body of work by social epidemiologists, particularly their critiques and commentaries on mainstream epidemiology, has occasioned much debate,[9] even earning the label of "the epidemiology wars" (Poole and Rothman 1998). Certainly there is much contestation within the epidemiological community over whether more, better, or radically different theoretical and methodological explications can strengthen the discipline. Simultaneously there is considerable deliberation about how epidemiology can achieve its ends of determining disease mechanisms and improving public health, and what kinds of scientific, methodological, and conceptual endeavors might accomplish this.

But the question remains: what will happen to laypeople's claims about how structural forces (such as racial oppression, material deprivation, and racial and gender segregation in the labor market) are direct contributors to cardiovascular risk as epidemiologists work to manipulate and translate them into hypotheses that they can test? Based on my interviews, it seems clear that at least some social epidemiologists feel an enormous personal and intellectual commitment to revolutionizing their discipline. From their perspective, they are not attempting to co-opt or appropriate a competing knowledge system. Rather than a "biologization" of social causes of health inequality, what they seek to effect is a "sociologization" of epidemiology.

However, like its larger discipline, social epidemiology is not a unitary set of beliefs and practices. Certainly some social epidemiologists are attempting to grapple with precisely the kinds of structural determinants of disease that laypeople point to as being consequential for their health. At the same time there are efforts to move to a population perspective that are still underwritten by individualistic and culturalist conceptions of human difference. Moreover, an assessment of epidemiological futures cannot rest on the commitments and desires of individual epidemiologists. It must instead analyze the organizational infrastructure of epidemiology, the larger sociocultural landscape of risk in which it is situated, and the stratified relations of power-knowledge that run through it. From this vantage point, significant obstacles stand in the way of social epidemiologists achieving their aims, however well-intentioned.

Chief among them are problems with scientific validity and issues of causal inference. Epidemiology has developed a well-understood hierarchy of research models and kinds of data based on the ability of various kinds of study designs to make scientifically supportable claims. At the apex of this hierarchy is the randomized controlled trial (see Marks 1997), in which exposure to some suspected causal factor is applied to a randomly selected "experimental" group of research participants while the "control" group receives some kind of placebo or alternate intervention. In contrast, observational data are viewed as problematic. Factors such as race and class status cannot randomly be assigned to individuals as can a drug or a low-fat diet; moreover, even if such factors could be assigned, they operate so interdependently with other factors that the experimental and control groups could not be "otherwise equal" (see, e.g., Kaufman and Cooper 1999). One's race, social class, and sex/gender permeate and indelibly shape so many other aspects of life potentially related to cardiovascular health that it would be impossible for groups of varying racial, socioeconomic, or sex designations to be conceptualized as "otherwise equal." Thus epidemiologists have little choice but to conduct observational studies on the effects of race and social class on cardiovascular health.

But social epidemiologists encounter other methodological quandaries with observational studies that have hampered their ability to state with confidence the conclusions of a study and its implications. First, such research depends on the reliable collection of myriad factors that influence cardiovascular health. Through the application of multivariate statistics, it is assumed that the contributions to the outcome of each variable — including race, socioeconomic status, and sex/gender — can be independently calculated. However, given the impossibility of disentangling the effects of race, class, and sex/gender from each other and from other aspects of life that may affect CVD incidence, the assumption that "independent" variables can be isolated from each other does not hold.

A second issue is that social epidemiological studies often involve the measurement of race, social class, and other complex social variables, as well as observations of health-related behaviors. However, such measures are plagued by questions of scientific validity (the extent to which the chosen methods actually measure what they set out to measure) and reliability (usually judged by the degree to which measurements are consistent when repeated across space and time). Social epidemiologists I interviewed who use or had considered using more novel means to measure social variables in their research recount the subtle pressures they feel to choose instead more standard measures and biological indicators perceived to be more reliable because they are viewed as more replicable.

Many social epidemiologists are thus "disciplined" by their colleagues, by epidemiological conventions, and by their own socialization as scientists to favor those risk factors and clinical markers that can be operationalized and measured physiologically. These are viewed as less contestable and more believable than potential risk factors such as racial or class discrimination. Within epidemiology, stable and quantifiable markers that reside in the biological body are constructed as more likely to be "really measuring what you want [them] to measure," as one epidemiologist puts it, and therefore less threatening to internal validity. Causal claims are thus seen as far more legitimate when they emerge from studies using biologically manifested measures. Epidemiological research on the causes of social inequalities in cardiovascular disease, then, often falls short of ideal standards of measurement and validity in this regard.

Continuing concerns about scientific validity, causal inference, and measurement error in social epidemiology often frustrate and confound the funding, conduct, and dissemination of such research.[10] Only with new institutional practices, such as establishing separate review committees dedicated to selecting behavioral and more novel research that characterizes much of the epidemiological research on social inequalities in CVD, can this systematic marginalization be reversed. Thus, rather than the question of whether or not lay knowledge claims are being ignored or co-opted by epidemiology, the crux of the matter is that attempts to incorporate social causes into scientific research run into existing conceptual, methodological, and institutional infrastructures that sustain the cultural prism, a behavioral orientation, and individualism. These in turn further stratify biomedicalization.

Conclusions

As we seek in the current era of biomedicalization to identify and treat more and more of our health risks, existing tendencies within epidemiology to neglect the social, structural, and historical contexts within which people live, work, get sick, and die are exacerbated. I have argued that the constructions of race, social class, and sex/gender that are prevalent in cardiovascular epidemiology are examples of stratified biomedicalization, where definitions of risk and pathology have subjected individuals and groups of particular races and classes to the selective surveillance of biomedicine. In epidemiology, risky behaviors have become tied to and embedded within constructions of race and social class and perceived as the natural and expected attributes *of some group*, rather than shaped by socially constructed power relations *between groups*. Such attributions then intensify the imperatives for individuals from racialized and classed groups perceived to be "risky" to comply with surveillance and behavioral mandates. In this fashion, individualized constructions of difference facilitate stratified notions of risk and pathology that contribute to the construction of some people, defined by their demographics and behaviors, as epidemiologically legitimate targets for biomedical intervention. Epidemiology, as "a 'normalizing' science" (Epstein 1996, 47), thus helps to underwrite processes of stratified biomedicalization through the explicit inclusion of racial and socioeconomic difference in the production of epidemiological knowledge, but in ways that strategically exclude dimensions of those differences that refer to structural forces.

In addition, by enforcing a definition of the problem that centers on cultural and behavioral characteristics seen as inherent to racial and class groups, traditional epidemiology embeds within its own logic dynamics of inequality. Thus health imperatives are systematically disproportionate in their effects, even if on their face they appear to be neutral with respect to race and class. This sedimentation of stratifying conceptions of social differences into the practices of epidemiology and into common-sense understandings of disease causation demonstrates the productive capacity of biopower. Epidemiological expertise achieves biomedicalization "not through the threat of violence or constraint, but by way of the persuasion inherent within its truth, the anxieties stimulated by its norms, and the attraction exercised by the images of life and self it offers to us" (Rose 1990, 10). Cardiovascular epidemiology as a regime of power-knowledge acts to invest power not so much in individuals as in the much more diffuse and pervasive acts of knowledge construction, surveillance, and regulation via behavioral norms and processes of self-governance. Rather than working to dominate citizen subjects, cardiovascular epidemiology and the health imperatives based on its claims instead act to educate and solicit them "into a kind of alliance between personal objectives and ambitions [regarding health, body, and behavior] and institutionally or socially prized goals or activities" (10). They identify new behaviors and habits that comprise "techniques of the self," that is, "the ways in which we are enabled, by means of the languages, criteria, and techniques offered to us, to act upon our bodies, souls, thoughts, and conduct in order to achieve happiness, wisdom, health, and fulfillment" (10–11). Epidemiological claims about difference thus mediate and shape relations between self and self, self and others, self and the state, and between communities and the state. In so doing, cardiovascular epidemiology often furthers processes of stratified biomedicalization and operates to reproduce inequalities in cardiovascular health.

In contrast, people of color with heart disease also have their own causal analyses of CVD. They conceptualize multiple social processes structured on such differences as sources of cardiovascular risk. Their narratives illustrate the uneven dimensions and consequences of biomedicalization. Laypeople's own uptake of proscriptions and prescriptions for health maintenance often occurs alongside their general dismissal of the individualized conceptions of risks and difference dominant in epidemiology. Thus despite the ubiquity of biomedicalization processes, some people with CVD still rely on the authority of their embodied and biographical knowledges.

Yet there is a lack of any direct engagement between lay and expert knowledges, notwithstanding the existence of competing constructions of race and class as disease mechanisms. Absent any arena in which laypeople and scientists can actually interact, or in which their knowledges are mutually interrogated, epidemiological science will continue to maintain its cultural authority. The widespread, chronic, and ordinary nature of heart disease has made it an accepted and normalized illness rather than a contested one. Thus, from a public standpoint, there are no hot-button questions, existing social movements, or clear community-level impacts from which a basis for mobilization around heart disease could be formed. While patient movements and key policymakers succeeded in the inclusion of racial and sex/gender differences and the subsequent establishment of niche standardization (Epstein 2007), the epidemiological meanings and conceptualizations of those differences have become exceedingly naturalized.

I have shown here that social forces, the content and conduct of science, and processes of biomedicalization are mutually shaped and constitute one another. Structural factors within the organization of epidemiology and its own disciplinary culture together create a situation in which devoting additional funding to investigating disparities in CVD — as critical as that is, and as important a political accomplishment as that would be — will not be enough. Under the existing conditions of production, further epidemiological research will continue to generate results that largely replicate what we already know about individual-level risks and health-related behaviors. Thus, to consider seriously the notion that health and illness are socially produced phenomena, stratified by racial and class inequality, requires fundamental changes to the social and political conditions in which expert knowledge is produced.

Notes

1 Cardiovascular diseases are diseases of the heart and blood vessel systems. One such disease is coronary heart disease (CHD), a condition where the coronary arteries that provide the heart tissue with blood become occluded and harden, resulting in insufficient blood supply to organs or tissue (ischemia). Another condition is high blood pressure, or hypertension, considered to be a risk factor for CHD as well as a cardiovascular disease in and of itself.

2 Unless otherwise noted, statistics are drawn from the American Heart Association (2001) and the National Center for Health Statistics (NCHS 1998). I refer here to various kinds of groups (e.g., "American Indians," "Hispanics") by the category labels used in the original sources.

3 All names are pseudonyms.

4 I use the term "culturalist" to refer to a particular idiom of "culture" as it is used to refer to behaviors and traits seen as intrinsic to ethnic groups. Thanks to an anonymous reviewer for suggesting this change.

5 All characterizations of participants' social class are their own.

6 In the eyes of epidemiologists, however, these standardized measures of class are not without their problems.

7 As Katz (1989, 16–35) points out, culture-of-poverty arguments were originally developed and articulated by intellectuals (e.g., Lewis 1959, 1966) who stressed the structural foundations and adaptive functions of a culture marked by lack of participation and integration into mainstream institutions, apathy, and hostility.

8 Here I rely especially on the discussion in Shim 2005.

9 See, for example, Diez-Roux 1998; Krieger 1994; McMichael 1995; Pearce 1996; Poole and Rothman 1998; Savitz 1997; Shy 1997; Susser 1998, 1989; and Taubes 1995.

10 As true of the other chapters in this volume, this is specific to the U.S. context. The status of social epidemiology in other countries, such as Canada, that are far more sympathetic to considering the social determinants of health, is very different. I thank an anonymous reviewer for pointing this out.

Sara Shostak

8 / Marking Populations and Persons at Risk

MOLECULAR EPIDEMIOLOGY AND
ENVIRONMENTAL HEALTH

This chapter describes the emergence of molecular biomarker research, its applications in environmental epidemiology, and its consequences for the biopolitics of environmental health and illness. Beginning in the 1980s and accelerating in the 1990s, molecular biomarkers became an integral part of epidemiologic research, in general, and environmental epidemiology, in particular. Epidemiologists use molecular biomarkers to assess individual exposure, internal dose, preclinical effects of exposure, and genetic susceptibilities to toxic substances (Perera and Weinstein 2000, 517; Schulte and Perera 1993, 7). Working with molecular biomarkers has allowed epidemiologists to look "inside the black box" of the human body, to refine the outcomes of interest in epidemiologic studies, and to expand levels of analysis. Environmental epidemiologists also have used biomarkers to elaborate the pathways leading from an exposure to an outcome of interest and to propose the molecularization of disease phenotypes.[1] Insofar as molecular biomarkers enable new forms of clinical and public health interventions, they provide an opportunity to examine the relationships between three key components of biomedicalization: transformations of biomedical knowledge production, the "technoscientization" of biomedical practices, and the elaboration of risk.

In the context of disease prevention, the practices of molecular epidemiology are connected inextricably with those of risk, "a family of ways of thinking and acting, involving calculations about probable futures in the present followed by interventions into the present in order to control that potential future" (Rose 2001, 7; see also Castel 1991; Lupton 1994/2003; Petersen and Lupton 1996; Rose 2007). In their research, epidemiologists use biomarkers to ascertain three types of risk: environmental risks (exposures), acquired risks (effects of exposures), and inherited genetic risks (intrinsic genetic susceptibilities). Concomitantly, molecular epidemiological research may be used to identify three potentially overlapping categories of persons at risk: people at risk by virtue of where they live, work, or play; people at risk as a result of the effects of prior environmental exposures; and people at risk as a result of inherited genetic susceptibilities, often in interaction with environmental exposures (Ottman 1996). By constituting new categories of people at risk, molecular epidemiological research creates new opportunities for biomedical surveillance and intervention, self-monitoring, and regimens of behavior change. In addition to treating disease, such interventions aim to prevent disease occurrence following exposure to environmental contaminants.

Although biomedicine has long played a prominent role in diagnosing and treating diseases caused by environmental exposures (Brown 2007), responsibility for preventing environmentally associated illness has historically been the mandate of government regulatory agencies, such as the Environmental Protection Agency (EPA) and the Food and Drug Administration (FDA). At these agencies the classification and regulation of environmental chemicals in the ambient environment (e.g., air, water, and soil) — rather than the classification and regulation of people — have constituted the dominant logics of control for protecting human health vis-à-vis the environment (Shostak and Rehel 2007). The explicit goal of many molecular epidemiologists is to improve exposure assessment and enable government agencies to more effectively regulate the ambient environment.

In seeking to develop better means of measuring exposures and their effects, molecular epidemiologists have extended the foci of environmental health research inside the human body and to the molecular level. They thereby have created, however inadvertently, the conditions of possibility for forms of governance that rely on the identification of persons at risk, biomedical intervention, and "technologies of the self" (Foucault 1988).

These individualized modes of governance become possible precisely because molecular biomarker techniques render the individual, as well as populations, a "work object" (Casper 1998) in environmental health research and a potential subject of regulation. As a consequence, the governance of environmental health and illness, though it remains a public policy concern and focus of state regulation, may also become "an individual goal, a social and moral responsibility, and a site for routine biomedical intervention" (Clarke et al. 2003, 171). Examining molecular epidemiology's contributions to the biomedicalization of environmental health and illness thus provides a unique vantage point for understanding the relationships between biomedicalization and other forms of governance in late modern society.

Biopolitics: Science and Social Order

In addressing questions of the contemporary relationship between knowing and governing one aspect of "life itself" — that is, the relationship between knowing and governing the effects of the environment on human health and illness — this chapter concerns "biopower." Foucault defined biopower as that which brings life and its mechanisms "into the realm of explicit calculation and [makes] knowledge-power an agent of transformation of human life" (Foucault 1978/1990, 143). Analytically, the concept of biopower challenges us to look at the ways in which scientific knowledge and sociopolitical orders are co-produced in relation to each other (e.g., Jasanoff 2007; Rabinow 2003; Reardon 2004).

"Biopower today" is constituted by "knowledge of vital life processes, power relations that take humans as living beings as their object, and the modes of subjectification through which subjects work on themselves qua living beings" (Rabinow and Rose 2003, 34). As such, understanding contemporary biopolitics requires analysis of the production and assemblage of practices, such as molecular epidemiology, that attempt "to intervene upon the vital characteristics of human existence — human beings, individually and collectively, as living creatures who are born, mature, inhabit a body that can be trained and augmented, and then sicken and die" (Rabinow and Rose 2003/2006, 3). This chapter examines especially forms of biopower which "in a quintessential Foucauldian sense, are no longer contained in the hospital, clinic, or even within the doctor-patient relationship" (Clarke et al. 2003, 172).

This chapter is oriented also to an ongoing conversation in science, technology, and medicine studies focused on the contexts and consequences of molecular genetic and genomic research (for reviews, see Clarke et al. 2009; Freese and Shostak 2009). I draw especially on the work of Novas and Rose (2000; see also Rose 2001, 2007), who argue that rather than generate fatalism, discourses of genetic risk instead "create . . . an obligation to act in the present in relation to the potential futures that now come into view" (2000, 486). Individuals in late modern society are already operating in "a political and ethical field in which individuals are increasingly obligated to form life strategies, to seek to maximize their life chances, to take actions or refrain from actions in order to increase the quality of their lives, and to act prudently in relation to themselves and to other"; indeed, these obligations and expectations increasingly define what it is to be an "ethical subject" (Collier and Lakoff 2005). By opening possible futures to the strategic consideration, choice, and action of individuals, molecular biomarker technologies further contribute to this mode of subjectivity. At the same time, attending to such possibilities in the context of environmental health and illness, especially given the extensive evidence that people of color and with fewer resources are more likely to be exposed to environmental hazards (Brulle and Pellow 2006), highlights the possibility that such forms of subjectivity may be highly stratified.

Finally, my research draws on symbolic interactionist perspectives on the production of scientific knowledge and technologies. In particular, I use an interactionist conceptualization of science as *work practices*. Methodologically, this means that empirical sociological analyses of "conceptual changes in science" can begin with an examination of "individual and collective changes in the ways scientists *organize their work*" (Fujimura 1988/1997, 97). Thus I explore the emergence of new forms of molecular disciplinarity by examining the ways in which assemblages of tools, work objects, experimental systems, concepts, and human subjects are brought into productive knowledge relations with each other (Clarke and Fujimura 1992; Messer-Davidow, Shumway, and Sylvan 1993; Murphy 2006). It is to such an analysis of molecular epidemiology that I now turn.[2]

Molecularizing Epidemiology:
Biomarkers and the Human Body

Beginning in the 1970s, epidemiologists commonly used molecular biological techniques to study particular disease vectors or "agents" of disease.[3] Indeed, the first use of the term "molecular epidemiology" was in 1973, in a publication titled "The Molecular Epidemiology of Influenza" (Kilbourne 1973). However, contemporary molecular epidemiologists generally use molecular biomarkers as indicators signaling events in *human* biological systems at the molecular or biochemical level (e.g., DNA adducts,[4] gene mutations, and chromosomal aberrations) (Perera and Weinstein 2000, 518).

Epidemiologists point to the publication in 1982 of a paper in the *Journal of Chronic Disease* by Frederica Perera and Bernard Weinstein as the beginning of a deliberate effort to bring molecular biological techniques to environmental epidemiology and to focus those techniques within the "black box" of the human body. The article was structured around two main arguments. First, the authors asserted that existing epidemiologic approaches were insufficient for detecting and evaluating environmental hazards and their role in causing human cancer. Second, they argued that molecular biological tools offered a means of transcending these limitations. In support of this argument, Perera and Weinstein identified four types of molecular biomarkers that could be used to improve measurement techniques in environmental epidemiologic studies.[5] Since then, molecular biomarkers and their applications have been developed by epidemiologists working not only in cancer epidemiology but also in cardiovascular, infectious disease, pulmonary, reproductive, genitourinary, neurologic, musculoskeletal, and genetic epidemiology (for an overview of molecular biomarker development in these fields, see Schulte and Perera 1993).

However, for several reasons, molecular epidemiology has been particularly prominent within environmental health research. First, the etiologic factors in most of the first "proof of principle" experiments were environmental chemicals, such as benzo(a)pyrene (Perera and Weinstein 1982). Environmental epidemiologists had long perceived that their subdiscipline was constrained by the very measurement issues that molecular biomarkers were designed to address. Thus environmental epidemiologists had a strong interest in developing molecular biomarkers as dosimeters of environmental exposure so as to minimize exposure misclassification in their research.

> I do pesticide research, and in pesticide research it's impossible . . . to get any kind of dosimeter unless you're using a biomarker. People just don't know what they're exposed to; they have no idea. They may know whether or not they use pesticides, but they have no idea what compounds they might be exposed to. And there's a lot of exposure that's involuntary, and they just don't know it at all. So if you're trying to do epidemiology of environmental pesticide exposure and you're not using biomarkers, you're going to get terrible exposure misclassifications. (Interview 5)

Environmental epidemiologists were also interested in developing susceptibility biomarkers as tools for identifying the modification of responses to environmental exposures. Consequently, departments of environmental health science have been the institutional homes of many leaders in the field of molecular epidemiology. As one university-based molecular epidemiologist commented, she is located in a department of environmental health sciences because "this is how it [molecular epidemiology] developed. . . . The people who got really interested in using these biomarkers were doing environmental health sciences research, looking at effects of environmental exposures" (interview 5). Finally, by virtue of their participation in the arena of the environmental health sciences, with its historical emphasis on animal models, environmental epidemiologists may have been especially interested in molecular biomarkers as tools enabling environmental health research to focus on the interior of the *human* body.

OPENING THE BLACK BOX OF THE HUMAN BODY

Perhaps the greatest shift made possible by molecular biomarkers comes from the efforts of molecular epidemiologists working in the environmental health sciences to look inside the human body and to focus research on the chain of events leading from exposures to outcomes. Before the advent of molecular epidemiology, environmental epidemiologists had largely been excluded from the risk factor paradigm that characterized post–World War II epidemiology in the United States (Susser and Susser 1996a, 1996b). Measurement issues, such as those described above, meant that though environmental epidemiologists might be able to identify large populations at risk (e.g., by virtue of massive environmental exposures), they faced tremendous difficulties in identifying subpopulations or individuals at risk; as a consequence, environmental epidemiology existed at the margins of the discipline as a whole.[6]

Molecular biomarkers offered environmental epidemiologists a means of more precisely identifying individual and subpopulation risks, thereby bringing their scope of practice in line with contemporary epidemiologic standards and norms. Additionally, molecular biomarkers were seen by environmental epidemiologists as enabling techniques that could *improve* on "risk factor" or "black box" epidemiology. Specifically, while risk factor epidemiology focuses on producing "risk ratios that relate exposure to outcome *with no elaboration of intervening pathways*" (Susser and Susser 1996a, 672; italics mine), molecular epidemiologists seek to use biomarkers both to identify risks and to elaborate precisely those pathways.

A focus inside the body had begun at the time that molecular biomarkers became popular tools in epidemiologic research. As one epidemiologist explained:

> Previously epidemiology mostly was just looking at some kind of exposure determinant and disease; that was it. So that's what the change to molecular epidemiology is about. Now, of course, even before this genetic revolution, we were already starting to open up some of those boxes . . . looking at contaminants in urine, hair, other fluids, that had . . . to do with chemistry and biochemistry. (Interview 6)

However, by using molecular biomarkers, epidemiologists hoped to gain an unprecedented level of access to an interior terrain at the molecular level: "Biomarkers essentially are very powerful tools because they . . . allow us to look into this kind of lead box" (interview 25). This "opening" has had manifold consequences for epidemiologic research.

First, molecular biomarkers have widened the scope of epidemiologic research that can be done with human subjects or with samples taken from human subjects. This expansion of research practice drew many scientists, and especially environmental epidemiologists, to molecular epidemiologic research. As an environmental epidemiologist recounted:

> What I was interested in is molecular mutagenesis—the basic mechanisms of molecular mutagenesis. . . . I had been working on bacterial models of mechanisms. . . . In the 1980s I switched to studying genetic

variations in humans. This is when molecular epidemiology was just getting started. The techniques were being developed, and I wanted to get into human studies. (Interview 26)

Additionally, that molecular biomarkers enabled scientists to pursue environmental health research using human (rather than animal) subjects attracted the attention and engagement of researchers from other fields, such as toxicology. As a toxicologist recounted, he was compelled by the possibility of using molecular biomarkers as a way to transcend the limitations of extrapolating from animal to human models in assessing the risk of environmental chemicals:

> By the end of the 1980s, I had become very disenchanted with animal experiments. . . . I was challenged at a committee meeting in the late '80s. . . . "What's the alternative?" And I thought that was a good question. So my research went into the direction of biological markers in the late '80s and the beginning of the '90s, with *the idea that we could study . . . and we'd be able to learn a lot more than we could from rats and mice with regard to risk.* So . . . what are people actually exposed to? How much are they exposed to? What does the dose response curve look like? Are there susceptible people? All this could be much better answered in people. (Interview 15; italics mine)

Indeed, some scientists described organization-wide shifts from animal to human studies as a result of the application of molecular biomarkers in environmental health research:

> It has allowed us to move a lot of our toxicology from animals into humans. In the division that I'm in, the Division of Occupational and Environmental Health, all of our studies are in humans now, whereas ten or fifteen years ago we still used a lot of animal studies. (Interview 12)

In short, environmental epidemiologists see molecular biomarkers as a more precise means of ascertaining the pathways that lead from the external environment (e.g., the air, water, and soil) to the internal environment of the human body. For example, one goal of molecular epidemiologic research is to correlate molecular biomarkers of internal dose and effect with measures of ambient exposure levels (e.g., the level of a chemical in the air or water). This has also been an important strategy in molecular epidemiologists' efforts to validate new molecular biomarkers, as such efforts rely, in part, on their ability to convincingly claim that internal molecular measurements have a reliable and valid relationship to previous measurement techniques (Shostak 2005; Timmermans and Berg 1997): "We're doing both [internal and external measurements]. And I think you need both because you need to relate it [the molecular biomarker] back to what's out there in the end" (interview 4). Molecular biomarkers thus allow epidemiologists to make new claims about the relationships between conditions external to and processes inside of the body.

Additionally, molecular biomarkers are hailed by environmental epidemiologists for their enabling research with outcomes *other than mortality*. As this epidemiologist noted, early environmental epidemiology was quite limited in its outcomes of interest:

> When you think of the early environmental epidemiology studies . . . they were really mortality studies. [These] mortality studies maybe had monitoring stations during the London fog or in northern Pennsylvania, where they have some idea of the magnitude of total particulates from some area stations, and have mortality as the outcome. (Interview 6)

In contrast, molecular biomarker measurements allow molecular epidemiologists to pursue multiple outcomes of hypothesized relevance to etiologic pathways. For example, molecular epidemiologic research may focus on the biological processes by which persons "handle" an environmental exposure:

> The biomarkers are telling you more than you can get through monitoring, and it's telling you something different. It is telling you not only was there exposure, but it is telling you something about how that individual processed, handled the exposure. (Interview 4)

A related transformation is the molecular epidemiologic focus on individual susceptibility, or what factors modify the events inside the body. Thus molecular biomarkers are used as tools for the study of environmental exposures, their molecular consequences, and the individual genetic characteristics that modify their effects. The availability of these markers as predictive, diagnostic, and biomonitoring technologies has extensive consequences vis-à-vis biomedicalization both within and beyond the clinic.

Measurements, Models, Effects:
Molecular Epidemiology and Disease Prevention

Molecular epidemiologists frame (Benford and Snow 2000) their research as an effort to understand and improve public health: "the goal of such research is practical: to apply relevant and valid biomarkers and to incorporate the resulting information into public health actions" (Christiani 1996, 921). This framing is consonant with the jurisdiction (Abbott 1988) of epidemiology more generally, as one of "the core sciences of public health" (Omenn 2000, 1). Molecular epidemiologists frequently refer to their potential contributions to disease prevention in their publications (Christiani 1996; Hemminki et al. 1996; Perera 1997; Schulte and Perera 1993), as well as at scientific meetings. For example, in a presentation at the annual meeting of the American Association for Cancer Research in 2002, Perera commented to a group of molecular epidemiologists that she was concerned that "we don't spend enough time translating our results" and challenged her colleagues to "lay out contexts and frameworks and their importance to prevention." As this discussion unfolded, there was broad agreement among the molecular epidemiologists present that "the goal of molecular cancer epidemiology is to develop *early warning systems* to prevent cancer onset" (field notes, AACR, 2002, emphasis mine).

Like molecular epidemiology itself, the translation of molecular biomarker research for use in disease prevention has been articulated beyond the original focus on cancer (Christiani 1996; Hemminki et al. 1996). In general, molecular epidemiologists argue that by improving knowledge about disease etiology and progression, molecular biomarkers will contribute to public health efforts to prevent and ameliorate adverse health outcomes following environmental exposures. The specific proposed contributions of molecular biomarkers to disease prevention include more precise forms of measurement, a reconceptualization of disease phenotypes, and new forms of intervention.

MOLECULAR MEASUREMENTS
From the perspective of molecular epidemiologists, a key contribution of molecular epidemiologic research to disease prevention lies in the potential of molecular biomarkers to address limitations of previous research methods. Such potential contributions to environmental epidemiologic

research include improving methods of quantifying the magnitude of risk posed by a substance in the environment, reducing the uncertainties produced by extrapolating from animal models to humans, alleviating the challenges posed by the latency period between exposure and clinical manifestation of disease, and providing more accurate dosimeters (Perera and Weinstein 1982, 581–84).

The demand and desire for molecular measurement are nowhere more apparent than in the case of molecular biomarkers of *exposure*. Exposure assessment techniques in epidemiology have been referred to as "the Achilles' heel" of the discipline, a point that advocates of molecular epidemiology have emphasized in their publications (Perera 1987, 887–88). Exposure assessment is a particular challenge to environmental epidemiology because most people are exposed to environmental chemicals without their knowledge and from multiple sources.[7]

> It's very, very hard to determine what environmental exposures are by questionnaire, because most people just don't know what particular compound they've been exposed to. We really needed to develop better dosimeters. So I got really interested in validating, in developing and validating biologic markers of exposure, especially for environmental epidemiology. (Interview 5)

As a result, the promise of molecular epidemiology to address these challenges was of great interest to epidemiologists:

> We started doing molecular epidemiology because the traditional exposure assessment was not good enough. So if I wanted to figure out how lead effects neurobehavioral function in adults . . . you would have to take a detailed occupational and environmental history to estimate what the person's exposure was, and I don't think that worked very well. The movement towards biomarkers . . . was to improve on our exposure and dose and risk assessment. (Interview 11)

Molecular biomarkers of exposure were seen by environmental epidemiologists as a means of reducing misclassifications and of minimizing false-negative results:

> I think one of the reasons, and maybe the most important reason, was to improve our exposure and dose assessment. Lots of studies had been

done relying on exposure measures that would report no association between the exposure assessment and the health outcome, and many of us were just worried that those techniques were just too error prone, and that would bias you towards a null finding. (Interview 20)

As one molecular epidemiologist commented, in the early days of molecular epidemiology, "the primary goal was measuring exposure itself" (interview 26). Indeed, in offering an alternative to extant methods of exposure assessment, molecular epidemiologists addressed significant challenges in the broader discipline of environmental epidemiology.

The contribution of molecular biomarkers of *effect* has been in identifying measurable molecular objects *inside* organs, tissues, and cells (e.g., carcinogen-DNA adducts, DNA breakage), which epidemiologists can now observe, measure, and use to map the pathways between environmental exposure and disease outcomes. As I detail in the following pages, molecular biomarkers of effect have been of particular interest to epidemiologists who study diseases characterized by complicated pathways and long temporal delays between exposure to suspected etiologic factors and clinical manifestation of disease. As one epidemiologist noted, "Previously, epidemiology was just looking at some kind of exposure determinant and disease; that was it, whereas now there are multiple factors, including host factors . . . that can now be assessed, which you could not assess before" (interview 6).

Molecular epidemiologists also use molecular biomarkers to measure what they conceptualize as three types of intrinsic susceptibility: (1) markers pertaining to enzymes that increase or decrease the ability of a chemical to interact with DNA, RNA, or proteins; (2) markers of genetic differences in the capacity of cells to repair DNA damage caused by environmental insult; and (3) preexisting inherited genetic conditions that increase the risk of disease (Eubanks 1994). For epidemiologists, molecular biomarkers of effect and susceptibility are complementary tools that are often used in tandem: "A lot of people are doing . . . work using biomarkers both of effect and susceptibility and looking at disease modulation" (interview 4). Together these new forms of molecular measurements and risk assessment have had significant consequences for epidemiological models of environmental health and illness and for the potential for individually focused biomedical intervention and treatment.

MOLECULAR MODELS OF DISEASE

In bringing molecular biomarkers to epidemiological research practices, molecular epidemiologists also have reshaped models of disease etiology and progression. First, molecular epidemiological practices "molecularize" disease phenotypes. As this epidemiologist commented, "We have used phenotype forever to diagnose disease. Now what we're doing is actually looking at that phenotype at the molecular level" (interview 20). The molecularization of disease phenotypes has profound implications for our understanding of health and illness, as such molecular disease phenotypes may exist in individuals who believe themselves to be healthy. Thus, in a process characteristic of biomedicalization, individuals with no experiential symptoms of illness are seen as "marked" at the molecular level as "becoming ill" or "at high risk" for adverse health outcomes.

The molecularization of disease phenotypes relies on new technoscientific practices, specifically the potential of molecular biomarkers to enable disease definition based on continuous and quantitative variables, rather than categorical variables. For example, as this molecular epidemiologist described, a case (that is, someone who has the outcome of interest) can now be defined with a continuous variable (e.g., number of deformed proteins) rather than a categorical one (e.g., normal versus pathological):

> [The] technologies that are developing are all lending themselves for quantitative measures. We're going to start to define cases as "you've got one hundred thousand deformed proteins," as opposed to . . . "you have emphysema." (Interview 20)

Molecular epidemiological models of disease replace "step function" models of environmental health and illness with fully normalizing models based on continuous gradients of quantifiable markers of disease. A step function model, as described by this molecular epidemiologist, would tell you: "So you're healthy, now you have hypertension, now you have advanced cardiovascular disease, now you have congestive heart failure, now you're dead" (interview 21). In contrast, a continuous model measures accumulation of molecular biomarkers and their associated *risks* over the life course. Therefore, from a molecular epidemiologic perspective, one may not be merely "healthy" or "ill"; rather, as this epidemiologist described: "What we would do is . . . use biomarkers as . . . measures of disease accumulation. . . . *You would be dealing with somebody in a variety of gradations of disease"* (interview 21; italics mine). Thus molecular epidemiological practices are used to define a continuum from health to illness with different quantities of markers of "disease accumulation" marking an individual's position and movement along the continuum. As a result, not only may disease be identified in presymptomatic individuals (Clarke et al. 2003; Conrad 2005) but new molecular disease definitions are instantiated. In this way, molecular biomarkers provide a vivid example of how innovations in scientific measurement techniques play a role in biomedicalization.

Molecular epidemiologists are excited about this reconceptualization of disease phenotypes because it may increase the period of time during which the risks of disease can be identified, categorized, prevented, treated, or managed; it expands biomedical options. For example, molecular biomarkers may allow scientists to better map, understand, and intervene in the "latency period," that is, the time after an exposure but prior to onset of any clinical symptoms of disease.[8] Before molecular epidemiology, latency was incorporated in study design as a part of efforts to ascertain *when* an exposure occurred. For example, if one was investigating the relationship between an environmental chemical and a form of cancer suspected to have a ten-year latency period, one would ask individuals with that cancer about their exposures approximately ten years before their cancer diagnosis. However, molecular biomarkers of effect have given epidemiologists something to measure and to monitor *during* the latency period between the occurrence of an exposure and its suspected effects:

> One area that I think is becoming really well developed is a better understanding of the idea of latency. Because, you see, latency is a complete artifact of epidemiology. It turns out that latency doesn't really exist. What they teach you in epidemiology is that you have to really figure in five or ten years of latency for solid tumors. . . . But it's not like nothing happened in those five or ten years. (Interview 7)

Thus latency may be reconceptualized as the period during which a person is becoming ill as he or she moves along the continuum from health to illness. Molecular epidemiologists frame the possibility of identifying the changes that occur during the latency period as a means of enabling interventions that may *prevent* the manifestation of symptomatic disease:

Rather than just looking at exposure to disease relationships, there is now [a way to look at] exposure and intermediate changes in perfectly healthy people that would perhaps go away if the exposure stopped, and then [disease] would not ever be manifested. So that pathway has been sort of stretched out. (Interview 20)

Consequently, molecular epidemiologists who employ biomarkers of effect can redefine latency as an opportunity for biomedical interventions rather than an obstacle in their research. This deconstruction of latency is especially significant for diseases like cancer, some of which have decades-long latency periods; one molecular epidemiologist told me that cancer was a disease that was "calling out for this kind of research." Put differently, applications of molecular biomarker techniques make a scientific category—latency—into a process amenable to biomedical intervention.

Finally, as noted, molecular epidemiologists may use biomarkers of *intrinsic* genetic susceptibility to identify persons who, by virtue of some aspect of their genetic inheritance, are at increased risk of illness after a given environmental exposure. As with molecular biomarkers of effect, biomarkers of susceptibility identify persons who, although currently free from symptoms of disease and absent of gross pathology, are nonetheless seen as not fully well or of normal health status. Like persons who carry biomarkers of effect, persons with identifiable biomarkers of intrinsic susceptibility may become subjects of self-surveillance and biomedical interventions aimed at reducing the risk that they will progress to symptomatic disease states. This knowledge constitutes, even as it is constituted by, biomedicalization.

EFFECTS

The implications of molecular biomarker research vary across disease prevention strategies. For example, many molecular epidemiologists are interested in using biomarkers of exposure to advocate for primary-prevention interventions oriented to the entire population (Christiani 1996). Primary prevention refers to the eradication or reduction of an exposure that has been identified as a risk factor associated with adverse health outcomes. Biomarkers support primary prevention in circumstances where the "data show that we have had disproportionately high exposure in the community and we need to do something about it" (interview 4). In this application, they provide an empirical basis for policies which would reduce exposure at the population level. The paradigmatic example of this approach to disease prevention is the mandatory removal of lead from gasoline and paints to prevent lead toxicity (and associated neurocognitive damage) in children (Christiani 1996).

A primary prevention approach assumes that exposures can be reduced or eliminated. Therefore, in the context of environmentally associated diseases, primary prevention is accomplished by regulating the entities responsible for the presence of chemicals in the environment. This form of governance emphasizes the responsibility of the state and its representatives for ensuring the health of the population: "The state retains the responsibility that it acquired in the eighteenth or nineteenth century—the precise timing varying across national contexts—to secure the general conditions of health" (Rose 2001, 6). Many environmental health and environmental justice groups advocate for the active and equitable fulfillment of this responsibility (Brulle and Pellow 2006; Shepard et al. 2002; Shostak 2004).

In contrast, a biomedically oriented strategy would use molecular biomarkers of effect to identify at-risk individuals and populations who are appropriate candidates for secondary prevention.[9] Secondary disease prevention occurs after an environmental exposure but before clinical manifestations of disease (i.e., during the latency period). It is directed toward people who feel healthy but are identified, using markers, as being on their way to, or at high risk for, environmentally associated disease(s). Such strategies often take advantage of the elucidation of disease pathways made possible by molecular biomarkers. For example, molecular epidemiologists have proposed that biomarkers be used to identify exposed individuals and populations who might benefit from various forms of clinical intervention. These forms of biomedical intervention may include surveillance of exposed individuals for early detection of disease, chemoprevention, and counseling regarding behavioral or lifestyle changes to interrupt the progression from exposure to illness.

> In essence, what we want to do [is] use that knowledge . . . in individually designed prevention. . . . So if you have an individual who is in the workplace and is exposed to an agent in the workplace that could be very toxic, if it's metabolized and converted to a more toxic metabolite, then you might intervene with another agent that would inhibit that enzyme. That's frequently done in clinical medicine. When you can-

not alter the exposure . . . this is where there have been . . . a number of nutritional-based, chemo-preventive-based, and so on, approaches. (Interview 20)

For example, research is under way to determine whether dietary antioxidants and micronutrients might provide chemoprevention for individuals at risk of environmentally and genetically associated cancers (Ames, Elson-Schwab, and Silver 2002). Research has suggested that dietary antioxidants might lessen the risk for lung cancer among persons who smoke tobacco (Christiani 1996; Garcia-Closas et al. 1997). In addition to identifying who might benefit from such clinical interventions, biomarkers can also be used to assess individuals' response to these interventions:

What drives my basic science in terms of developing these biomarkers is to really use them as tools in preventive interventions, which have to be individually based. . . . We use these markers to look at trajectory change in individuals and in response to intervention. In other words, if an individual is exposed, we need to know that exposure, so then we can intervene and drive the effective consequences of that exposure down. (Interview 21)

The language of "individually based" interventions is quite striking here. In essence, it marks the shift from disease prevention strategies that operate at the community level (e.g., regulatory policy) to explicitly biomedical strategies that operate at the individual level. To the extent that the individual emerges as a key site of monitoring, intervention, and regulation, environmental governance is biomedicalized.

A third and related use of molecular biomarkers in disease prevention focuses on the use of biomarkers of genetic susceptibility to identify "risky individuals" (Rose 2001) and to intervene to reduce their risk. However, even molecular epidemiologists focused on issues of susceptibility differ about how disease ought to be prevented for "the person genetically at risk." Underlying these differences are issues regarding the locus of responsibility for health and illness. For example, many molecular epidemiologists argue that the only way to protect susceptible populations from environmental exposures is for the state to use public policy to regulate the sources of those exposures to maintain them at levels that will protect the most susceptible individuals:

So what you have to do is reduce exposures, to protect those who may be somewhat more susceptible because of heritability, and in the process everyone is protected. (Interview 6)

These epidemiologists argue that given the community setting of most environmental exposures, genetic screening and behavior modification as disease prevention strategies are, at best, impracticable:

There's no way you can screen people and have them move out of their communities. I think that's just over the top, ethically, and probably legally challengeable too. I think they have to put extra controls on the [power plant] stacks to get the level to nondetectable. (Interview 6)

In contrast, some public health practitioners advocate a genetic counseling and testing approach as a means of identifying individuals who might be "motivated to take special steps, beyond those taken to protect everyone" (Omenn 1991). Finally, still other molecular epidemiologists are interested in "molecular therapeutics" as disease prevention strategies (interview 7). They propose using information on the genetic basis of susceptibility to environmental exposures to identify molecular targets for chemoprevention. Again, this is a secondary prevention strategy, aimed at preventing disease onset following an environmental exposure.

By and large, these different approaches to disease prevention — and these different forms of biopower — exist rather comfortably alongside each other. This range of variation reflects historical tensions and ongoing debates within the larger fields of epidemiology, public health, and biomedicine; it is widely accepted among epidemiologists, as well as the clinicians and policymakers who are potential end-users of molecular biomarker research. Therefore molecular epidemiology may contribute simultaneously to both public policy and biomedical strategies for disease prevention.[10]

Conclusion: Molecular Epidemiology and Biomedicalization

Molecular biomarkers enable environmental epidemiologists to identify persons and subpopulations at increased risk for developing environmentally associated diseases. Molecular epidemiologic techniques focus on three different sorts — and sources — of risk. Molecular biomarkers of exposure identify risks posed by substances in the environment. Molecular biomarkers of effect identify the effects of exposure to substances in the en-

vironment. Molecular biomarkers of intrinsic genetic susceptibility identify risks resulting from inherited genetic traits, given an environmental exposure. As such, molecular epidemiologic practice may direct state public
health policy and biomedical interventions to three different populations
of persons at risk: (1) persons at risk as a result of the environment in which
they currently live, work, or play; (2) persons at risk as a result of previous
environmental exposures; (3) persons at risk as a result of inherited genetic
susceptibilities and their interactions with environmental exposures. That
a wide variety of strategies, or combination of strategies, may be deployed
to assess and address the risks marked in these populations is characteristic of the contemporary moment, in which biopolitics focuses precisely on
"the specific strategies and contestations over problematizations of collective human vitality, morbidity and mortality, over the forms of knowledge,
regimes of authority, and practices of intervention that are desirable, legitimate and efficacious" (Rabinow and Rose 2003/2006, 4).

Indeed, my argument is not that molecular epidemiology is determin-
istic of a particular mode of knowing and governing the relationship(s)
between human health and the environment. Rather, molecular epidemiology enables the biomedicalization of environmental health and illness
to occur *alongside* the ongoing (if perhaps diminishing and ever stratified
[Brulle and Pellow 2006; Evans and Kantrowitz 2006]) state regulation of
the ambient environment. Thus regulation of "external nature" (i.e., the
world around us) may co-occur with efforts to harness and transform "internal nature" (i.e., the world within us) (Clarke et al. 2003, 164). By molecularizing disease phenotypes and identifying new categories of persons
as "at risk" or "becoming ill," molecular epidemiologic research opens new
possibilities for biomedical and personal practices that intervene in the
pathways from environmental etiologic factors to symptomatic illness.

Insofar as the governance of environmental health, once wholly a matter of public policy and state regulation, becomes also "an individual goal,
a social and moral responsibility, and a site for routine biomedical intervention" (Clarke et al. 2003, 171), it is biomedicalized. This resonates with
Epstein's eloquent observation that the contemporary moment is characterized not only by the politicization of biomedicine but also by the biomedicalization of governance (2007, 18). Thus, in addition to elucidating
how technoscientific innovations transform medical practices—a hallmark of biomedicalization—analysis of molecular epidemiology high-

lights emerging relationships between technoscience, biomedicalization, and broader biopolitical domains.

Notes

1 By "molecularize," I mean to visualize them at the submicroscopic level — between 10^{-6} and 10^{-7} (de Chadarevian and Kamminga 1998; Kay 1993b).

2 This analysis draws on data from a multisited ethnographic project on disciplinary emergence in the environmental health sciences conducted, in two stages, from September 2000 through September 2002 and from September 2003 through July 2004. The primary data on which this paper is based come from in-depth qualitative interviews ($n = 59$ and $n = 26$, respectively, from each wave of data collection) with scientists, policymakers, and activists. Respondents were initially identified via a comprehensive literature review, which allowed me to map the arena of concern (Clarke 2005, 137–38), and then by snowball sampling. Consistent with the principles of grounded theory, I used purposive sampling once themes began to emerge from my coding of interview data (Charmaz 2006). I also conducted participant observation at a variety of scientific conferences, meetings, and symposiums and at the National Institute of Environmental Health Sciences (NIEHS), first as an intern in the NIEHS Program in Environmental Health Policy and Ethics (summer 2002), and then in 2003–4, when I returned to the NIEHS (for approximately one week of every month), as a fellow in the Office of NIH History. Data from both interviews and ethnographic observation were uploaded into ATLAS. ti, where they were coded and analyzed using the principles of grounded theory (Charmaz 2006; Strauss 1987). In order to protect the confidentiality of respondents, I identify each speaker only by an interview number.

3 A variety of biologic markers had been used for many years as tools in biomedical and epidemiologic research (Schulte 1993, 3–44). For example, immunologists have long relied on white cell counts and antibody titers as indicators of infection (Schulte 1993, 19). Likewise, medical studies had begun to incorporate biochemical markers as early as the 1930s (Kohler 1982). More recently, cardiovascular epidemiologists relied on blood pressure, serum cholesterol, and lipids as markers in their studies of cardiovascular disease (Shields and Harris 1991).

4 Adducts occur when a chemical or its metabolites bind covalently to cellular DNA, thereby potentially inducing mutations in critical cellular genes. Chemical-protein adducts can also be measured as surrogates for DNA damage (Perera and Weinstein 2000).

5 Molecular biomarkers of internal dose indicate the actual level of a compound within the body or specific tissues and compartments. Molecular biomarkers of biologically effective dose indicate the amount of a compound that has reacted with specific cellular macromolecules. Molecular biomarkers of effect identify early biological effects resulting from exposure. Molecular biomarkers of susceptibility are used to identify individual and population differences in susceptibility

to adverse outcomes following environmental exposures (Perera and Weinstein 1982).

6   I thank Ezra Susser for helping me to understand these dynamics, and for his very insightful comments on an early draft of this manuscript.

7   Even exposures that require individual agency, such as the number of alcoholic drinks one consumes per day, are difficult to ascertain.

8   The latency period varies across individuals and is sensitive to many parameters, including, for example, the type of exposure, the type of outcomes, nutrition, presence of comorbidity, and intervening events.

9   In the United States, where at least forty million persons do not have health insurance, access to biomedical resources, and thus access to forms of secondary prevention, varies significantly. As such, it is unsurprising that access to biomedical resources (and financial resources with which to pay for health services) is a frequent redress sought by persons bringing "toxic tort" litigation against industry and the state. Molecular biomarkers of effect may thus be doubly relevant, as persons who believe that they have been harmed by exposure to environmental chemicals may use biomarkers of effect to substantiate their claims. Some environmental advocacy groups are optimistic about these possibilities. For example, in their position paper "The Crisis in Chemicals," the U.K.-based environmental group Friends of the Earth predicts that "the biomedical revolution" will result in an increased ability to identify and measure the damaging effects of exposures, and as a result, affected individuals "will be better able to mount liability cases," and "industry will be faced by people, armed with scientific evidence who can show damage from exposure to chemicals" (Warhurst 2000, 9). Though the Friends of the Earth position paper is largely speculative, it may well be a harbinger of a time in the near future when "biopolitics . . . is waged about molecules, amongst molecules, and where the molecules themselves are at stake" (Rose 2001, 17).

10 Many observers of the emergence of molecular techniques in epidemiology have compared it to the advent of germ theory during the nineteenth century (Loomis and Wing 1990; Susser 1998; Susser and Susser 1996a, 1996b; Vandenbroucke 1988). This historical comparison has been used by some to argue that, like the miasma theory of disease, which preceded germ theory and over time was largely effaced by it, social and environmental explanations of disease (and associated interventions) will wane as molecular biomarker techniques become standard epidemiologic practices (Vandenbroucke 1988). However, more commonly, these authors argue that just as neither miasma theory (and environmental interventions) nor germ theory (and medical interventions) could effectively address the full range of etiologies important to human health and illness, neither a focus solely at the molecular nor one at the environmental level will be able to provide an adequate basis for the field of public health (Loomis and Wing 1990; Susser 1998; Susser and Susser 1996a, 1996b); the future of environmental health, they contend, will therefore incorporate a variety of levels of analysis and intervention.

Jonathan Kahn

9 / Surrogate Markers and Surrogate Marketing in Biomedicine

THE REGULATORY ETIOLOGY AND COMMERCIAL PROGRESSION OF "ETHNIC" DRUG DEVELOPMENT

On June 23, 2005, the U.S. Food and Drug Administration formally approved the heart failure drug BiDil to treat heart failure in "self-identified black patients" (FDA 2005). Widely hailed throughout the media and professional journals as the first "ethnic" drug, BiDil has become a focal point for debates over the appropriate use of racial categories in biomedical research, drug development, and clinical practice. It is also emerging as a new model of how to exploit race in the marketplace by capitalizing on the racial identity of minority populations and leveraging the disproportionate risk of adverse health outcomes they suffer into a cheaper, more efficient way to gain FDA approval for drugs.

BiDil is noteworthy because it is the first race-specific drug ever approved by the FDA. It has been touted by the FDA and others as a significant step toward the promised era of personalized pharmacogenomic therapies. Upon closer examination, however, BiDil's story is far more complex. Its emergence as a racially marked drug reveals a multifaceted interplay of legal, commercial, and technoscientific interventions that both exemplify and problematize the dynamic of biomedicalization (Clarke et al. 2003).

BiDil is marketed by NitroMed, a hitherto small Massachusetts biotech company with no other products currently on the market. BiDil does
indeed appear to significantly help many people suffering from heart
failure—a debilitating and ultimately fatal disease afflicting several million Americans. There is no scientific evidence, however, that race has any-
thing to do with how BiDil works. This is for the simple reason that the
data supporting NitroMed's submission to the FDA were based on a clinical trial (A-HeFT: the African-American Heart Failure Trial) that enrolled
only "self-identified" African Americans (Kahn 2005; Taylor et al. 2004).
With no comparison population, no legitimate claims can be sustained
that BiDil works differently or better in African Americans than in anyone
else.

The FDA, however, accepted NitroMed's argument that because the trial
population was African American, then the drug should be labeled as indicated only for African Americans. This sends the troubling and unsubstan-
tiated message that the subject population's race was somehow a relevant
biological variable in assessing the safety and efficacy of BiDil. Ominously,
it also gives the federal government imprimatur to use race as, in effect, a
genetic category. Moreover, most drugs on the market today were tested almost exclusively in overwhelmingly white male populations. But we do not
call these "white" drugs—nor should we. Rather, the operating assumption
for approving these drugs was that for the purposes of assessing drug safety
and efficacy, the unmarked racial category of "white" was coextensive with
the category "human being." In approving BiDil as a drug only for African
Americans, the FDA has also implicitly adopted an assumption that the
category "black" is somehow less fully representative of humanity than the
category "white."

The process through which BiDil became a race-specific drug elucidates
the power of legal and regulatory regimes in processes of biomedicalization. Legislation such as the Bayh-Dole Act provides the backdrop to the
emergence of BiDil. Enacted in the 1980s, Bayh-Dole allowed researchers
to patent inventions developed with federal funds. It has provided substantial incentives to the formation of new university-industry collaborations that have accelerated to commodification of research as proprietary
knowledge. BiDil is a product of such collaborations. The first studies of
the drugs that make up BiDil were conducted in the 1980s by university researchers in cooperation with the U.S. Veterans Administration. The first

patent to the BiDil drugs as a method to treat heart failure emerged from these studies in 1989 (Kahn 2004).

At this point, however, BiDil was not yet a racialized drug. To the contrary, the 1989 patent covers the use of BiDil in the general population, without any mention of race. The move to see race as relevant was framed by a diverse set of regulatory and legal considerations that reveal the complex imbrication of BiDil in an array of dynamic processes of biomedicalization. BiDil's first step toward racialization came in 1997, through a regulatory intervention, when the FDA rejected the first application to approve BiDil as a drug to treat heart failure in the general population. It was only at this point that BiDil researchers returned to their data from the 1980s and started analyzing it by race.

This move, however, must also be understood in relation to other federal regulatory interventions that were transforming the production and distribution of biomedical knowledge from outside the medical profession. An array of federal mandates in the 1990s directed federally funded researchers and those seeking federal drug approval to produce, collect, and analyze biomedical data according to categories of race and ethnicity promulgated by the Office of Management and Budget in its Directive 15, "Standards for Maintaining, Collecting, and Presenting Federal Data on Race and Ethnicity" (U.S. OMB 1997). These federal mandates directed researchers to produce, collect, and use racially coded data. They provided an incentive, external to professionally controlled medical knowledge production, to reshape researchers' approach to interpreting research data along racial lines.

[handwritten margin note: a push for researchers to incorporate racial categories in research.]

Merely having and seeing racially coded data, however, was not enough to transform BiDil into a race-specific drug. Only when combined with the new commercial incentives presented by patent law, as enabled by the Bayh-Dole Act, did the BiDil researchers begin to make the full turn toward racializing the technoscientific object that was BiDil. The incentive to combine the two lay in the fact that the original, non-race-specific patent to BiDil was set to expire in 2007. By recasting BiDil as a racial drug, the researchers were able to obtain a new race-specific patent to BiDil—for use in African Americans—which lasts until 2020 (Kahn 2004).

But again, the race-specific patent alone was not sufficient to ensure BiDil's ultimate success before the FDA. The true innovation of the new racial patent did not involve the drug itself. Rather, the newly racialized

BiDil succeeded because the patent provided a basis for exploiting bio-medicalization's customization of bodies in the new field of pharmaco-genomics. The entire marketing model for BiDil, including its presentation to the FDA for approval, has been framed by an appeal to personalized medicines that are tailored to particular individuals or groups. This model is premised on an underlying molecularization of bodies as a function of genetic variations that affect responses to particular drugs. BiDil exploits this model but does not exemplify it. In many respects, BiDil is a highly traditional pharmaceutical intervention, comprising two generics widely used before the modern era of biotechnology. No one yet understands how BiDil works at the molecular level. Indeed, no data support any claim that the drug works differently or better in African Americans than in anyone else. Nonetheless, NitroMed has capitalized on the wave of enthusiasm for "personally tailored" pharmacogenomic therapies to position its product within a matrix of biomedical discourses and practices. Its success here is evident in the FDA press release announcing the approval of BiDil, which characterized the drug as "representing a step toward the promise of personalized medicine" (FDA 2005).

NitroMed has also exploited existing models of ethnic niche marketing to promote its product. In this regard, BiDil does not *exemplify* biomedicalization so much as it *exploits* it to serve commercial ends. And yet, while NitroMed is primarily exploiting biomedicalization, its race-specific marketing of BiDil is also having the unintended consequence of contributing to the biomedicalization of race as a genetic category. In this regard, the entire process of BiDil's conception, development, and marketing has contributed to the molecularization of racial identities, a distinctive dynamic of biomedicalization that is transforming race into a disembodied function of genetically based statistical correlations.

BiDil's Origins

How did we get to this point?[1] If we go back to its origins, we find that BiDil did not begin as an ethnic drug. Rather, it became ethnic over time and through a complex array of legal, commercial, and medical interventions that transformed the drug's identity. Over the past twenty years, a revolution has occurred in heart failure treatment with the development of a wide range of pharmaceutical interventions to improve both the quality of life and longevity of people suffering from heart failure. One of the earliest breakthroughs came in the 1980s with the first Vasodilator Heart Failure Trial (V-HeFT I). This trial lasted from 1980 to 1985. It was led by Dr. Jay Cohn of the University of Minnesota and involved cardiologists from around the country working together with the U.S. Veterans Administration. The trials found that patients receiving a combination of two vasodilators called hydralazine and isosorbide dinitrate (H/I) seemed to have a lower rate of mortality. These generic drugs — H/I — would later become BiDil.

The V-HeFT I trial was soon followed by V-HeFT II, which lasted from 1986 to 1989. This trial compared the efficacy of the H/I combination against the drug enalapril, an ACE inhibitor. It found an even more pronounced beneficial effect on mortality in the enalapril group, establishing ACE inhibitors as a front-line therapy for heart failure (Cohn et al. 1991). ACE inhibitors, however, have not totally supplanted H/I because not everyone responds well to them and some others cannot tolerate the side effects.

The V-HeFT investigators did not build the trials around race or ethnicity. They enrolled both black and white patients, but in the published reports of the trials' successes, they did not break down the data by race. Rather, they presented H/I (the BiDil drugs) as generally efficacious in the population at large without regard to race.

The Legal Construction of BiDil

The role of law as a player in the emergence of BiDil as an ethnic drug begins most immediately in 1980, more or less coincidentally with the initiation of V-HeFT I. In that year President Carter signed into law two pieces of legislation that would come to transform relations between industry and academic researchers.[2] The first, the Stevenson-Wydler Technology Transfer Act (15 U.S.C. § 3701 [1994]),[3] encouraged interaction and cooperation among government laboratories, universities, big industries, and small businesses. The second, the Bayh-Dole Patent and Trademark Laws Amendment (35 U.S.C. §§ 200–212 [1994]), allowed institutions conducting research with federal funds, such as universities, to retain the intellectual property rights to their discoveries. It is in this context that the research findings of V-HeFT, produced in cooperation with the U.S. Veterans Administration, could be commercialized through patent and trademark law. Thus were the lead cardiologists in the V-HeFT trials, Jay Cohn and Peter Carson, later able to obtain intellectual property rights in BiDil-related patents and enter into deals with the likes of NitroMed to commercialize the discoveries made through the V-HeFT trials.

The first intervention of patent law in the development of BiDil, however, was negative and restrictive rather than productive. After the successful completion of V-HeFT II in 1989, the next logical step would have been to conduct a trial that explored the combined effects of ACE inhibitors plus H/I. Cohn himself pushed for such a trial and openly bemoaned the lack of corporate support to enable him and other cardiologists to go forward (Cohn 1991). The key reason, as Cohn later noted, was because hydralazine and isosorbide dinitrate were both generic drugs. In the absence of intellectual property rights to the therapeutic compound, corporate support for further tests involving the BiDil drugs would not be forthcoming.[4] Thus, years before BiDil was ever taken before the FDA for approval as a new drug, the lack of relevant intellectual property value seemed likely to condemn hydralazine and isosorbide dinitrate to obscurity as treatments for heart failure.

Cohn revived the prospects of BiDil by obtaining a patent, in 1989, on a "method of reducing mortality associated with congestive heart failure using hydralazine and isosorbide dinitrate" (U.S. Pat. 4,868,179) and then by developing BiDil as a new drug—being a combination of H/I in single-dose form. BiDil was a breakthrough of convenience—it made it easier to use and dispense the drug—but it was not itself a new therapy. Again, at this point it was still a drug for everyone. So it was that Medco, a North Carolina biotech firm, first acquired the rights to BiDil in the early 1990s and started investing time and money in conducting bioequivalence tests and developing marketing strategies in preparation for submitting its New Drug Application (NDA) to the FDA in 1996.

There was a measure of convenience to BiDil, but that alone was not sufficient to drive its development or, it turned out, to obtain FDA approval. One consultant to the FDA panel that ultimately rejected the BiDil New Drug Application in 1997 noted that the two generic component drugs of BiDil "are there" for anyone to use for heart failure. The FDA's denial of the BiDil application would not change that. Rather, he observed, "the practical impact of the FDA not approving this combination today is that there won't be an economic incentive for the sponsor to get out and provide educational material for a lot of doctors to know how to use the drugs best" (FDA 1997, 210).

The truly "convenient" breakthrough for BiDil, therefore, was not simply the combination of two generic drugs into one; it was the development of new intellectual property rights whose value was contingent on FDA approval of the new drug. With a patentable therapy in hand, drug companies would have an incentive to educate physicians and market the new drug — thus changing the behavior of both doctors and patients. Patent law (and to a lesser extent trademark law, which allowed for added brand-name value in the marking of BiDil®) thus provided a critical impetus toward the creation of BiDil. In contrast to the classic justification for patents as incentives to develop new products, intellectual property rights here instead provided an incentive for developing a new marketing strategy based on an existing therapy. Moreover, given that the two drugs that made up BiDil were already available as generics, this also indicates how patent law may also distort a market, potentially obscuring less expensive generic alternatives that have the same therapeutic value.

The FDA rejected this first NDA in 1997 because it found the retrospective analysis of data from the V-HeFT trial was insufficiently powered to meet the regulatory criteria of statistical significance. It is important to note that the FDA advisory committee reviewing the drug did not think BiDil didn't work. To the contrary, many of the doctors on the panel were generally convinced of the drug's clinical efficacy. They turned down the application because V-HeFT trials were not designed as new-drug trials, so the data they produced could not meet the regulatory criteria of statistical significance required for new-drug approval (Kahn 2004, 14–15).

After the FDA rejection in 1997, the value of the intellectual property rights to BiDil plummeted along with Medco's stock. The rights reverted to Cohn, and Medco exited the story of BiDil's development. It was at this point that Cohn, together with Carson and others, went back to the V-HeFT data and broke it out by race. In 1999, Carson was lead author on the first article arguing for a race-based differential in response to H/I treatment based on a retrospective analysis of the V-HeFT data — in particular the nearly fifteen-year-old data from V-HeFT I (Carson et al. 1999).

The intervention of the federal regulatory system to deny the NDA marks the turning point in BiDil's journey toward ethnicity. The regulatory action taken by the advisory committee impelled the BiDil researchers to reconceptualize their drug along racial lines to get a "second bite" at the apple of FDA approval. After the publication of Carson's article, the value of the intellectual property rights to BiDil rebounded—not because of any changes to the underlying molecular structure or biological effects of BiDil as a drug but through the reanalysis of the old V-HeFT data along racial lines.

NitroMed acquired the intellectual property rights to BiDil in September 1999 (NitroMed 1999)—the same month Carson published his paper on purported racial differences in response to the drug during the 1980s trials. In the hands of its new corporate handlers, together with their public relations consultants, BiDil was soon reborn as an ethnic drug. Hearkening back to the comment by the FDA panel consultant, the subsequent spate of publicity attending the inauguration of A-HeFT marks how the renewed value of the patent to BiDil provided an incentive for NitroMed to educate doctors and the public about the nature and value of this "new" drug for African Americans.

In the next logical extension of patent rights into the process of creating an ethnic drug, Cohn and Carson jointly filed for a new BiDil-related patent on September 8, 2000. Titled "Methods of Treating and Preventing Congestive Heart Failure with Hydralazine Compounds and Isosorbide Dinitrate or Isosorbide Mononitrate," the patent appears much the same as Cohn's original 1989 patent. On closer inspection, however, the abstract to the patent specifies that the "present invention provides methods for treating and preventing mortality associated with heart failure in an *African American* patient" (U.S. Patent 6,465,463; italics mine).

The issuance of the new patent is commercially important because the original patent is set to expire in 2007. The new race-based patent will not expire until 2020. Significantly (and rather astonishingly), in issuing the second patent, the U.S. Patent and Trademark Office found that Cohn's first method-of-use patent for BiDil did not constitute "prior art" with respect to the new patent application. Rather, it found the application's race-specific method of treatment to be a "nonobvious" extension of the earlier concept and hence patentable (NitroMed 2003, 11). In this chapter in BiDil's development, patent law did not spur the invention of a new drug but rather prompted the reinvention of an existing therapy as ethnic.

In 2001, NitroMed approached the FDA with its proposal to obtain race-specific approval for BiDil. The FDA responded with a letter stating that

BiDil might be "approvable" pending the successful completion of a race-specific confirmatory drug trial. With the FDA letter in hand, NitroMed was able to raise over $30 million in venture capital during the nadir of the dot.com bust in 2001 to initiate A-HeFT, the African-American Heart Failure Trial. A-HeFT enrolled only "self-identified African American" subjects. After a slow start, 1,050 subjects were ultimately enrolled. The trial was halted early, in July 2004, by NitroMed's Data Safety Monitoring Board because it found a striking degree of efficacy in early results finding that BiDil reduced mortality by some 43 percent (Taylor et al. 2004). Nitro-Med's stock price more than doubled on the announcement (TSC Staff 2004). Then, on June 23, 2005, the FDA granted final race-specific approval to BiDil as a drug to treat heart failure in "blacks" (FDA 2005).

It is a great irony that the results of the A-HeFT trial say nothing about whether BiDil works differently or better in African Americans than anyone else because the trial enrolled *only* African Americans. Why then did NitroMed design A-HeFT to enroll only African Americans? It is not necessarily inherently unreasonable to hypothesize that a particular drug works differently, on average, between ethnic or racial groups. The only logical way to test this hypothesis scientifically would be to enroll two or more such groups for comparison. The problem with this approach is, if the drug shows efficacy in the comparison population (as the A-HeFT researchers generally expected it would), then NitroMed would have run the risk of getting FDA approval for use of the drug in the general population, without regard to race. Why would this have been a problem? Of the two key patents to BiDil that NitroMed has licensed from Cohn, the non-race-specific one expires in 2007; the second, race-specific patent, however, lasts until 2020. Under these circumstances, it made sense for NitroMed to design a trial to enroll only African Americans because this maximized its chances of getting race-specific approval of the drug from the FDA — a good commercial basis for trial design, but not good science. The trial design thus exemplifies a dynamic of biomedicalization insofar as it reflects the imbrication of clinical practice in diverse legal, regulatory, and commercial processes.

The role of the federal legal and regulatory system in producing BiDil as an ethnic drug is especially important because it lends the imprimatur of the state to the use of race as a biological category. Between the FDA's letter commenting on the ultimate approvability of BiDil as a race-specific drug and the PTO's issuance of the patent for using H/I in African American patients, powerful federal agencies have legitimized the use of race as a marker for biological difference. To the extent that institutions of the state, such as the PTO or the FDA, come to mark certain biological conditions as "racial," race may become a surrogate not only for medical research but also for a wide array of legally sanctioned discrimination.

[handwritten marginalia: Scientific discrimination]

Developments after A-HeFT

Bearing this story in mind, I now turn to developments that have transpired since the A-HeFT trials were concluded in July 2004. In particular, I explore the strategic reification of race in three related contexts: first, identifying past and current manipulations of statistical data to make it appear as if the race-specific character of BiDil's development was driven more by medicine than by commerce; second, considering how the appearance of difference created by such manipulations then becomes conflated with genetics; and third, broadening the analysis to examine how particular themes in debates about race and drug development connect to strategies that use genetics to recharacterize race-specific health disparities — caused by structural inequality or discrimination — as mere differences, rooted in genetics and personal choice. The resulting focus on "difference" prioritizes market mechanisms over state intervention to redress persistent problems of social and economic inequality.

Statistical Manipulation

The legal and commercial interventions driving BiDil's move to the market have played a role in reconfiguring race from an embodied experience of real people walking in real time through physical space into a statistical concept based on molecular correlations and frequencies. Traditional legal schemes of assigning racial identity in the United States were based on the concept of "hypodescent," where "one drop of blood" was sufficient to consign a living person to the marked racial category of "black" (Nobles 2000, 135–36). In the new statistically informed legal schemes of biotechnology patents, it is "one drop of correlation" that is providing the basis for making claims that mark out one therapy from another on the basis of race.

From its outset, the development of BiDil has been strategically framed by statistics concerning race and disease. These statistics, however, have been misleading and sometimes simply wrong. The media reports following NitroMed's announcement of the race-specific A-HeFT trials in early 2001 almost uniformly repeated NitroMed's own assertion that African Americans died from heart failure at a rate twice that of white Americans. Proponents of BiDil used the statistical disparity to buttress claims that heart failure was somehow a "different disease" (Yancy 2002) in African Americans and needed to be addressed at the genetic level.

The 2:1 mortality statistic, however, was wrong—egregiously wrong. Current data from the CDC show the difference to be approximately 1.08:1 (Kahn 2003). Yet, through a curious series of events, involving miscitations to a decade-old study and the too-ready acceptance of assertions of racial difference, the 2:1 mortality statistic was taken up widely throughout the media (via NitroMed's press releases) and professional journals. Of particular significance and influence was an article by Dries et al. in the *New England Journal of Medicine*, which asserted that "the population-based mortality rate from congestive heart failure is 1.8 times as high in black men as for white men and 2.4 times as high for black women as white women" (Dries et al. 1999, 609). These figures, it turns out, were derived from an editorial published in 1987 by Richard Gillum of the National Center for Health Statistics. Gillum's article did use these figures, but with an important qualification. He specified explicitly that they applied only to "persons aged 35 to 74 years." Leaving out this age-specific qualification is a major problem on its own, but Gillum also noted in the same editorial that the ratio of black to white mortality in persons over seventy-five approached 1:1. Moreover, this data was based on information from 1981 (Gillum 1987). For current information that included overall mortality from all age ranges, a simple visit to the Centers for Disease Control statistical information website, http://wonder.cdc.gov, revealed the actual ratio to be approximately 1:08 to 1 (Kahn 2003, 474).

An article on these statistical missteps was published in late 2003 (Kahn 2003). By mid-2004, it appeared that NitroMed and the doctors involved with A-HeFT changed their rhetoric—somewhat. They merely asserted that African Americans had a "higher rate" of mortality than the "corresponding" white population (see, e.g., NitroMed 2004). Nonetheless, as NitroMed prepared to bring BiDil before the FDA for approval in 2005, it put forth a newly configured statistic of race and disease. On January 11,

2005, NitroMed issued a press release asserting that "African Americans between the ages of 45 and 64 are 2.5 times more likely to die from heart failure than Caucasians in the same age range" (NitroMed 2005). Unlike the previous 2:1 statistic, this new statistic is technically accurate. Nitro-Med fails to mention, however, that only about 6 percent of overall mortality from heart failure occurs between forty-five and sixty-four. About 93 percent of mortality occurs after age sixty-five, and in that group there is almost no difference in age-adjusted mortality rates between blacks and whites. Indeed, the crude death rate for blacks is actually lower than that for whites.[5]

Why this investment in creating a major racial difference where none exists? One can only ask: if you have a medical interest in the underlying etiology of a disease, do you look at a subgroup where 6 percent of mortality occurs or at one in which 93 percent of mortality occurs? If, however, you have a commercial interest in convincing the FDA and capital markets that a legitimate basis exists for approving a race-specific drug, then showing a huge racial difference becomes central to marketing your product— but, of course, you do not mention that your subgroup represents only 6 percent of overall mortality.

Ironically, while the statistics help NitroMed cross the FDA regulatory threshold, their very inaccuracy simultaneously keeps open an arena for the play of the traditional concept of race as an embodied concept. This is important because it is the embodied consumer who will be purchasing and taking the drug. This points up the Janus-faced feature of BiDil in relation to biomedicalization—both looking forward to new molecularized concepts of identity and looking back to individualized embodied identities in the marketplace.

In the emerging field of pharmacogenomics, where drug companies are hoping to tailor therapies ever more closely to the genetic profile of individuals or groups of consumers, identifying racial or ethnic correlations with disease is becoming big business. As an announcement in 2005 for the industry conference "Multicultural Pharmaceutical Market Development and Outreach" put it:

> The unprecedented growth in ethnic populations across various regions in the United States opens doors to a wide array of new market opportunities for healthcare and pharmaceutical companies. Untapped consumers for both new and proven therapeutics sold as prescription or

over-the-counter products represent a total population of almost 80 million whose combined purchasing power by 2009 is estimated $2.5 trillion. With the onslaught of generics, pricing battles and DTC competition, reaching out effectively to America's emerging majority is a clear road to brand building and market growth for U.S. pharmaceutical and healthcare companies. (Strategic Research Institute 2005a)

The drive to develop race-specific therapies is not subtle, and NitroMed's A-HeFT model of race-specific trials is also on its way to becoming a new market paradigm. Thus Waine Kong, the CEO of the Association of Black Cardiologists, which cosponsored A-HeFT, was one of four featured "thought leaders" giving a keynote address on BiDil to this same multicultural pharmaceutical marketing conference. The sponsoring website urges attendees to "find out how NitroMed partnered with the Black Cardiologists Association to conduct this study [A-HeFT] and understand the opportunities and implications for drug manufacturers, disease management, clinical trials and health care companies" (Strategic Research Institute 2005b). Similarly, NitroMed's chief medical officer Manuel Worcel was a featured speaker at the 2005 Bio-IT World Conference and Expo in Boston, where he gave a presentation on A-HeFT as part of the section "Advances in Genomic Medicine" (Davies 2005). Additionally, a report titled "Cardiovascular Marketing: Budgets, Staffing, and Strategy," from Cutting Edge Information (which bills itself as "the world's largest market research resource"), features BiDil as a teaser to sell the report, which retails for $5,995 (Cutting Edge Information 2004a). As one senior analyst at Cutting Edge put it: "If trials prove successful, and drug responses prove different based on ethnicity, drug companies will certainly have new avenues for the discovery, development, and marketing of medications" (Cutting Edge Information 2004b).

In webcast presentations to the twenty-third annual J. P. Morgan Healthcare Conference and the UBS Global Life Sciences Conference, NitroMed's CEO Michael Loberg discussed the company's marketing strategy for BiDil. As part of the rollout for BiDil, the company has hired 195 sales representatives through Publicis exclusively to sell NitroMed products. NitroMed has been using this sales force to focus on doctors who are providing cardiovascular and metabolic care to African Americans, and it is especially interested in "specializing in the African American cardiovascular marketplace" (Loberg 2005a, 2005b).

In the context of pharmacogenomic marketing, it is important to consider that when a drug such as BiDil gets produced, researchers understand that it works at the molecular level, affecting, for example, levels of nitric oxide in the blood. AHeFT.org has one web page titled "A-HeFT and Genomic Medicine," which notes that the A-HeFT researchers "theorize that race may serve as a marker for multifactorial variations in endothelial dysfunction" that affect nitric oxide levels. "Finding effective medical therapy," the page continues, "for the subset of heart failure patients with these genetic differences may be a step toward personalized therapy based on pharmacogenomics" (AHeFT.org 2005).

It is here that NitroMed is most adeptly exploiting a phenomenon of biomedicalization, in the form of the promise of pharmacogenomics, while in fact being grounded in a more traditional medical model. A distinctive genetic component to BiDil's efficacy has not been established. Even if it were, the technology does not yet exist to allow NitroMed effectively to market BiDil to the biological group of individuals who have a particular genetic polymorphism that may lead to lower levels of nitric oxide. NitroMed, however, has actually embraced this limitation and turned it to the company's advantage.

In 2004, NitroMed hired Vigilante, a subsidiary of the multinational PR firm Publicis, to help market BiDil to the social group known as African Americans. NitroMed vice president of marketing B. J. Jones described Vigilante as "a leader in the field of advertising and marketing to the African American, minority, and urban communities." The firm also handled the giveaway of a fleet of Pontiac automobiles to audience members on *The Oprah Winfrey Show* (Holmes 2004). Targeting a racial audience is perceived as necessary because at this point the resources or technology does not exist to scan every individual's genetic profile. Instead one must market the product to a particular social group that is hypothesized to have a higher prevalence of a relevant genetic variation. It is far easier to target African Americans than to identify a market of particular individuals who happen to respond well to BiDil because of their genetic makeup regardless of race. NitroMed has used the Carson article purporting to identify a racial difference in response to BiDil to create a market based on a social group. Medical researchers may say they are using race as a surrogate to get at biology in drug development, but corporations are using biology as a surrogate to get at race in drug marketing.

In this context, commercial imperatives can drive drug companies to seek out and emphasize racial difference such that it becomes conflated with genetic difference. Thus, for example, in 2003 we witnessed the spectacle of VaxGen, on the heels of its failed trials to prove the efficacy of an AIDS vaccine, trying to revive its commercial prospects by claiming that a retrospective analysis of the results seemed to indicate a beneficial impact on African Americans (Pollack and Altman 2003). In February 2003, Vax-Gen announced the results of the first-ever efficacy trial of an AIDS vaccine. The overall findings were that the vaccine failed to protect against infection with the virus that causes the disease. The VaxGen researchers claimed to be surprised by the findings, but they were also undeterred. Like the BiDil researchers before them, they decided, post hoc, to break the results out by race and claimed that a retrospective analysis of the data revealed "significant efficacy in 66.8 percent of blacks, Asians, and people of mixed race, and 78.3% in blacks alone" (Cohen 2003b). One headline for a Gannett News wire service story obligingly took up VaxGen's spin on the results to report, "AIDS Vaccine Protects Asians, Blacks," with the subhead "AIDSVAX seems ineffective in whites and Hispanics. Results may be good news for HIV-plagued Africa" (Gannett 2003). VaxGen's race-based claims, however, were quickly shot down by the medical and scientific community as based on a deeply flawed, even tortured, reading of the data (see, e.g., Gannett 2003; Cohen 2003a; Pollack 2003), but not before VaxGen stock value momentarily rallied—giving rise to a series of class-action lawsuits for stock manipulation (Stanford 2005). As one HIV specialist at Emory University School of Medicine put it, "It was a desperate act by a company that was trying to save a failed product. . . . If they really cared about racial and ethnic differences, they would have structured a very different trial" (Elliott 2004). Nonetheless, in January 2004 a VaxGen spokeswoman said the company would like to do a trial focused on an African American study population to settle the question but does not have the funding to do so (Elliott 2004). This is the BiDil model at work.

More recently, on the same day in November 2004 that the A-HeFT principal investigator Anne Taylor announced their successful results at the Scientific Sessions of the American Heart Association, Keith Ferdinand (a cardiologist and co-investigator on A-HeFT) announced separate positive results for a different race-specific trial—ARIES: African American Rosuvastatin Investigation of Efficacy and Safety (AstraZeneca 2004). ARIES is a postmarketing trial of Rosuvastatin, whose trade name is CRESTOR, a cholesterol-lowering statin marketed by AstraZeneca, which also sponsored the ARIES trial. In AstraZeneca's press release presenting the ARIES data, Ferdinand emphasized that the trial showed both the safety and efficacy of CRESTOR in African Americans (AstraZeneca 2004). Later Ferdinand reiterated that the ARIES trial should "add to physicians' comfort level. It's an additional study to show that Crestor is more effective and safe" (Gardner 2004).

The emphasis on safety might seem superfluous given that CRESTOR had already been approved by the FDA in 2003. The approval, however, came amid some controversy concerning the drug's safety. In March 2004, Dr. Sidney Wolf, the director of Public Citizen's Health Research Group, petitioned the FDA to withdraw CRESTOR from the market for safety reasons. His petition noted that two major U.S. insurance companies had refused to reimburse for the drug (Wolfe 2004). In November 2004, David Graham, an FDA whistle-blower, also mentioned CRESTOR as one of the drugs the FDA should remove from the market (Zwillich 2004). In December 2004 the FDA actually issued a warning letter to AstraZeneca concerning a misleading advertisement it had placed asserting that the FDA had concluded that "the [safety] concerns [about CRESTOR] that have been raised have no medical or scientific basis" (FDA 2004). In short, CRESTOR's safety is of central concern to AstraZeneca's marketing strategy for the drug. AstraZeneca thus has a powerful incentive to produce findings of CRESTOR's safety. And one way to do that is through race-specific subgroup trials. Beyond ARIES, AstraZeneca is also sponsoring two other race- and ethnic-specific CRESTOR trials: the IRIS trial (Investigation of Rosuvastatin in South-Asian Subjects) and the STARSHIP trial (Study Assessing Rosuvastatin in the Hispanic Population) (AstraZeneca 2004). AstraZeneca is mobilizing race in relation to a pharmacogenomic model of individualized medicine to address technoscientific concerns over the safety of its valuable drug. As with BiDil, here race both produces and is produced by the commodity value of a drug.

Beyond BiDil: Twenty-nine Drugs

The dynamic relation between markets and statistical manipulation has recently moved beyond BiDil to support larger claims about the legitimacy of developing race-specific drugs. In a special supplement on race and genetics in November 2004, *Nature Genetics* published an article by Sarah Tate and David Goldstein titled "Will Tomorrow's Medicines Work for Everyone?" Among other things, the article identified "29 medicines (or combinations of medicines) [that] have been claimed, in peer-reviewed scientific or medical journals, to have differences in either safety or, more commonly, efficacy among racial or ethnic groups" (Tate and Goldstein 2004, 34). This number was immediately taken up throughout the media and in certain professional contexts as providing further evidence of sup- posedly real biological differences among races. Moreover, reports of these striking results were almost invariably paired with a discussion of the near-contemporaneous formal announcement of the A-HeFT results for BiDil. Thus, for example, after discussing BiDil, an article in the *Los Ange- les Times* referred to "a report in the journal *Nature Genetics* last month [that] listed 29 drugs that are *known* to have different efficacies in the two races" (Maugh 2004; italics mine). Similarly, a *Times* of London article as- serted that "only last week, *Nature Genetics* revealed research from Univer- sity College London *showing* that 29 medicines have safety or efficacy pro- files that vary between ethnic or racial groups" (Ahuja 2004; italics mine). And a *New York Times* editorial titled "Toward the First Racial Medicine" began with a discussion of BiDil and went on to note that "by one count, some 29 medicines show evidence of being safer or more effective in one racial group or another, suggesting that more targeted medicines may be coming" (Editorial 2004). The consistent linking of BiDil to the "twenty- nine medicines" reflects an underlying assumption that there is some real difference underlying the response to these drugs. The "real" difference must ultimately be read as genetic. These diverse sources thus reflect the popular media taking up an implicitly molecularized conception of race as a function of statistical correlation. Racial difference here is not conceived of in terms of phenotypic body type or of cultural practices and perfor- mances. Rather, it is being marked in terms of presumed differential re- action of chemical entities at the molecular level.

Yet in their rush to embrace a genetic basis to race, these stories totally misrepresent the findings and assertions of the Tate and Goldstein piece. Remember first that Tate and Goldstein asserted that these twenty-nine medicines have only been "claimed" to have racial differences in safety or efficacy. They go on in the next sentence to assert, "But these claims are universally controversial, and there is *no consensus* on how important race or ethnicity is in determining drug response" (Tate and Goldstein 2004,

34). If one took the trouble actually to read their analysis of the claims, one would see that of the twenty-nine medicines, Tate and Goldstein considered only four to provide evidence of a genetic causation being related to the differential drug response, and only an additional nine to provide evidence that "the association has a reasonable underlying physiological basis." For the remaining sixteen medicines, Tate and Goldstein found either no demonstration of a physiological basis to any observed difference, or possibly false positive claims. Moreover, of the thirteen medicines with some supporting evidence of racial difference, three were ACE inhibitors — whose claims of racial difference have been hotly contested in the professional literature — and one was BiDil (36–37). All of the thirteen dealt with hypertension, and the International Society on Hypertension in Blacks has issued guidelines arguing against race-specific treatment of hypertension (Moore 2003; ISHIB 2005).

The mischaracterization of the Tate and Goldstein article also matters because promoting race-specific drugs can lead to a misallocation of healthcare resources. This is not to advocate "colorblind" medicine. To the contrary, there are very real health disparities in the country that correlate with race. African Americans suffer a disproportionate burden of a number of diseases, including hypertension and diabetes. Like heart failure, these are complex conditions caused by an array of environmental, social, and economic, as well as genetic, factors. Central among these is the fact that African Americans experience discrimination, both in society at large and in the healthcare system specifically. The question, once one identifies these disparities in health outcomes, is how to address the underlying causes. Of course, outcomes can have multiple causes, both social and genetic. But health disparities are not caused by an absence of "black" drugs. As studies by the Institute of Medicine, among others, make clear, such disparities are caused by social discrimination and economic inequality. The problem with marketing race-specific drugs is that it becomes easier to ignore the social realities and focus on the molecules.

This very dynamic appears to be driving the use, or rather misuse, of the "twenty-nine medicines" statistic in more expert and often more conservative circles. I refer here specifically to John Entine and Sally Satel, fellows at the American Enterprise Institute (AEI). Both have gained a good deal of notoriety for their popular works on race and genetics: Entine for his book *Taboo: Why Black Athletes Dominate Sports and Why We Are Afraid* to *Talk About It* (2001) and Satel for, among other writings, a prominent *New York Times Magazine* article titled "I Am a Racially Profiling Doctor" (2002). Entine framed a recent AEI symposium on BiDil by noting, "Only last month, the prestigious journal *Nature Genetics* reported that at least 29 medicines have so far been identified that are either safer or more effective in certain populations because of genetic differences between those population groups" (Entine 2004). Satel echoed Entine's move in a more qualified manner later in the AEI symposium when she asserted that "generally, when we're talking about BiDil and things like that, it's skin color as a marker for genetic heritage" (Satel 2004b). A month later, she repeated Entine's claim about the "twenty-nine medicines" and genetics almost word for word in an article for the conservative Manhattan Institute titled "Race and Medicine Can Mix without Prejudice: How the Story of BiDil Illuminates the Future of Medicine" (Satel 2004a). Not only do Entine and Satel elide any reference to Tate and Goldstein's qualifying analysis, but they also extend the purported connection between race and drug response into the realm of genetics. BiDil provides the starting point for this move toward identifying race with genetic difference — a difference that the A-HeFT investigators themselves do *not* make.

Here, then, is another critical moment of reification. By connecting BiDil to the manipulated "twenty-nine medicines" statistic, Satel and Entine cast BiDil as the poster drug for the future of addressing racial difference in medicine — much as the corporate analysts cast it as the new paradigm for multicultural pharmaceutical marketing. Entine's and Satel's message is that race and genetics correlate closely enough to provide the basis not only for general medical practice but for addressing specific health disparities — remember, these discussions are also indirectly being framed by the misleading 2.5:1 heart failure mortality statistic. The related message is that the correlation also provides the basis for market-driven pharmaceutical development to produce new drugs such as BiDil to address these differences.

This is where reification in the context of medical practice intersects with broader strategies regarding commerce and the politics of difference. At work here is an appropriation of race as reified in the BiDil story to serve larger political agendas aimed at transmuting health disparities rooted in social and economic inequality into mere health differences — rooted in biology and genetics. Attempts to address social disparity generally implicate the power of the state or other nonmarket institutions consciously to intervene in both the allocation of resources and the sanctioning of racist practices. In contrast, attempts to address genetic "difference" may be located at the level of the molecule and targeted by pharmaceuticals developed and dispensed through the purportedly impersonal forces of the market.

Implicit in the logic of conservatives such as Satel and Entine, who use BiDil to characterize disparate health outcomes in terms of genetics, is an argument for privatizing efforts to address what are currently characterized as health disparities. This goes far toward explaining why free-market conservative organizations such as the American Enterprise Institute and the Manhattan Institute have taken such an interest in BiDil and Tate's and Goldstein's article.

This observation was driven home by the recent publication of two articles, one by Satel and the other by the University of Chicago law and economics professor Richard Epstein in a special issue of *Perspectives in Biology and Medicine*. Both pieces attack the Institute of Medicine's 2002 report, *Unequal Treatment: Confronting Racial and Ethnic Disparities in Health Care*, which chronicled an array of health disparities and connected them directly to social and economic issues of equity, access, and racism. Epstein posits that "the leap from disparity to discrimination is not, on balance, established" (Epstein 2005, S29), thereby rendering disparity the functional equivalent of mere difference. Satel and her coauthor Jonathan Klick complain that the IOM report was "too quick to diagnose bias" and object that "many medical schools, health philanthropies, policymakers, and politicians are proceeding as if 'bias' were an established fact. In other words, they consider part of the solution to the disparities problem to be located in the arena of race politics" (Satel and Klick 2005, S22). As an alternative, Satel and Klick argue that "understanding health disparities as an economic problem tied to issues of access to quality care and health literacy, rather than as a civil rights problem born of overt or unconscious bias on the part of physicians, is a more efficient and rational way to address the problem of differential health outcomes" (S23). In contrasting race politics with economics, Satel prioritizes private action operating in the market over affirmative government-sponsored action as the preferred mechanism of response to inequality.

In their respective articles, Satel and Epstein provide intellectual cover to revive a previous and far more egregious attempt by former U.S. Health and Human Services secretary Tommy Thompson to transmute the IOM's focus on disparity into difference. In December 2003 the Department of Health and Human Services issued a report on health disparities, supposedly based on the IOM report. The HHS report, however, dismissed the "implication" that racial differences in care "result in adverse health outcomes" (Bloche 2004, 1569). It turns out that top officials told HHS researchers to drop their initial conclusion that racial disparities are "pervasive in our health care system" and to delete or recharacterize findings of "disparity" as mere evidence of healthcare "differences" (Bloche 2004, 1569). Thus, for example, an earlier version of the report mentioned the term "disparity" thirty times in the "key findings" section, but the final report mentioned it only twice and left the term undefined (Vedantam 2004). HHS officials accompanied this push toward "difference" with a directive to emphasize "the importance of . . . personal responsibility" for health outcomes (Bloche 2004, 1569). Ultimately, Thompson had to backtrack when word of the manipulation of the report was leaked by concerned HHS staff. Nonetheless, Satel and Epstein effectively take up where Thompson left off—emphasizing personal choice and market forces rather than racism or inequality as the basis of health differences among races.

The embrace of BiDil and the repudiation of the IOM report are not unrelated. Together they constitute a strategic move to locate the responsibility for health disparities in the individual through biology or "personal choice" rather than society. The implicit goals are first to undermine calls for further state action to address the underlying racism that leads to disparities, and second thereby to privatize the move to address health disparities by leaving it to market forces—such as drug development along the model of BiDil and the "twenty-nine medicines"—to address the issue. In the world of law, it is a way of saying that disparate impact is due not to discrimination but to "natural" forces and so is nobody's "fault"—and therefore requires no conscious effort at redress.

Conclusion

The story of BiDil presents a cautionary tale of how America's first racially marked drug was both produced by and exploited processes of biomedicalization in the drive to invest new commodity value in a product that is already available in generic form. BiDil became racially marked not through the independent operations of clinical investigation but through the dynamic interplay of legal, regulatory, and commercial consideration that produced racialized data, shaped the clinical trial design, and framed later marketing strategies for the drug.

It is perhaps the crowning irony of BiDil's story that the very act of revealing the fallacy of its race-specific marking may redound to NitroMed's commercial benefit. To the extent that doctors come to understand that the A-HeFT data say nothing about whether BiDil works differently or better in African Americans, they will be willing to prescribe the drug to anyone they think may benefit from it—regardless of race. This is known as off-label prescribing, and it is perfectly legal. As doctors prescribe to non-African Americans, the market for BiDil grows, and NitroMed makes more money. This is a good thing insofar as more people with heart failure will get what appears to be an efficacious therapy. But it also rewards NitroMed for structuring and capitalizing on processes that have commodified, reified, and molecularized race in highly problematic ways.

BiDil clearly raises concerns over how commercial imperatives provide incentives to reify race in a manner that could lead to new forms of discrimination. The drug, however, is part of a much larger dynamic of reification in which the purported reality of race as genetic is used to obscure the social reality of racism. This dynamic ranges from the appropriation and distortion of the report on the "twenty-nine medicines" to attacks on the IOM report. To the extent that this dynamic succeeds in reductively reconfiguring health (and other types of) disparity in terms of genetic difference, it casts personal responsibility and the market as the appropriate arenas for addressing differential outcomes. It also undermines the rationale for deliberate state or institutional interventions to address discrimination. For all the legitimate concerns that the genomics revolution might lead to new forms of discrimination, we must also be alert to the potential appropriation of genetics to obscure or justify existing inequalities.

Notes

1 For an extended discussion of this story, see Kahn 2004.
2 Sheldon Krimsky notes: "The new federal initiatives on technology transfer and academic-industry-government collaborations were responsible for a marked rise in university patents. In 1980, American university patents represented one percent of all U.S. origin patents. By 1990, the figure rose to 2.4 percent. Within that decade,

the number of applications for patents on NIH-sponsored inventions increased by nearly 300%" (Krimsky 1999, 22).

3   In particular, the act encourages the transfer of technology developed in federal laboratories to the private sector for further development through Cooperative Research and Development Agreements (CRADAS). In some instances, this involves the transfer of legal rights, such as the assignment of patent title to a contractor or the licensing of a government-owned patent to a private firm. In other cases, the transfer endeavor involves the informal movement of information, knowledge, and skills through person-to-person interaction.

4   Reviewing the course of the V-HeFT trials, Cohn noted, "The natural evolution of V-HeFT would have mandated that the vasodilator regimen [to be combined with enalapril in V-HeFT III] would be the combination of the hydralazine and isosorbide dinitrate, which has been so effective in V-HeFT I and V-HeFT II. Unfortunately, the need for financial support has made it necessary that the vasodilator be an agent with potential commercial interest. Thus, a calcium antagonist has been substituted in V-HeFT III for the hydralazine nitrate combination, and it will be felodipine—a calcium antagonist with considerable vasoselectivity" (Cohn 1993, 2–3). Elsewhere Cohn also states: "One of the problems with advocating non-ACE vasodilators in treatment of the post-infarct period relates to the inadequacy of the database on these drugs. Since hydralazine and isosorbide dinitrate are generic agents, there has been no effort on the part of a pharmaceutical company to mount large-scale trials or to develop an NDA for drug approval. In contrast, the ACE inhibitors have been heavily marketed and their use for infarct related heart failure appears to be growing rapidly" (Cohn 1994, 20).

5   To obtain this information, I visited the CDC's Wonder website at http://wonder .cdc.gov (accessed January 11, 2005). The percentages are derived from queries for information concerning compressed mortality by race, age adjusted for ages 45 to 64 and ages 65 and above. In the over-65 age group, the crude death rate for blacks is 142.9 per 100,000; for whites it is 153.3 per 100,000.

▷▷▷PART III / FOCUS ON ENHANCEMENT

Jennifer R. Fishman

10 / The Making of Viagra

THE BIOMEDICALIZATION OF
SEXUAL DYSFUNCTION

This chapter examines the emergence of Viagra as a treatment for erectile dysfunction within the context of the biomedicalization of impotence over the last half of the twentieth century in the United States, including the categorical transformation of the male sexual disorder "impotence" into "erectile dysfunction" (ED) and from a psychological condition to a biomedical disorder. Through this shift, impotence was not only medicalized in the traditional sense but *biomedicalized*, in that it was simultaneously transformed into an exemplary "lifestyle" condition. It is a key example of how medical discourses and treatments can now legitimately be used to treat a life-limiting (as opposed to life-threatening) condition, through a ratcheting up of what is considered normal and desirable and the risk factors that might impede one's quality of life. I build on the premise that biomedicalization is not a radical break from the processes that preceded it, but often represents "medicalization *and*"—that is, medicalization formed through traditional medical channels plus new channels that incorporate biomedicine's turn to lifestyle and to the molecularization and organicization of the human condition.

Historically, ideas about impotence and then erectile dysfunction accumulated over time, creating a ripe environment for Viagra's widespread acceptance as a credible treatment for men's sexual problems. Concomitant with the medicalization and organicization of erectile (dys)function has been a continued molecularization of the body, in which men's erections and sexual arousal became refined, reduced, and delimited to the functioning of neurotransmitter pathways. Locating Viagra's development in the decades before its commercial release is important for a number of reasons. It demystifies the rhetoric of Viagra as a "wonder drug," enacting overnight transformations for male sexuality. This analysis seeks to rupture mystifying notions of the "spontaneous" and "novel" implicit in much of the discourse of biomedical innovation.[1] Viagra developed within the context of long-standing efforts by medical professionals and others to place male sexuality under the medical gaze.

For this reason, I do not ask how Viagra as an emergent technology has transformed our understanding of impotence, but rather pose the inverted question: what is the historical context that enabled Viagra to be considered a viable and legitimate treatment for men's sexual problems? Viagra is a culminating event in a longer trajectory of scientific work on the biochemistry of erections. In fact, acceptance of Viagra as a way to treat impotence depended in part on a shift in knowledge claims about the physiological etiology of the problem of erectile dysfunction and the broader acceptance of this biomedical shift. In this way, the biomedicalization of impotence depended in part on the molecular turn in the life sciences and the interweaving of science and (pharmaceutical) technology. Furthermore, Viagra's success relied on networks, discourses, and organizations in the sciences that emerged in the decades preceding its development. This historicized understanding of Viagra's emergence reconfigures it as part of a nexus of biomedicalizing developments within the production of scientific and cultural knowledge about male sexuality in the United States.

The Pathologization of Impotence

In the early to mid-twentieth century, impotence was explained primarily as a biological event that accompanied aging in men (Marshall and Katz 2002). Unlike contemporary theories of erectile dysfunction that uncouple impotence from the "normal" aging process, in the early 1900s, decline in sexual function for men was considered part and parcel of the normal aging process. Men had a choice of either accepting and adjusting to these changes or consuming morally and medically questionable elixirs or nostrums such as Brown-Sequard's "Pohl's Spermine Preparations."[2] With the emergence of gerontology and sexology as fields of study in the early twentieth century, discourses about "positive aging" began to emerge, touting scientific developments as the path to successful aging, through ideals of vitality, activity, autonomy, and well-being. These ideals still persist in contemporary understandings of erectile dysfunction. However, discourses about positive aging were stymied in part due to claims in gerontology and sexology that made sexual decline an element of middle age. To develop a "legitimately positive discourse on sexual decline," impotence had to be transformed from an organic disorder into a psychological one (Marshall and Katz 2002, 52).

Psychological explanations for impotence were considered as early as the 1920s. For example, the psychoanalyst William Stekel, a colleague of Freud's, published *Impotence in the Male* in 1927 (cited in Marshall and Katz 2002). By midcentury, psychological causes of impotence pervaded the literature. However, this understanding was restricted to *younger* men with impotence. As Marshall and Katz (2002, 53) summarize, "impotence was still largely regarded as a disease of the young, and a condition of the old." Therefore psychotherapy was recommended (and considered successful) for men under forty-five, while impotence in men over forty-five was still considered a normal part of the physical decline of aging.

"Successful Aging" and the Sexual Imperative

The psychological paradigm, however, quickly spread to include aging men as well, with explanations for impotence ranging from "anxiety" over their supposed loss of sexual function to their "fear" of loss of potency. From the 1960s through the 1980s, ideas about positive aging and positive sexuality took hold, and men were often told that psychological factors were primarily responsible for loss of sexual function and that to *cease* having sex would hasten aging itself. Sexological reports[3] emerging in the mid-twentieth century contributed to this conception of both the importance of one's continued sexual activity over the life course and the normalization of sexuality itself. The Kinsey Report (1953) and Masters and Johnson's books (1970, 1966) normalized sexuality by demystifying and naturalizing it for an American audience. Through discourses about the "naturalness" of sexual activity, Masters and Johnson (1970) further adhered to a notion of a universal sexuality (Tiefer 1995), which implies not only a homogeneity of men's and women's experiences but also an imperative to be sexual. These techniques are normalizing, not in the sense that they produce bodies or objects that conform to a particular type but in the sense that they create a standard model against which objects and actions are judged (Ewald 1990). The work of Masters and Johnson and others such as Kinsey (1953) and Hite (1976) each mentioned that sexual activity was enjoyed throughout the life course. Intending to promote the acceptance of older people as sexual beings, these researchers have also contributed to what amounts to a mandate for sexual activity for everyone. If one does not measure up to these norms, one may feel inadequate and abnormal. When aligned with the "successful aging" movement in gerontology, sexual activity becomes a necessary and healthy component of the "good life" as one gets older, though it may be curtailed through psychological blocks stemming from fear and anxiety.

Despite the physiological emphasis of the Masters and Johnson human sexual response cycle, William Masters himself thought that most impotence cases have psychogenic etiologies, as indicated by the modes of behavioral therapy at his institute. This became the widely accepted view of impotence. During the 1970s, a statistic began to circulate in much of the psychological and popular literature on impotence: "At least 90 percent of impotence" is psychologically based. However, the exact origins of this statistic are unclear, and it appears not to have been generated by any empirical study (Bancroft 1982). Untethered from whatever roots it had in empirical research, this statistic traveled across a number of sources, from psychology and sex therapy books and journals (e.g., Kaplan 1974) and medical articles (Spark, White, and Connolly 1980) to self-help books (e.g., Butler and Davis 1988) and mainstream magazine articles. Because it decoupled impotence from the normal aging process, establishing impotence as a psychological condition proved to be an important step in paving the way for Viagra's eventual acceptance as a legitimate treatment for impotence. Impotence was no longer considered a component of aging but rather was seen as an avoidable and even preventable condition created by psychological blocks.

Erections for Life

In the two decades before Viagra's release (ca. 1980–98), impotence was already deep in the process of being transformed once again, this time from a psychological condition to a physiological disorder. During this era, the medical specialty of urology came to consider impotence a problem within its disciplinary purview and also developed new techniques and technologies for its diagnosis and treatment. This transformation occurred through a number of both coordinated and serendipitous events during the 1980s and 1990s. These culminated in 1998 with the release of Viagra to treat "erectile dysfunction," thereby solidifying its conceptualization as an organic condition, controllable through pharmacologic means.

Beginning in the 1980s, debates about the etiology of impotence began to intensify as urologists interested in developing new medical treatments sought to determine its underlying causes. Part of the insistence on the psychogenic causes of impotence in the 1960s and 1970s stemmed from the fact that psychotherapeutic techniques were the only ones available to impotent men and their partners, with the marked exception of penile prostheses, which were available but not widely used throughout the second half of the twentieth century.[4] Understandings of the underlying etiology of impotence began to shift due to urologists' trial-and-error approach to new medical treatments and as failure rates of psychotherapeutic approaches began to emerge (e.g., Spark, White, and Connolly 1980). As is common in the production of scientific knowledge, new techniques and instruments can be crucial elements in clearing the path for an emergent science and scientific understanding (Canguilhem 1978, 1992; Latour 1987). In fact, ideas about the underlying physiological bases of impotence were then and continue to be co-constituted with new techniques for diagnosis and treatment.

Although sex therapy dominated the discourse and clinical treatment of impotence throughout the 1970s and 1980s, during this time urologists also began to interest themselves in the problem from a medical and physiological perspective. Former Kinsey Institute director John Bancroft, a psychiatrist by training and a prominent sexologist with over thirty years of experience in the field, explains this development through urologists' self-interest in maintaining their profession and livelihood:

> Urology had something of a crisis. . . . Urology is a surgical specialty, and was being squeezed out by non-medical methods and management. So, urologists were turning to other types of problems to earn their living. Erectile dysfunction, or whatever you prefer to call it, was one that attracted their attention. (Personal interview, October 1, 2001, lines 37–45)

Leonore Tiefer (1994) similarly linked urologists' attraction to sexual dysfunction as a subspecialty to the capacity to diversify their outpatient and inpatient services by clinically treating a patient population that was not in the strictest sense chronically ill, diseased, or likely to die from this condition. This does bespeak medicalization in the classical sense, as the expansion of medical jurisdiction by physicians to extend their power and social control (see, e.g., Zola 1972; Illich 1976/1982; Conrad 1992). In fact, this was occurring during what was the heyday of medicalization, both in terms of the process itself and in terms of the scholars who were describing and analyzing it.

At first, surgical solutions were offered to men with impotence by way of penile implants, an inflatable prosthesis of semirigid rods inserted into the penile shaft. These treatments were (and still are) invasive and not without significant risk of malfunction and infection. They were hardly considered ideal solutions by either physicians or patients. However, their availability contributed to the acceptability of remedies for erectile dysfunction, which paved the way for Viagra's legitimacy and acceptability. Urologists thereby succeeded in establishing themselves as having the expertise and authority to treat impotence and conducted an increasing amount of biomedical research on the physiological mechanisms of erections. Such treatments also made the etiology of erectile dysfunction irrelevant, since they worked regardless of the cause of impotence (e.g., psychological causes, iatrogenic effects of surgery or other treatments, or a medical condition). Mechanical treatments, which quite literally could give men erections on demand, also made sexual desire irrelevant for sexual performance. Implants created the ability for a man to have an erection irrespective of his state of desire or arousal.

Despite the seeming irrelevance of sexual desire for sexual activity, this conceptualization further delineated and divided processes of desire from processes of arousal. Instead of constructing the impotent man as desireless, discourses about implant use took male desire as already given. An erection in this context was therefore not a bodily response to a psychological state of being or feeling; rather, the mind and body were considered distinct and separate entities. This idea ran counter to psychological theories of impotence that posited that the means of treating impotence was to treat the mind first and the body would follow. In fact, with the advent of new medical technologies such as implants, cause and effect were inverted. While before, the common wisdom was that fear, anxiety, depression, and so on could act on the body to produce impotence, now the emphasis was on the psychological distress *caused* by impotence. In turn, implants, through their mechanical means of repairing erections, could alleviate psychological problems: fix the body and the mind will follow. This switch marked a turning point in the reconfiguration of impotence as a "lifestyle" problem, wherein treating impotence could benefit and even enhance one's "quality of life."[5] Further, the implication was that impotence itself is a mechanical problem that needs a mechanical solution.

The Organicization of Impotence

Throughout the 1980s and 1990s, urologists continued to study the physiology of erections and to search for pharmacological treatments. Over time, the notion that "over 90 percent of impotence is psychologically based" began to shift to a more equitable distribution: eventually urologists and the mass media claimed that about half of all impotence cases had an *organic* basis (Tiefer 1986). A number of marker events signaled this turning tide, as urologists begin to usurp what was previously the territory of sex therapists in both treating and defining impotence. In 1980 the endocrinologist Richard Spark and his colleagues published an article in *JAMA* titled "Impotence Is Not Always Psychogenic" (Spark, White, and Connolly 1980). He argued that contrary to the claim that most cases of impotence are psychogenic, there are likely many cases of impotence that have their origins in hormonal abnormalities. He reported the results of a study of a sample of 105 "impotent" patients who had each had their serum testosterone levels checked;[6] 35 percent had hormonal abnormalities.[7] The article's title became something of a battle cry for clinicians and medical researchers in the following decades.

A second event reached legendary proportions in the sexual dysfunction lore. It occurred in 1983 at the American Urological Association meetings when Giles Brindley, a urologist, injected his own penis with phenoxybenzamine,[8] a pharmacologic solution that gave him an erection. He then presented his erect penis as specimen, walking the aisles of the conference room letting others inspect and touch it. This was often referred to as the watershed moment in the shift from impotence as a psychological to a physiological condition. In an early case of "reverse engineering" in this field, producing an erection by chemical means shed light on the physiological mechanisms of erections, codifying a *lack* of erections (that is, impotence) as a physiological condition. In other words, if an erection could be biochemically induced, then when one is unable to have an erection, there must be a biochemical problem. Because the penile injection used by Brindley was a "vasoactive" substance, meaning that it opened up blood vessels allowing greater blood flow, erections were reinterpreted as a function of blood flow. Like implants, injections such as these could produce erections regardless of the etiology of the impotence (e.g., Ghanem et al. 1998), and regardless of whether or not a man had any difficulty getting an erection "on his own." Working within the same paradigm as penile implants, penile injections induced erections irrespective of a man's "feelings" of desire or arousal.

Some scholars considered these innovations a confirmation and reification of the phallocentrism of male sexuality, wherein constructions of male sexuality focused exclusively on penile erection for sexual pleasure (Tiefer 1994). However, while appreciated by clinicians and male users for their "reliability" (Incrocci, Hop, and Slob 1996), penile injections, implants, and vacuum pumps were not considered ideal treatments by patients or urologists precisely because of the "artificialness" of the erections produced in this manner (Hatzichristou 2002; Leungwattanakji, Flynn, and Hellstrom 2001; Rowland et al. 1999). As discussed later, Viagra provided a nice foil to earlier innovations in that it was capable of producing an erection through a noninvasive pill that created erections through "natural" processes.

Throughout the 1980s and 1990s, various techniques and devices were developed and implemented to differentiate diagnoses of organic versus psychogenic impotence. A medical device called the RigiScan was developed by Timm Medical Technologies Inc. to measure penile rigidity. Still widely used today, it consists of a recording device and two loops that are placed around the base and tip of the penis. The loops tighten periodically (usually every fifteen or thirty seconds) and a recording of the rigidity of the penis is calculated. The device can be operated by patients in the laboratory or in at-home settings. It is often used to measure "nocturnal tumescence," meaning that a man wears the device while he is sleeping, where it measures the degree of the rigidity of the penis at thirty-second intervals over the course of the night. It is thought that men who are able to get nocturnal erections likely have impotence with a psychogenic origin because these men do not seem to have physiological difficulties.

Through such research, the demarcation between organic and psychogenic etiology of impotence became more distinct during this period, as researchers believed that all erections (i.e., those that emerge from sexual stimuli and those that do not) were the same. That is, if a man produced nocturnal erections, then he should be physiologically capable of erections for sexual activity. Once again this medical model elides the question of desire or relegates it to the psychological sphere.

A second diagnostic technique developed during this time was duplex ultrasonography.[9] As medical science came to understand erectile function through theories of vascularization and blood flow, doctors used the duplex ultrasound to evaluate the status of penile arteries. Ultrasound could detect arterial blockages that might prevent erectile functioning. Since its introduction in 1985, duplex ultrasonography has become the most commonly used clinical test for ascertaining the vascular status of patients with erectile dysfunction (ED) in order to differentially diagnose it (Steel et al. 2003).

During the 1980s, while working on the physiology of erections and potential treatments, a group of urologists also made a concerted effort to reconstruct impotence as a medical rather than psychological condition (Katz and Marshall 2003; Tiefer 1994). As detailed by Tiefer (1994), in 1982 this group of urologists held an informal meeting that resulted in the formation of the International Society for Impotence Research. The first World Meeting on Impotence was held in 1984. A few years later, three urologists published an overview of this emergent field in the *New England Journal of Medicine* (Krane, Goldstein, and DeTejada 1989). In 1989 the International Society for Impotence Research began publishing its own medical journal, the *International Journal of Impotence Research*, whose editorial board primarily consisted (and continues to consist) of urologists.

In 1992 this shift of impotence from a psychological to a biomedical disorder was stabilized and codified by a "consensus development conference" sponsored by the National Institutes of Health. This also marked the beginnings of the transformation of terminology from impotence to "erectile dysfunction" (ED). A number of experts in the field of impotence research gathered in Bethesda for a two-day public session to present their own work in areas relevant to impotence research, followed by an open public discussion (NIH Consensus Development Panel on Impotence 1993). A separate panel then met behind closed doors to develop a consensus statement based on the previous days' discussions. The planning committee for the conference consisted entirely of medical doctors and NIH employees. The experts who presented were mostly (75 percent) physicians.

Reports of the consensus conference appeared in at least two venues. The first was in the *International Journal of Impotence Research* in 1992 (National Institutes of Health 1992), and the second later appeared in *JAMA* (NIH Consensus Development Panel on Impotence 1993). Although the *JAMA* document was titled "Impotence," the first paragraph said that although *impotence* has been traditionally used to signify the inability of the male to attain and maintain an erection, this use had led to confusion within scientific investigations because of its lack of specificity. It continued: "This, together with its pejorative implications, suggests that the more precise term *erectile dysfunction* be used instead" (83; italics in original). The article defined erectile dysfunction as "the inability to attain and/or maintain penile erection sufficient for satisfactory sexual performance" (89). The renaming process codified men's sexual problems as medical conditions, encouraging further medical research to understand male sexual function. The chronology of events indicates that this was not simply a terminological change but an indication of medical claims-making about the organic and physiological etiology of the disorder itself.

In retrospect, this seems an obvious and purposeful way to reframe impotence within medical parlance, through focusing on the functionality of a specific organ. Surprisingly, however, this change was first recommended by psychologists themselves. Leonore Tiefer, a psychologist by training and a critic of "the medicalization of male sexuality" (Tiefer 1986), claimed that it was *psychologists* who preferred the terms "erectile disorder" and "erectile dysfunction" and urologists who promoted their own claims through the continual use of the word "impotence" (Tiefer 1994). The term "dysfunction" itself, in existence since at least the 1970s (as in the terminology of "sexual dysfunction"), was initially used by sex therapists (e.g., LoPiccolo 1978). Apropos the discussion about biomedicalization, the *pathologization* of sexuality, as epitomized by the "dysfunction" nomenclature, is neither coincident nor coterminous with the *medicalization* of sexuality. In fact, at the 1992 consensus development conference, Tiefer, speaking about nomenclature, was the one to suggest that "erectile dysfunction" replace the term "impotence," which she felt was pejorative and confusing (Tiefer 1995). She advocated use of "erectile dysfunction" as a strategy for the demedicalization and destigmatization of impotence. Given Tiefer's stance on the dangers of medicalization, it was ironic that she provided nomenclature that later served to help medicalize the condition. In ways that appear predictable in retrospect, erectile dysfunction, as later heavily promoted by Pfizer vis-à-vis Viagra, in many ways legitimated impotence as a biomedical condition, wresting it away from its much older psychoanalytic and psychological associations. The new term capitalized on its ability to destigmatize and sanitize the condition. Furthermore, it focused attention on erections themselves as the source of the problem, rather than, as most psychologists would prefer, a less local and more holistic dysfunction around sexuality broadly conceived.

Accompanying the organicization of erectile dysfunction, the consensus statement also concretized and institutionalized the emergent idea that ED is not a normal occurrence as men age. In a section on "prevention," the report said that "although erectile dysfunction increases progressively with age, it is not an inevitable consequence of aging" (NIH Consensus Development Panel on Impotence 1993, 85). As in the broader discourse of aging in contemporary biomedicine, aging itself need not be accompanied by any disease or even limitations, only an increasing coalescence of risk factors (Bury 2000; Cohen 1991; Estes and Binney 1989; Katz and Marshall 2002; Marshall and Katz 2003).

This is indeed one of the hallmarks of biomedicalization, which was occurring alongside the simple expansion of the jurisdiction of urologists. With ED, such risk factors (which often "accompany" aging but are not integral to it) included diabetes mellitus, hypertension, vascular disease, high cholesterol, depression, and alcohol ingestion (NIH Consensus Development Panel on Impotence 1993, 85). There was not only an expansion in the jurisdiction of the medical profession but further an expansion of those to be considered under its purview. Although here we see how aging itself is in some ways *demedicalized* (not considered a disorder or disease in and of itself), we simultaneously see a biomedical expansion into all the conditions thought to influence aging. To achieve the state of healthiness now expected for everyone as they grow older, more of the concomitant components of aging are placed within biomedicine's realm.

In some ways, then, the notion of impotence has come full circle. The conception of impotence at the beginning of the twentieth century was as an inevitable organic process that occurs as men age. After the shift to understanding impotence as a psychological condition, once again we have returned to reconceptualizing it as a physiological and once again organic process. Only this time around, erectile dysfunction has been decoupled from the natural aging process and thereby pathologized and subsequently medicalized. Although the processes of pathologization and medicalization are often related, they are not necessarily coterminous or coincident. In the case of erectile dysfunction, impotence first became pathologized through psychological interventions and changing norms about the benefits and prevalence of sexual activity in older men. The idea that men continued to be sexual as they age made those who were not or could not be abnormal, thereby making impotence pathological. It was not until later, in the 1990s, that we witnessed the complete biomedicalization of impotence, whereby it was considered not only pathological but also a physiological problem that requires a biomedical solution (NIH Consensus Development Panel 1993, 85).

The NIH Consensus Statement was both a result of, and instigator for, medical attention to the problem of erectile dysfunction. Urologists today reflect on the work conducted throughout the 1990s on the physiology of erectile function as a foundational example of how medical research can advance our knowledge about the inner workings of the human body. For example, one urologist began an article on Viagra's clinical success for treating erectile dysfunction as follows: "In the past 15 y[ears], erectile dysfunction (ED) has witnessed an enormous growth in scientific interest, which has been translated impressively to the development of several treatment options" (Hatzichristou 2002, S43). A urologist who began studying erectile function in the early 1980s said similarly:

> You know, over the last fifteen to twenty years I think . . . researchers made a lot of progress in terms of understanding what is a normal physiology, what happens during an erection, what is the pathophysiology for why impotence occurs, why diabetes causes impotence, why high cholesterol causes impotence. You know, I think there has been a lot of progress made in better understanding. (Interview R20, December 28, 2001, lines 74–82)

With continued focus on the physiology of the penis, attention shifted to studying its function as an organ in much the same way that a medical researcher might study the kidney or the heart; it became devoid of most of the remnants of the psychological components of erections.

The Commodification of Erections

Research conducted during the 1990s produced a theory of the physiology of erection that depended on a complex interplay between vascular and neurological events. In 1992 an article published in the *New England Journal of Medicine* reported the results of a study that demonstrated the role of nitric oxide as a mediator of the smooth-muscle relaxation of the corpus cavernosa of the penis (Rajfer et al. 1992).[10] The study's lead author, Jacob Rajfer, is a urologist, and the last author listed, Louis Ignarro, is a pharmacologist.[11] Both were (and remain) professors at UCLA. Ignarro had spent the previous fifteen years studying the potential effects of nitric oxide on cardiovascular tissue and had published his first article in this area in 1979. Along with two other American researchers at different academic institutions, he was awarded the Nobel Prize in Medicine in 1998 (the same year as Viagra's release) for their work in identifying nitric oxide as a messenger signal of the cardiovascular system, inducing relaxation of the smooth muscle of blood vessels. In the 1990s Ignarro and Rajfer conducted some of the research identifying nitric oxide as the neurotransmitter responsible for smooth-muscle dilation in the penis, enabling erections. Nitric oxide is released by both the cavernosal nerves and the epithelial cells of the corpus cavernosum.

Finding the mechanism that induces erections lent further credence to the idea that erectile response, and male sexual arousal more generally, could best be understood as a biochemical, mechanical, and physiological phenomenon. In fact, the most promising means of developing new drugs to create seemingly natural erections with fewer side effects was through investigation of the physiological pathways of erections.

The history of the development of Viagra, then, is in many ways typical of other pharmaceutical technologies. While like Rajfer and Ignarro, academic and other research scientists often conduct basic scientific research, it is private biochemical and pharmaceutical companies that apply this research to create profitable products (Gaudillière and Löwy 1998). Pfizer was testing a compound called sildenafil for the treatment of coronary angina. The trials failed to demonstrate the effectiveness of sildenafil for this purpose. However, because of the research conducted on the physiological mechanisms of erections, sildenafil could be rehabilitated as a treatment for erectile dysfunction. Once male clinical trial subjects (in the coronary angina study) reported that they were experiencing better erections with sildenafil, Pfizer was able to capitalize quickly on this knowledge and on the concurrent basic science research already conducted on the physiology of erections to develop a successful drug.

Therefore, perhaps more important than the scientific developments leading up to the biomedicalization of impotence were the research networks, organizations, and overall transformation of biomedicine's conceptualization of the nature of impotence constructed in the two decades before the compound that came to be known as Viagra. Not only did the concerted efforts of urologists and others contribute to the science of Viagra, but moreover they contributed to a shift in the social and cultural climate. This made Viagra a credible, legitimate, and successful commodity for both medical practitioners and consumers more broadly in treating what is now known as erectile dysfunction. Developments throughout the last quarter of the twentieth century reconstituted impotence as a physiological phenomenon and a legitimate medical condition separate and apart from the normal process of aging. It was this environment that enabled the emergence of Viagra and foretold its success.

The Profitability of Erections

In the first five years of its availability on the market, Viagra netted approximately $7.4 billion in total sales for Pfizer Inc. It is estimated that over twenty million men worldwide have received prescriptions for the drug. Accurately labeled a blockbuster drug, Viagra was colossally successful from the moment of its release; in its first three months of availability in the United States, $411 million worth of Viagra was sold, and 2.7 million prescriptions were written for the drug.

Although I note the ways in which biomedicine itself contributed to the biomedicalization of erectile dysfunction in the years leading up to Viagra's release, it is equally important to document and consider how Viagra itself has perpetuated this trend. In what has elsewhere been called "disease mongering," we have witnessed how Pfizer, the maker of Viagra, has attempted to expand the potential Viagra-taking population to include more and more people (Lexchin 2006; Moynihan and Henry 2006).

Through marketing campaigns that feature younger and younger spokespeople, Pfizer has expanded the idea of the typical Viagra user. When

Bob Dole was enlisted as Pfizer's first spokesman for Viagra in 1998, he was seventy-five years old. The most recent celebrity Viagra spokesman is Rafael Palmeiro, a Major League Baseball player for the Texas Rangers. He is Cuban and was thirty-seven years old when he became a Viagra spokesman in 2002.[12]

A further shift is taking place in marketing to the ideal Viagra consumer. Whereas Bob Dole "educated" the public about erectile dysfunction, Palmeiro doesn't mention a medical condition at all. In the television and print advertisements, he says simply, "I take Viagra. Let's just say it works for me." Unlike the ads featuring Bob Dole, Palmeiro does not talk about erectile dysfunction or label himself with that diagnosis. In fact, the shift in Viagra advertising campaigns and appeals went from linking Viagra with erectile dysfunction to linking it with *erectile difficulties*, a milder form of ED.[13] This is a means of enrolling more potential users of Viagra, extending it to those who might suffer from occasional erectile problems rather than have ED as a chronic problem. By saying simply that "Viagra works" for him, Palmeiro is purposely vague. It is not clear whether he takes Viagra because he in some way "needs" it, or if he takes it to enhance his sexual performance, creating harder and longer-lasting erections, so that he can be read as in some way *boasting* about his sexual performance.

The expansion of Viagra to treat *female* sexual dysfunction (FSD) has been documented elsewhere as another case of biomedicalization and disease mongering (Fishman 2003; Tiefer 2006). It is a case of biomedicalization worthy of its own chapter, given that there are as yet no FDA-approved drugs to treat FSD, yet it still maintains its cachet as a disorder in need of a drug.

Conclusions

The scientific and cultural understandings of erectile dysfunction over the twenty-five years before Viagra's release illustrate not only the medicalization of impotence but also the ways in which this process contributed and ultimately gave rise to its subsequent biomedicalization. First, the psychologization of impotence in the 1960s and 1970s pathologized impotence, such that it was considered no longer a component of normal aging but a dysfunction that required treatment. Second, the organicization of impotence beginning in the 1980s relocated impotence from a problem of the mind to a problem of the body, focusing attention on the penis as an organ. As the changing medical model likened erectile dysfunction to other physiological disorders, such as heart disease or diabetes, researchers and healthcare professionals came to see ED as a condition needing a biomedical fix, divorced from the psychological stigma of previous eras.

When taken together, these ideas transformed impotence into a "lifestyle" condition. Accompanied by a shift from mind to matter, impotence was neither inevitable for aging men nor indicative of larger psychological problems: it became simply a condition to be treated, not because it is life threatening but because it is life limiting. The biomedicalization of the problem as "erectile dysfunction" legitimated it as one in need of a biomedical solution. Erectile dysfunction, then, needed to capitalize on both its status as a medical condition in need of treatment and a lifestyle condition that deserved treatment because of the hindrances it placed on one's quality of life. The "erectile difficulties" label that emerged in the post-Viagra period exemplifies not only the expansion of our ideas about normal aging but also a shift to biomedicalizing everyday living.

Without the medicalization of impotence, the FDA could not have approved Viagra as a prescription drug to treat erectile dysfunction. Urologists' intentions to claim jurisdictional ownership of impotence and the intense focus on uncovering the physiology of erections led not only to a medicalization but to a molecularization of impotence. This was instrumental in paving the way for Viagra's eventual breakthrough and immense success on the market.

Most significantly, then, the years leading to the FDA approval of Viagra are an exemplar of the combined processes of medicalization and biomedicalization over the course of the last thirty years of the twentieth century. Medicalization of impotence worked through the channels that we grew accustomed to talking about: the professional expansion of physicians, the pathologization of a previously "normal" condition, and a changing in the standards of what was considered normal, especially as we grow older. However, in the latter part of the twentieth century, other components were added to the equation, altering the medicalization process and shifting it to what we have called biomedicalization. As this chapter documents, these include a new emphasis on improved lifestyle as a worthy application for biomedical innovations; a focus on risk factors rather than on a disease itself; a molecular understanding of a treatment that seems to validate the condition it was designed to treat; and new forms of commodification and marketing of a disease by pharmaceutical companies with marketing campaigns that appeal to a new form of consumer and patient. As these took shape and were cumulative over time, medicalization was never, nor could it be, severed from these processes.

Notes

1 For an analysis of the deployment of novelty in biomedicine, see King 2001.

2 Brown-Sequard may be best known for his early work with identifying sex hormones through his now famous experiments using animal gland injections in other animals and humans. See Clarke 1998.

3 By "sexology," I refer here not only to the discipline dedicated to the study of sex but also to the notion that sexuality is an area worthy of *scientific* pursuit.

4 Tiefer (1986) estimates that "hundreds of thousands of men had received implants" by the mid-1980s, which seems minuscule when compared to Pfizer's statistics of the twenty million men who had received Viagra prescriptions by the end of 2002 (Pfizer 2003). Yet Tiefer also points out the difficulty in estimating the number of penile implant operations conducted because they are implanted primarily by private practitioners.

5 In studies to test the effectiveness of erectile dysfunction treatments, "quality of life" became a meaningful outcome variable in and of itself (Althof et al. 2002; Litwin, Nied, and Dhanani 1998; Seidman et al. 2002; Woodward et al. 2002).

6 In this article, impotence was determined according to the same standards that Masters and Johnson used at the time: a man was defined as impotent if "erectile failure" occurred in more than 25 percent of the attempts at intercourse (Spark, White, and Connolly 1980).

7 The methodological problems with this article are numerous, although not particularly relevant to the discussion at hand. Because of the article's prominent placement in the medical literature and its decisive sentence title, the argument carried much weight in propelling a reconfiguration of impotence as a medical problem.

8 Phenoxybenzamine is a drug classified as an antihypertensive and an alpha adrenergic blocker. The drug is currently available in capsules only, not for injection.

9 For a history and analysis of duplex ultrasonography, see Mol 2002.

10 The corpus cavernosa are the two chambers of smooth muscle that run alongside the urethra in the penis.

11 Unlike social science journal articles, in medical journal articles it is often the case that the principal investigator of a study is listed last.

12 Although Palmeiro's appearance in the Viagra ads for a "performance drug" raised few eyebrows, he has come under more serious scrutiny for accusations that he uses another performance drug—steroids.

13 In fact, this seems to suggest a prophylactic use for Viagra. Viagra must be taken one hour before desired sexual activity, such that if a man was concerned about his performance — that is, if he *on occasion* has erectile difficulties — he might take Viagra as insurance (Potts et al. 2003). In fact, some urologists have suggested that men consider taking Viagra as a prophylactic measure before they experience any erectile problems, in the belief that it might help prevent ED in the future (field notes, November 16, 2000).

Natalie Boero

11 / Bypassing Blame

BARIATRIC SURGERY AND THE CASE OF BIOMEDICAL FAILURE

Increasingly the term "epidemic" is being used in the media, medical journals, and health policy literature to describe the prevalence of fatness in the United States. Skyrocketing rates of obesity among all groups of Americans, in particular children, the poor, and minorities, have become a major public health concern. Indeed, it is difficult to open a newspaper or magazine without in some way encountering a discussion of the "expanding American waistline" and the health problems and risks associated therewith.

The "obesity epidemic" incorporates both traditional processes of medicalization as well as newer processes of biomedicalization.[1] In this chapter, I examine medicalization and biomedicalization through surgical interventions for weight loss.[2] I use the case of weight loss surgery to analyze both the biomedicalization of obesity and, more significantly, the relationship between biomedical failure and moral discourses of health and illness.[3] I argue that the case of failed weight loss surgeries allows us to see that processes of biomedicalization do not simply replace medicalization and discourses of morality vis-à-vis health; rather, biomedicalization draws on and breaks with these discourses in employing the complex language of personal responsibility for health, risk, and the success or failure of biomedicine.

My research draws on data from three primary sources. First, I derived ethnographic data from participant observation in seven informational seminars on weight loss surgery, five support group meetings, a weekend-long national convention of the group Obesity Help, and observation of a variety of online message boards and chat rooms for those who have had or are interested in weight loss surgery. Second, I conducted ten in-depth semistructured interviews with postoperative weight loss surgery patients. Third, I did a textual analysis of informational literature from a variety of weight loss surgery programs and *Obesity Help Magazine*.[4]

Obesity as a Postmodern Epidemic

Given the reporting on statistics of death, expense, and contagion attributable to obesity and the sense of panic they have inspired, one would expect to see the streets of American cities littered with the dead bodies of fat people. However, the obesity epidemic is not a traditional epidemic of contagion and mass death. Rather, it is what I call a "postmodern epidemic,"[5] one in which unevenly and situationally medicalized phenomena lacking a clear pathological base are cast in the language and moral panic of more traditional epidemics. The moral panic central to the postmodern epidemic is, like other moral panics, as much an expression of outrage against "deviant" groups as it is about unadulterated fear (Goode and Ben-Yehuda 1994). Therefore, obesity as a disease or condition is only part of the concern in this epidemic. Those at risk of becoming obese are as central to the epidemic as those who actually are obese, again making obesity a clear example of processes of biomedicalization.

This focus on risk is the first of three central characteristics of a postmodern epidemic. What makes the obesity epidemic unique is that we are all at risk for obesity; what varies is our degree of risk. Indeed, in an era of personal responsibility for health, one no longer need manifest any concrete symptoms to be considered at risk for any given disease (Clarke et al. 2003). Both this concern with risk and the divorcing of the concept of an epidemic from a traditional biomedical model are the true hallmarks of a postmodern epidemic like obesity. The second central characteristic of postmodern epidemics is a shift in thinking about public health issues away from the public and onto the individual. That is, while a focus on specific populations remains significant and the consequences of these epidemics are understood as widespread, the solutions to public health problems are seen as increasingly microlevel (Clarke et al. 2003; Foucault 1973). Third, and most centrally, postmodern epidemics no longer rely on the presence of a contagious and potentially deadly biological entity as their defining characteristic.

Traditional epidemics[6] like cholera and influenza and postmodern epidemics like obesity, teenage pregnancy, youth violence, and drug use involve a rapid spread of fear and calls for vigilance (Glassner 2000; Luker 1996; Rosenberg 1992, 1962). Fear comes to characterize the obesity epidemic early. On December 5, 1994, an obesity researcher was quoted in the *New York Times* as having said, "We're frightened right now because obesity is an epidemic that has made all of us wake up." In the case of obesity, an epidemic whose biological basis is questionable, this fear and sense of chaos cannot be fueled by the existence or even "spread" of fatness alone; there must also be a shift in the signification of fatness and fat bodies. Paula Treichler, in her work on the AIDS epidemic, suggests that along with biomedical epidemics there often exists a "parallel epidemic of meanings, definitions, and attributions," what she terms the "epidemic of signification" (1999, 19). The idea of an epidemic of signification is indispensable when looking at an "epidemic" like obesity. Indeed, this sense of chaos is a central part of constructing obesity as an epidemic and works with the designation of fat bodies as out of control and threatening (Bordo 1993; Spitzack 1990).

The post–World War II period marks the intensification of what Jeffrey Sobal calls the "medicalization of obesity." With a cultural aesthetic of slimness already in place, Sobal suggests that during this period the moral model of fatness shifted to a medical model in which obesity was designated as a disease to be treated through medical intervention (Sobal 1995; Schwartz 1986). This fell in line with a general trend toward the expansion of medical categories to include undesirable or stigmatizable differences and had the consequence of normalizing medical treatment of these newly medicalized conditions (Conrad and Schneider 1980a/1992; Zola 1972; Goffman 1963). In this period the measurement of overweight became significant with the introduction of ideal height and weight charts by the Metropolitan Life Insurance Company. Methods for the measurement and classification of body weight have changed a great deal since the 1950s, yet the normative and "scientific" measurement of weight remains central to discussions of weight and weight loss (Gaesser 2002; Sobal 1995; Schwartz 1986).

Early medical treatments for obesity initially focused on drugs (most commonly methamphetamines). By the mid-1950s, "diet doctors" frequently provided these "diet pills" to adolescent girls as well as adults and required regular office visits as part of treatment. Subsequently treatments included jaw wiring and, by the 1970s, intestinal bypass surgeries to treat "extreme" cases of obesity (Stearns 1997; Sobal 1995; Riessman 1983). What distinguishes this early medicalization of obesity from the current obesity epidemic is today's focus on public health and risk. This is consonant with newer processes of biomedicalization, one characteristic of which is "the extension of medical jurisdiction over health itself in addition to illness, disease, and injury" (Clarke et al. 2003, 162). Historically, obesity was viewed as bad for one's health and a sign of weakness or moral lassitude. Yet only in recent years has this concern for the health impacts of overweight and obesity been parlayed into the public health crisis of an American obesity epidemic.

So how can we begin to understand the designation of obesity as an epidemic and public health crisis spreading most rapidly among children, minorities, and the poor? Some have suggested that the focus on obesity as an epidemic is indicative of its medicalization (Sobal 1999, 1995). This medicalization is a process that does not break from earlier moral and religious models of body size; rather, it cumulatively incorporates them in various ways. According to Sobal, while one model of obesity may dominate, "most of the time the degree of medicalization and demedicalization has only been partial, with several models competing simultaneously in their claims to typify obesity" (Sobal 1995, 84; Best 1989). Medicalization is a process in which competing definitions and remnants of previous definitions intersect with the process of defining a given phenomenon in medical terms (Conrad and Schneider 1980a/1992). What is needed, however, is a better understanding of how these factors actually impact the degree to which something is medicalized (Conrad 1992) or biomedicalized. Unpacking the language, tensions, production, and contradictions of the obesity epidemic reveals how the moral, medical, and biomedical models of body size reinforce each other and thus truly situate obesity as an epidemic, and enables an analysis of the material and cultural consequences of this designation. Our cultural understanding of obesity, its causes, and its consequences has a profound impact on the techniques designed to "treat" it.

To be sure, weight loss surgery is but one intervention in the obesity epidemic. More traditional behavior modification and fitness programs have been reframed to address the epidemic. For example, in 2003 Weight

Watchers, the largest commercial diet program in the world, teamed with the American Heart Association to promote the first "Great American Weigh-In."[7] This event was designed both to bring awareness of the personal and public health "dangers" of overweight and obesity and to promote Weight Watchers as a way to lose weight and get healthy. In addition, research and development of new weight loss drugs, slowed in the mid-1990s by the fen-phen scandals (Mundy 2001), has expanded dramatically in recent years (Campos 2004). Attention to epidemic childhood obesity and calls for public health and legislative interventions into things like school nutrition and soft drink consumption can be heard almost daily. However, weight loss surgery is unique among these interventions not only because it permanently changes one's anatomy but because it is the most dramatic, expensive, and rapidly spreading obesity treatment available.

The Obesity Epidemic and Biomedicalization

The biomedicalization of the obesity epidemic centering on technoscientific solutions is nowhere more evident than in the development, popularity, and urgency surrounding the development of bariatric or intestinal bypass surgeries.[8] There are several varieties of weight loss surgeries, but all in some way involve sealing off or removing most of the stomach to limit food intake, or bypassing parts of the intestines to prevent food absorption. Designed to treat the most extreme cases of "morbid" obesity,[9] in the face of an epidemic, the surgeries are becoming more and more common. The American Society of Bariatric Surgeons (ASBS)[10] estimates that in 2004 more than 140,600 of these surgeries were performed in the United States alone.[11] This is up from 16,200 in 1992 and 36,700 in 2000. The ASBS estimates that over 80 percent of these surgeries are performed on women.[12]

The most common type of weight loss surgery is the Roux-en-Y gastric bypass, frequently referred to as RNY. In this procedure, the stomach is reduced to approximately 2 percent of its normal size by creating a small pouch or "new stomach." In addition, three to four feet of the small intestine is then bypassed, and the "new stomach" is connected below the bypassed intestine segment, at which point digestion is allowed to begin. This procedure facilitates weight loss in two ways. First, as the new stomach holds only one to two ounces, there is an extreme reduction in the amount of food a person can eat. Second, since a large portion of the small intestine is bypassed, fewer calories and nutrients are able to be absorbed; therefore the RNY is both a "restrictive" and "malabsorptive" procedure.[13] Preoperative patients are warned that the procedure has many potential side effects and complications ranging from blood clots, bleeding, hernias, infections, ulcers, and chronic anemia to constipation, hair loss, vomiting, and weight gain.[14] The frequency with which any of these side effects occur is unclear due to lack of data, but the number of patients who experience more serious complications, while likely significantly larger than in other elective surgeries, is relatively small in relation to the large percentage of patients who experience more common complications like vomiting, hair loss, and fatigue.

The popularity of weight loss surgery can be attributed to several factors. The most significant appear to be the visibility of celebrity weight loss surgeries like those of the singer Carnie Wilson and the TV personality Al Roker, the sense of urgency that goes along with designating obesity an epidemic, and general despair over the high failure rates of traditional diets and lifestyle changes. The prevalence of celebrities having weight loss surgery has grown in the last few years, and those like Carnie Wilson, once an outspoken advocate of size acceptance, now declare themselves to be advocates for morbidly obese persons seeking a surgical solution. Wilson even had her surgery broadcast live over the Internet to raise awareness about the seriousness of morbid obesity.[15] Not only do these celebrities talk about their surgeries in mainstream media, but specialty publications on weight loss surgery often feature interviews and photo essays about them and their surgery experiences. One of my interviewees, Alan, told me that though he had heard of weight loss surgery, it was watching Al Roker talk about his own experience in a television interview that inspired him to begin to research weight loss surgery and talk to his doctor about having it. Others, like Kathy, are skeptical of the focus on celebrity surgeries:

> I think weight loss surgery is a great thing, but I also think it is dangerous to glamorize it with the focus on celebrities who have done it. I mean, Carnie Wilson has had tons of plastic surgery and probably has a personal trainer too. Your average weight loss surgery person has all kinds of flabby skin and doesn't have the money to get it cut off.

Though others shared Kathy's sentiments, most people saw these celebrities as role models and a way to bring weight loss surgery into the mainstream.

Equally central to the popularization of surgical options for weight loss is the sense of urgency that accompanies the designation of obesity as an epidemic. This sense of urgency is furthered by the moral imperative for those "affected" or "at risk" to "act now." This sense of urgency is conveyed in many ways, through the media, in the medical field, in public health, and most significantly in informational materials for the surgeries. We see an example of this in a weight loss surgery brochure from a bariatric clinic in Los Angeles. The overview story is titled "Surgery for Severe Obesity: Drastic Treatment for a 21st Century Epidemic."[16] In the story, the surgeon details the conditions thought to be associated with obesity and gives statistics on the economic cost of treating obesity and obesity-related conditions. While estimates of the public health costs of obesity vary by source and calculation method, most are in the tens or hundreds of millions of dollars. A brochure for a different weight loss surgery program states in boldface on the first page, "Both the medical profession and the general public are recognizing the fact that obesity kills."

The presumed catastrophic impact of obesity on individual health is also cited as a reason for surgery. In one informational seminar, Joy, a middle-aged white woman one year out of gastric bypass surgery, explains to seminar attendees that in fact, when she had her surgery, she had none of the "obesity-related co-morbidities" that many surgery patients have. In fact, she had no health problems at all. For Joy, her general good health was her main reason for having the surgery. She tells us: "It is easier for your body to handle surgery when it is healthier, and at the rate I was going, it was just a matter of time before I had high blood pressure, diabetes, and all that good stuff." Indeed, every surgeon I heard speak at an informational seminar or support group said that people who are healthier at the time of surgery have more success and fewer major complications. Ironically, to qualify to have the surgery covered by insurance, patients must have a certain number of "co-morbidities" or health problems presumed to be related to their weight. One interview told me that she and her doctor "basically made up co-morbidities," reporting that she was "*pre*arthritic" and "*pre*-hypertensive." The interviewee quickly followed this statement by telling me that she was "certain that had I stayed fat, I would have had all of those conditions anyway." Surgeons prefer to do the surgeries on people who, while meeting the BMI threshold for surgery, lack the co-morbidities required for insurance approval. According to one surgeon, this would both be easier for patients and "lower the complication rates" associated with the surgeries.

Though it may seem odd to perform major intestinal surgery on healthy fat people, it is the *risk* assumed to be inherent in obesity that becomes the justification for such surgeries. It is taken for granted by surgeons and patients alike that though an obese person may not have any obesity-related health problems at the time of surgery, it is a virtual certainty that without the surgery they would develop them.

Finally, *all* weight loss surgery patients have tried traditional diets based on caloric restriction and exercise,[17] and everyone I spoke with cited these diet failures as a reason they sought out weight loss surgery. Charmaine told me:

> At some point, you have taken enough pills, counted enough calories, and drunk enough nasty shakes that you know that it is not going to work. I mean, I tried *everything*; in the 1970s I even did a diet where I had to drink the urine of pregnant women. Isn't that disgusting? Eventually you learn that these things don't work, and the people who sell them just want to make a buck. I guess the surgeons do too, but at least surgery works.

Like Charmaine, many others I spoke with rightfully criticized the profit motives of the diet industry. Ironically, it is to these same techniques that weight loss surgery patients must often return to maintain the weight loss they experience in the first year after surgery.

Surgical Weight Loss and Biomedicalization

The social construction of the obesity epidemic is contingent on traditional medicalization processes and newer processes of biomedicalization, as well as older yet constantly shifting moral discourses of body size. A theory of biomedicalization allows for understanding how medicalization and moralism work together rather than looking at one or the other. Though all three of these discourses and processes are central to the designation of obesity as epidemic, nowhere is the biomedicalization of the obesity epidemic more evident than in the case of surgical treatments for weight loss.

The biomedicalization of obesity in the case of weight loss surgeries orbits around the interaction of these biomedicalization processes: the creation of new bodies through surgical techniques, the creation of a new community through online and face-to-face support groups, the development of a new market for specialized products, new techniques of disseminating information about the surgeries, and the co-opting of fat-acceptance language by surgeons and surgery advocates.

These processes interact with and buttress each other such that beyond the common development of a "new" bodily form, it is difficult to trace just how these processes co-constitute each other. The radical change in body size facilitated by weight loss surgeries and the irreversible change in intestinal structure characteristic of the most popular weight loss surgeries create new bodily forms that require specific medical and biomedical, as well as social, considerations. At the Obesity Help convention, attendees were frequently reminded that they "will never be normal again; we now have different plumbing." These new bodily forms also create a new market for everything from new forms of plastic surgeries to remove post-weight-loss "redundant" skin,[18] nutritional supplements and specialty foods, to beaded medic-alert bracelets and weight loss surgery scrapbooks in which one can chronicle one's weight loss journey. A convention organizer encouraged attendees to have a new professional photograph taken by one of the photographers at the convention by suggesting, "Isn't it time for a new picture? Didn't you always avoid having your picture taken before?" This statement also taps into the sense of a common past among weight loss surgery patients, something I discuss in more detail later.

These new bodily forms and the resulting new market also draw from and extend the sense of community that has developed among those who have had or are considering weight loss surgery, as well as their friends and families. Bariatric surgeons are remarkably absent from this community. Throughout the surgery process, patients may meet with their surgeons only a handful of times.[19] One woman I spoke to met her surgeon only once before surgery and once after surgery, as those were the only visits covered by her insurance. She did not seem to have a problem with this, because for her "the surgeons do the surgery, but only someone else who has gone through surgery can understand what it's like. I go to the doctor for my regular blood work,[20] but for advice I go to the message boards." This is a pattern among many people I have spoken with; once they are four to six months post-op, they rely primarily on their peers for support and medical and nutritional advice related to their surgery.

Though most of this support takes place online or over the phone, ObesityHelp.com members often travel to meet each other individually or at Obesity Help events like the convention I attended. Though the website has over 250,000 registered members, it is a tightknit community, and members respond to requests for help and advice rapidly. Susan, a fifty-five-year-old white woman who had RNY surgery in 2003 and has since lost over one hundred pounds, gave me an example of this. At the time, Susan's insurance company wanted her to have the surgery at a specialty bariatric clinic over five hundred miles from her home. The surgery resulted in major complications that kept her away from home and in the hospital for over two months. During this time, Susan's daughter-in-law posted a message to a board for that area, and almost instantly Susan was receiving visitors she had never met, but who had all had weight loss surgery and wanted to support her.

The weight loss surgery community is built around two central commonalities among its members: the experience of living life as a very fat person and the experience of having had or desiring to have weight loss surgery. The common experience of having lived life as a fat person in a fat-phobic society is drawn on by surgeons and surgery advocates in encouraging people to have weight loss surgery. Surgeons often cite the social, economic, and medical discrimination experienced by most surgery candidates as one of the most compelling reasons to have surgery. Indeed, my interviewees and people I have spoken with informally at weight loss surgery seminars and events all cite size discrimination as one of their reasons for deciding to have the surgery. Kate described it this way:

> It's like, all of a sudden you can sit in a movie theater seat again, or not get a seat belt extender on the plane. I think thin people take that stuff for granted, but it is really hard to be in the world like that. It is hard to never fit, to have people assume things about you because you're fat. I would be a liar if I said all that wasn't a big consideration going into this.

Most have experienced weight-related discrimination for a significant portion of their lives and hope that surgery and the potential resulting weight loss can alleviate some of this suffering. It appears to be particularly potent to hear an acknowledgment of this discrimination from a medical professional, as many people who choose weight loss surgery have had extremely bad interactions with the medical profession in the past. A number of people I spoke with cited their weight loss surgeon as the first doctor they had ever consulted who did not lecture them on their weight or seem disgusted by their bodies. Maya, a thirty-seven-year-old Caucasian woman, told me her bariatric surgeon was "the first doctor in my life who touched my body and didn't seem grossed out." In addition, both surgeons and patients expressed a hope that weight loss surgery could help them to appear "normal" in society and to live their lives without constantly running into barriers both literal and figurative. The social benefits of surgery are emphasized in informational materials. One pamphlet from a bariatric surgeon and member of the ASBS shows anatomical diagrams of two weight loss surgery procedures, yet under these diagrams is written in large print "Patients no longer face the social stigma or the many indignities attached to obesity." This focus on the alleviation of the stigma associated with obesity represents a partial co-optation of size-acceptance rhetoric on weight-based discrimination.

Another common experience shared by patients in their presurgery lives (and sometimes postsurgery as well) is the repeated failure at other weight loss attempts. While much of society seems to blame fat people for their weight and inability to lose it in spite of well-known statistics on diet failure rates (Fraser 1998), weight loss surgeons and others in bariatric surgery programs emphasize that people are not fat because they are lazy or undisciplined or because they simply eat too much. Informational literature from surgery programs and information given out at informational seminars often emphasizes that surgery is an option precisely because willpower has little or nothing to do with people's weight loss failures. Surgery is necessary because traditional diets don't work and because a large part of an individual's obesity can be attributed to genetics, not behavior. A speaker at the Obesity Help convention echoed a sentiment I heard at all the informational seminars I attended: "If all we had to do was eat less and exercise to lose weight, then no one would have weight loss surgery." Thus diets fail the vast majority of the time because they assume that obesity is simply an issue of food choices and personal behavior. However, according to weight loss surgery advocates, this misconception accounts for most diet failure and makes surgical intervention necessary.

This focus on taking the blame for obesity off fat people themselves is significant for two reasons. First, as I will show in the case of "failed" weight loss surgeries, this more biomedicalized discourse of body size and weight is often overshadowed by a reassertion of traditional diet discourse and moral understandings of diet failure. This biomedicalized approach to obesity is also significant because in large part it represents a co-optation of the political discourse of weight espoused and developed by the fat-acceptance movement (Sobal 1999, 1995). Indeed, this co-optation is one of the processes that make weight loss surgeries a particularly clear site from which to understand the biomedicalization of obesity.

Second, the attempt by surgeons to move obesity from the realm of moral failure to the realm of the biomedical represents the co-optation of the language of body size long used by fat activists. In his article on the size-acceptance movement, Sobal (1999, 1995) states that the core assumption of the size-acceptance movement is that fatness is not a moral issue and is not one best suited for medical treatment. Rather, for the size-acceptance movement, fatness is a political issue, and it is social, not individual, change that needs to happen to end discrimination against fat people (Wann 1998; Poulton 1997; Millman 1981). To be sure, the biomedical focus of surgeons does not exactly mirror the political approach of the size-acceptance movement, particularly vis-à-vis medical treatment of obesity. However, the shared desire to take the focus off moral failure on the part of fat people represents a co-optation of a central piece of the fat-acceptance message to make potential patients more comfortable with a profession and procedures that might otherwise be reminiscent of their all-too-common experience of medical discrimination. The core difference between the biomedical and political efforts at moving away from the doctrine of moral lassitude in fat people is that for the surgeons, the answer to the problem is internal and involves the permanent surgical alteration of the body, whereas for the size-acceptance community the answer lies in accepting oneself as a fat person and working to change a fat-phobic society.

The second axis of the weight loss surgery community is the actual experience of having undergone weight loss surgery. There is a sense that the unique experience of being a formerly fat person living in a surgically modified body creates both a body of experiential knowledge and somatic needs and characteristics that only those who have had bariatric surgery share. Indeed, the physical challenges brought by weight loss surgery are unique and many. Postoperative (post-op) patients must follow specific rules and regimens to avoid serious complications and to ensure that they experience maximum weight loss in the first nine to twelve months after surgery.[21] My interviews and participant observation show that after their typical one-month post-op visit with their surgeon, most patients get their nutritional and healthcare information from each other and Internet resources. The obesity surgery message boards are often filled with questions about how to deal with "dumping,"[22] hair loss, nutritional supplementation, and exercise from people who have not yet asked these questions of their surgeons or primary-care physicians.

Postoperative patients, or "losers,"[23] often serve as "angels," or primary surgery support, for preoperative patients. While surgeons and the weight loss surgery community alike emphasize that each person will have a varied experience with the surgery, the angel is presumed to have an experiential knowledge of the process that only someone who has gone through the surgery can have. This assumption, along with the frequent desire to avoid one's primary-care physician at all costs, also leads people to do most of their presurgery research online through resources like ObesityHelp.com before they ever even ask their doctors for a surgery referral. This democratization of information is yet another way in which the obesity epidemic integrates processes of biomedicalization (Clarke et al. 2003).

Biomedical Success or Individual Failure?

The official definition of weight loss surgery success as set forth by the American Society of Bariatric Surgeons defines success as a patient losing and maintaining a loss of 50 percent of his or her excess body weight over five years. Thus, according to the Body Mass Index (BMI)[24] definitions of overweight and obese, a person who had surgery at a height of 5'7" and 350 pounds would be considered a weight loss surgery success if she or he lost and maintained a loss of 100 pounds over a five-year period, even though she or he would remain classified as "severely obese" and still within the BMI range of those *eligible* for surgery. However, it appears that weight loss surgery patients themselves would not tend to think of this example as a success, because at 250 pounds they would hardly have achieved the "normalcy" they had hoped the surgery would give them. On a message board for failed weight loss surgery, one woman who weighed 460 pounds before her RNY surgery and had lost "only" 130 pounds sixteen months postsurgery wrote:

> I did not have surgery to weigh 330 pounds. It bothers me a lot to know that my goal weight will still qualify me for WLS. But I have no idea if I will ever even get close to that. Sure, I am glad that I have lost 130 lbs.

But in 16 months that is not what I expected. I do my exercise, I enjoy it. I never expected to be able to weigh 120 like that stupid chart says I should, but I at least wanted something closer to a normal weight than I am now.

This frustration is not unexpected given that most advertisements for weight loss surgery show patients who have clearly lost significantly more than 50 percent of their excess weight and in many cases appear to have already undergone plastic surgery to remove excess skin. It is often these people who are invited by their surgeons to speak at weight loss surgery seminars. Typically, at least one post-op patient from the surgery program being promoted is asked to share his or her story with seminar attendees. The speakers selected by program doctors are usually those who have achieved the most dramatic results, not those who have had the most typical results.

Lenny was the speaker at an informational seminar to promote a new bariatric clinic at a large suburban hospital. Lenny is a forty-five-year-old single white male who, at the time of the seminar, was eighteen months post-op from an RNY procedure. Dressed in form-fitting black jeans and a tight black T-shirt, Lenny looked like the type of guy who had been going to the gym his whole life. In fact, eighteen months earlier, Lenny had weighed close to 400 pounds and had never regularly exercised in his life. Now, at 180 pounds, Lenny works out at a gym at least two hours every day and often works overtime at his physically demanding job as a warehouse worker. He wears compression garments while he works out to help minimize the sagging skin that frequently develops in those who lose weight so rapidly and is proud to say that he has not had and does not plan to have any plastic surgery to remove excess skin.[25] When Lenny is asked what he eats in a typical day, he tells us that for breakfast he eats two scrambled egg whites and water. For lunch he eats a half cup of white rice with vegetables and soy sauce, and for dinner he eats half a skinless chicken breast and more steamed vegetables. To make sure he is getting his protein, he also drinks a special protein drink at some point during the day. He tells us that he expects that this will be his daily menu for the rest of his life and that he doesn't mind because he feels better than ever and the surgery has saved him from the certain early death he was facing due to his morbid obesity. Everyone attending the seminar seems very impressed with Lenny, and the doctors talk about how proud they are of him and his dedication. One of the surgeons calls him "a poster child for weight loss surgery." A thin man sitting toward the back of the room with his wife (who is considering surgery) asks the doctor if Lenny is representative of most post-op weight loss surgery patients. The doctor says no, most people do not lose as large a percentage of their weight as Lenny did, and most people will need to have surgical skin removal at some point after their weight loss slows. She says that they chose Lenny to speak because he is an example of "the best of what is possible through weight loss surgery."

Surgeons will often give potential patients the clinical definition of weight loss surgery success described earlier, yet the visual images of success are often people who have lost a far higher percentage of weight than the average person can expect to lose. It is also the case that post-op speakers at support groups and information sessions and featured in *Obesity Help Magazine* tend to be between nine and thirty-six months post-op, the period in which most people's weight loss peaks.[26]

There is no more agreement on what constitutes failed weight loss surgery than there is on weight loss surgery success. The central medical rationale for having weight loss surgery is to cure, manage, or prevent conditions like adult-onset diabetes, sleep apnea, and hypertension (to name but a few) that are assumed to be caused by obesity. However, among weight loss surgery patients, an assessment of surgical success or failure does not seem to rest on the alleviation or abatement of these co-morbidities. For most of the people I talked to and whose stories I read on the weight loss surgery message boards, like the woman I quoted earlier, the success of the surgery seems to rest on having lost enough weight to be able to consider oneself "normal."

Most of the online weight loss surgery chat rooms, listservs, and message boards tend to deal with more of the "positive" aspects of weight loss surgery, especially the message boards at ObesityHelp.com. Yet spaces exist for people to tell stories of complications, weight loss surgery failure, and weight loss surgery regrets. ObesityHelp.com hosts two such forums, the "Weight Loss Surgery Regrets" message board and the "Second Time Around/Weight Loss Surgery Failure" message board.[27] Most of the people who use these two boards have already had surgery, though occasionally someone still considering the surgery will post asking for information about the downside of weight loss surgery. Most of the people who participate in these two lists seem to know that they are outsiders in the weight loss surgery community, both because their surgeries, for whatever reason, were not successful, and because they are willing to voice their disillusionment with the surgery. One woman who felt that her surgeon had misrepresented the extreme diet and exercise modifications she would have to make after surgery posted the following:

> You know, I don't even go to my support groups anymore because I found myself being ignored. . . . I found that I am too blunt, too controversial, and the truth be told, [neither] doctors nor their staff want people like me speaking up at those groups. I understand why, it's a business and if I start telling it like it is, from my perspective, from my shoes, it deters people.

Many on these lists expressed their sense of alienation when attending support groups. Many said that the support groups are basically "preaching to the choir," and some expressed concern for people who feel that if they have had any problems as a result of the surgery, they cannot speak up. One woman gave her impression of the support groups: "Everyone seems all smiles and success stories. I'm happy for them, but I also know somebody in those rooms is having a hard time and is afraid to express any regret about having this surgery." There are perhaps many reasons why criticism of the surgery is so discouraged by the weight loss surgery community. Clearly, as the previous quote states, bariatric surgery is a business, and hearing negative experiences may steer potential patients away. Second, many of my interviewees and people I met at events feel defensive about the surgery, feel it is negatively portrayed in the media and that they must correct for that negative image. Third, regret may consciously or unconsciously feel like a futile emotion in those who have had an irreversible elective surgery. This is important because this avoidance of criticism is inextricably connected to how people understand and explain surgical failure.

In the absence of clear surgical error, weight loss surgery failure, most often defined as the failure to lose at least 50 percent of one's excess body weight or weight regain after surgery, is explained as a result of patient noncompliance, specifically patients' beginning to eat more than two years after surgery and subsequently "stretching" their pouches. This pouch stretching results in an ability to take in more food and thus gain weight. In a breakout session at the Obesity Help convention, a surgeon told the audience that most often weight regain in weight loss surgery patients is due to patients' not continuing to follow the post-op rules of eating once they are past their peak weight loss stage.

According to this surgeon, the rules of eating that patients must follow for the long-term success of weight loss surgery are as follows: first, "Eat no more than one bite of food every ten minutes"; second, "Use a stopwatch when eating"; and third, "Do not drink liquid half an hour before a meal, with a meal, or for one hour after finishing a meal. Drink no more than one ounce of liquid every five minutes."[28] The surgeon tells us that most of the patients he knows who fail do so because they eat and drink too fast. He tells us:

> There are no scientific studies on this, but if you talk to patients, and they are really being honest, they will tell you that they will eat a half a sandwich in fifteen minutes. . . . We [surgeons] give you this tool, but you have to ask yourself, "How do I use the tool given to me?" This surgery takes a lot of discipline and learning.

Another surgeon speaking at the convention warned, "If you drink a lot of shakes and high-calorie liquids, then we don't have a surgery for you." A bariatric nurse speaking on the subject of long-term pouch care told us, "Now you have a tool that if you treat it right will last for life. The problem when it doesn't work is that you have the same bad habits that you did before you got it."

What these quotes show is that although weight loss surgeons frequently tell patients before surgery that dieting doesn't work, and that their weight isn't their fault, behaviors associated with traditional dieting become the dividing line between those who are able to maintain long-term weight loss and those who regain.[29] These quotes also show that like the dieter who claims to not be overeating yet still gains weight, weight loss surgery patients who claim to be following their post-op regimen yet gain or do not lose weight are seen as less than honest with themselves and with their doctors. On the "failed weight loss surgery" message board, two women expressed their frustration at this assumption:

> I am concerned because now I can eat a whole single hamburger with a bun in one sitting.[30] When I told my doctor this, he just said, "well, just because you can do it doesn't mean you should." Please save me from skinny doctors!

In a separate thread on the same message board, another woman said:

> I was 248 pounds when I had this surgery in late 2003. 8 months later I weigh 174. It has been a full time job trying to get this weight off and now I have almost given up. My doctor says I am "out eating" the surgery, but I know I am not. No one seems to want to talk to you when you are a WLS failure; they only want to hear the good things.

These quotes show that while the success of surgery is attributed to surgical skill, patient compliance, and biomedical innovation, doctors attribute failure to the inability or unwillingness of individual patients to change their behaviors to facilitate weight loss. The assumed lack of motivation or willpower associated with fat people that doctors explicitly criticize in weight loss surgery seminars later becomes the comfortably familiar way in which to explain weight gain or failure to lose in "noncompliant" post-op patients. Doctors' tendency to revert to individual explanations for surgical failure is not simply hypocritical or contradictory; it is a further example of the centrality of moral understandings to biomedical discourses of weight and weight loss.

The specter of the emotional eater also looms large in attempts to explain weight loss surgery failure. The emotional eater, almost invariably presumed to be female (Orbach 1978; De Vault 1991; Zimberg 1993), is a figure often invoked to explain weight loss surgery failure. In support groups and at the Obesity Help convention, people talked about emotional eating as one of the most common ways to "eat your way around the tool of surgery." At one postoperative support group I attended, the discussion turned to weight regain and how to avoid it. The surgeon leading the group said that "surgery can take away the calories, but it can't take away the drive to overeat." He says that this is a problem he sees most often in his female patients.

> My male patients do better with the surgery than women do. I think this is because they [women] are emotional eaters and men are not. Men eat because they like to eat; I can fix that, but I can't fix emotional eaters.

Many postoperative patients seem to agree with this sentiment. One woman speaking at the convention told the audience that when she regained weight several years after surgery, "My surgery didn't fail me; I failed my surgery." She went on to explain that though she could never eat much in one sitting, she would try to deal with her emotions by "grazing" and thereby gained back the weight that the surgery had helped her lose.[31] This again highlights that it is patients, not surgeons, who are held responsible for the success or failure of weight loss surgery, and the "emotional woman" is the cultural trope that is most easily identified as problematic.

Though surgeons and advocates of weight loss surgery cite well-known statistics on the high failure rates of traditional diets to justify the need for bariatric surgery, when patients are two or more years post-op and their bodies have adjusted to the caloric restriction inherent in most weight loss surgeries, it is exactly such traditional dieting that is required to maintain weight lost in the first two years after surgery. Many of the people I spoke to have done just this. In one support group, Marlene, a middle-aged white woman who had RNY surgery in 2003, tells the group that she has started going to Weight Watchers "for emotional support." Marlene has lost 105 pounds in sixteen months, but she knows that "the hard part is still ahead of me" and that going to Weight Watchers meetings can help her maintain her weight once she reaches her goal. Currently she cannot actively participate in the program, as she is unable to eat even close to the minimum number of calories required by Weight Watchers, but she does "listen to people and get tips for later."

Still others have returned to pharmaceutical methods to maintain or supplement their weight loss. In response to a message board question about how to keep off weight lost as a result of surgery, a woman who had RNY surgery eighteen months earlier described her own method:

> I am taking prescription diet pills to curb my hunger and I haven't gained any weight but I haven't lost any either in 6 months. . . . I have lost 120 pounds and not a pound more. I do exercise, drink water and get in 90 grams of protein a day. I also watch my calories and keep then below 1,200 a day. I feel like I never even had the surgery and I am just back to dieting again.

Others shared similar techniques, such as over-the-counter diet pills, liquid diets, and fasting. Still others have developed anorexia or bulimia connected with their efforts to stay thin after surgery. In a session on eating disorders at the Obesity Help convention, the leader, a therapist who works with pre- and postoperative weight loss surgery patients, asked the audience, "How many of you have eaten something you knew would make you throw up just to avoid gaining weight?" Several hands went up, and the therapist said, "While it is good to be scared of weight gain, being obsessed with it is unhealthy." However, for many, it is unclear where the line between "healthy" fear and "unhealthy" obsession should be drawn. What is clear is that most people, whether they are critical of dieting or not, must return to traditional dieting behavior if they are to have a chance at maintaining the weight lost due to surgery.

These reversions to dieting and to the individualized understandings of success and failure that go along with it directly contrast the biomedicalized, "value-neutral" perspective on weight and weight loss that patients hear from their surgeons and others before they have surgery. However, what this research shows is that earlier and arguably more familiar discourses of weight and moral failure are accessible and drawn on to explain the failure of specialized biomedical interventions.

Conclusion

Weight loss surgery as an extreme intervention into the contemporary American obesity epidemic brings into high relief the relationship between processes of biomedicalization and moral discourses of bodies and body size. I have explored this relationship in two ways. First I have shown the ways in which processes of biomedicalization can be seen in the case of surgical weight loss. Second, and most significantly, I have shown how in the case of failed weight loss surgeries, doctors, patients, and others in the weight loss surgery community are quick to return to explanations of weight gain that hinge on familiar understandings of fat people as weak-willed, indulgent, lazy, and generally noncompliant patients. I have shown that even as surgeons co-opt and adapt the messages of the fat-acceptance movement as they promote weight loss surgery, they are quick to return to individual and highly gendered explanations of weight regain and the failure to lose weight after weight loss surgery. This shift goes largely unquestioned for the reasons I have discussed—market interests, patient defensiveness, and cognitive dissonance, among others. However, all these reasons are plausible because moral discourses of body size have become so familiar that they are easily used by patients, doctors, and many others in the weight loss surgery community and industry to make sense of biomedical failure without questioning the legitimacy or techniques of biomedicine.

Others (Clarke et al. 2003; Sobal 1999, 1995; Conrad 1992) have emphasized that moral discourses of health, illness, and risk continue to exist even as processes of medicalization and biomedicalization gain prominence. What I have done is to provide one example of how moral discourses of health continue to hold sway and remain useful even as biomedicine and technoscience expand ever further into the multilayered world of human experience.

This chapter raises another, larger question, namely, how can understanding the case of biomedical failure add to a theory of biomedicalization in an era in which the definition of an epidemic now extends far beyond the realm of mass contagion and death? I have begun here to address this question, but it demands further exploration through study of the obesity epidemic as well as other social phenomena subject to classification as epidemics.

Notes

Special thanks to Raka Ray, Dawne Moon, Barrie Thorne, Adele Clarke, Kristin Barker, Leslie Bell, Mark Harris, C. J. Pascoe, and the editors for their suggestions and guidance on various drafts of this chapter.

1 Clarke et al. (2003, 162) define biomedicalization as "the increasingly complex, multisited, multidirectional processes of medicalization that today are being both extended and reconstituted through the emergent social forms of a highly and increasingly technoscientific biomedicine."

2 The interplay of these two sets of processes along with other understandings of body size (moral, psychological, and political) is complex and beyond the scope of the chapter. This complex array of interactions and meanings is more fully examined in the larger project of which the present chapter is a part. The larger project of which this research is a part looks to the social construction of obesity specifically *as an epidemic*. I do not deny a rise in average weights and potential health problems *associated* with obesity; rather, I look specifically at the ways in which obesity is constructed not only as a social problem but as an *epidemic*. As others have pointed out, this separation of objective and subjective aspects of social problems is especially important at moments of apparent "crisis" — precisely the point at which they are most likely to be unreflexively collapsed (Treichler 1999; Epstein 1995).

3 There is a range of fatness, some of which is clearly more acceptable than others. In this chapter, however, given my attention to surgical weight loss, I focus on what some see as the extremes of fatness. Given the increased attention to all degrees of overweight and obesity, I argue that the discourse of weight used to talk

about surgical weight loss is in many ways an extension of the discourses of weight and weight loss found in more general cultural concerns about overweight and obesity.

4   All the support groups and informational seminars I attended were free and open to the public. The Obesity Help convention was also open to the public with payment of a $60 registration fee. As I discuss later, the Internet is a main avenue of information, support, and community for those who have had or are interested in weight loss surgery. *Obesity Help Magazine* is published six times a year by the organization ObesityHelp.com. Both the magazine and the website are promoted as "your gateway to the weight loss surgery community." ObesityHelp.com is a for-profit Internet community started in 1998. ObesityHelp.com offers online resources for weight loss surgery patients, those considering surgery, surgeons, primary-care physicians, insurers, and more. The organization currently claims more than 250,000 members, most of them postoperative weight loss surgery patients.

5   The concept of a postmodern epidemic is further theorized in the larger project on the social construction of the obesity epidemic of which the present paper is a part (Boero 2007, 2009).

6   A traditional definition of an epidemic is generally the presence of infectious disease and death "prevalent among a people or a community at a special time and produced by some special causes not generally present in the affected locality" (Rosenberg 1992, 78; Hatty and Hatty 1999).

7   This event was modeled after the American Lung Association's annual Great American Smokeout, a day on which smokers are encouraged to seek help in quitting smoking.

8   While I do not argue that describing the current prevalence of obesity as an epidemic has caused the increase in bariatric surgeries in recent years, I do argue that using the term "epidemic" contributes to the creation of a cultural climate in which drastic interventions become more acceptable.

9   Morbid obesity—sometimes called "clinically severe obesity"—is defined as being one hundred pounds or more over ideal body weight or having a BMI of 40 or higher.

10 The ASBS is recognized by the American College of Surgeons and is a specialty surgical society in the Specialty and Service section of the American Medical Association (http://www.asbs.com).

11 Many people whose insurance will not cover the cost of weight loss surgery will go to another country, most often Mexico, to have the procedure done at a lower cost. While no data are available for the number of such surgeries, there are several listservs, message boards, and even travel companies that cater to such patients. For more on plastic surgery tourism, see Gilman 2001.

12 According to the ASBS, there are not yet any accurate statistics on the racial and ethnic backgrounds of weight loss surgery patients. However, unlike plastic surgeries, most RNY procedures are paid for by private insurance, Medicare, or

Medicaid, thus potentially diversifying the class, racial, and ethnic composition of those who have or will have weight loss surgery.

13 It is difficult to find outcome data on weight loss surgeries, as there are not yet any controlled clinical studies of the surgeries (see, for example, the October 2004 special issue of *JAMA*). It is also the case that the studies that are available do not include long-term (five years or more) data on surgery outcomes.

14 Kaiser Permanente, "Patient Information Booklet: Gastric Bypass Surgery" (2003).

15 Carnie Wilson, interview, *Obesity Help Magazine*, no. 5 (2004).

16 Many bariatric surgeons and materials promoting weight loss surgery make an argument for surgery that draws on the language of epidemic obesity. What is needed is more research on how patients themselves negotiate the language of epidemic obesity as they understand and describe their own weight-related experiences.

17 For surgery to be approved by insurance companies as well as doctors, patients must be able to document having tried and failed at traditional weight loss methods.

18 The American Society of Plastic Surgeons estimates that in 2003 more than 52,000 plastic surgeries were performed on postoperative bariatric surgery patients. The most popular procedures for bariatric patients are tummy tucks, panniculectomy (removal of excess abdominal skin), breast reductions, breast lifts, lower-body lifts, and upper-arm lifts (http://www.plasticsurgery.org).

19 In hospitals with specific bariatric surgery programs, patients are far more likely to be seen by a bariatric nurse than by their actual surgeon.

20 Most surgeons advise that weight loss surgery patients have extensive blood work done every six months post-op to check for nutritional deficiencies, such as anemia and vitamin B_{12} deficiency, common in RNY patients.

21 It is generally accepted that the first nine to twelve months following gastric bypass surgery are the "weight loss window," the period when the vast majority of postoperative weight loss occurs.

22 "Dumping syndrome" is experienced by most RNY post-ops. The syndrome causes weakness, nausea, sweats, vomiting, diarrhea, cramps, and fainting. Dumping is most often caused by eating high-sugar foods but can be caused by a wide variety of foods depending on the individual ("Understanding Obesity Surgery," informational pamphlet published by the Stay Well Company, 2001). People I spoke with often welcomed dumping as a type of aversion therapy, and some expressed missing it when, after two or more years post-op, they no longer experienced dumping.

23 The term "loser" is often used by weight loss surgery patients to jokingly refer to themselves postsurgery.

24 BMI is calculated by dividing weight (in kilograms) by height (in meters) squared. See http://www.nhlbi.nih.gov/index.html.

25 Compression garments are popular among post-op patients, especially those who

have not yet had or cannot afford to have plastic surgery to remove "redundant skin." The garments range from compression panties to full-body stockings that cover the body from the ankles to the wrists and neck. There are even compression masks to wear while sleeping. Many of these garments are produced and marketed specifically to the weight loss surgery community. One woman I spoke to at the Obesity Help convention referred to her body stocking as "lipo in a box."

26 In my research, I never encountered a speaker at a convention, support group, or information session who was more than five years post-op.

27 "Second time around" refers to patients who are going through weight loss surgery revisions, a new surgery to rebuild the pouch and to attempt to narrow the stoma from the first surgery. Such procedures are often difficult to get approved by insurance because they carry added risks, and many doctors will not perform such surgeries due to the increased risk of complications and death.

28 This suggestion is particularly difficult to follow. Weight loss surgery patients are required to drink a minimum of sixty-four ounces of water daily to prevent dehydration that can result from bypassing parts of the intestines. However, at the rate of one ounce every five minutes and not drinking with meals or for a time period before or after meals, patients would need to drink one ounce of fluid every five minutes for five and a half hours (not including mealtimes) every day just to take in the minimum amount of liquid recommended.

29 Because there are no scientific studies that evaluate the results of weight loss surgery in those who are five years or more post-op, there is no way to accurately know rates of regain. The surgeon described estimates (he did not cite any specific data) that after five years approximately 30 percent of RNY patients will have regained most, all, or more than their lost weight.

30 In a person who has RNY surgery and whose pouch was not made "too large" or whose pouch has not stretched, being able to eat a whole small hamburger would be unusual.

31 "Grazing" is a term used in the weight loss surgery community for the eating of small bits of food all day long, rather than sticking to the three small meals one is supposed to eat. Often, when people are having trouble with postsurgery weight gain, the first question asked of them by their surgeon and their peers is "Are you grazing?"

Jennifer Ruth Fosket

12 / Breast Cancer Risk as Disease

BIOMEDICALIZING RISK

One of the sites at which biomedicalization is most evident in everyday life is in the heightened attention to risk and its transformation into a treatable health problem — a diseaselike state in itself. With the expansion of technoscientific tools of biomedical surveillance into ever greater areas of people's bodies and lives, being "at risk" becomes an increasingly common diagnosis, with its own set of proscriptions, prescriptions, and treatments. In this chapter, I explore this aspect of biomedicalization as it emerges in the arena of pharmaceuticals aimed at treating breast cancer risk: chemoprevention. Chemoprevention is the practice of ingesting pharmaceuticals or nutraceuticals to prevent disease. While chemoprevention in biomedicine is not new — it is a familiar practice in heart disease, birth control, and other areas — chemoprevention for breast cancer achieved legitimate status only in 1998 with the FDA's approval of the drug tamoxifen for the reduction of risk of breast cancer. By intensifying and capitalizing on slippages between risk and prevention, health and disease, chemoprevention provides a stark exemplar of the biomedicalization of *prevention*.

Within biomedicalization, it is no longer necessary to manifest symptoms to be considered either ill or even "at risk." With the "problematization of the normal" and the rise of what Armstrong (1995) calls "surveillance medicine," *everyone* is implicated in the process of eventually "becoming ill" (Petersen 1997). Beginning in the twentieth century, it was no longer assumed that normal was an unproblematic state of being. Rather, at all times, everyone is potentially *not* normal, inhabiting tenuous spaces between illness and health. "It is no longer the symptom or sign pointing tantalisingly at the hidden pathological truth of disease, but the risk factor opening up a space of future illness potential" (Armstrong 1995, 400). Surveillance medicine ushered in the concept of risk and risk factors. Once the idea that everyone is potentially ill becomes part of common discourse, the next step is to determine what the signs or symptoms of that potential illness might be — risk factors. The tasks of specifying risk factors that might lead to future illnesses then become part of health research.

Robert Castel helps map theoretical spaces to explore the complex networks of risks implicated in chemoprevention technologies. He writes, "A risk does not arise from the presence of particular precise danger embodied in a concrete individual or group. It is the effect of a combination of abstract factors which render more or less *probable* the occurrence of undesirable modes of behavior" (1991, 287; italics mine). In Castel's scheme, these factors are standardized into packages from which experts calculate the riskiness of particular embodied subjects. As articulated by Castel, the relations between individual persons and risk factors are complicated and highly unstable.

Despite this instability, within the current patterns of biomedicalization, risk factors are nevertheless increasingly conceptualized as treatable health problems. Indeed, within the realm of breast cancer, widespread controversy remains as to what causes the disease, what bundle of factors may increase one's risk of getting it. Yet pharmaceuticals have emerged that are marketed for their ability to treat breast cancer *risk*. In this chapter, I analyze the emergence of chemoprevention and the shift it has marked toward conceptualizing breast cancer prevention as the treatment of breast cancer risk with pharmaceuticals rather than the identification and elimination of cancer-causing agents.

Risk as a Treatable Health Problem

What marks chemoprevention as a uniquely biomedicalized phenomenon is the way in which it conceptualizes risk as a treatable health problem. Within the schema of chemoprevention, breast cancer risk becomes something detectable, diagnosable, and treatable with pharmaceuticals. This occurs within a milieu in which biomedicalization has already transformed the meaning of drugs from something one takes to *restore health* when one is ill to something that has a legitimate place in the daily lives of healthy people to *preserve health*. Chemoprevention for breast cancer thus joins other drugs aimed at preventing other health problems (e.g., hypertension, high blood cholesterol) in a world where being medicated is increasingly the norm. Biomedicalization as a cultural force shapes the way chemoprevention has been able to emerge when and how it has. Other factors are also important. First, chemoprevention for breast cancer enters the scene at a time when little else exists to prevent the disease and uncertainty dominates the identification of breast cancer's risks.

Uncertain Risk and Piecemeal Prevention

Knowledge of breast cancer causation is so uncertain that as many as 75 percent of breast cancer cases are estimated to occur in women with no known risk factors except increasing age (Madigan et al. 1995). Several unique aspects of cancer per se make it particularly problematic to construct stable, much less certain, etiological knowledge. Knowledge about direct one-to-one cause-effect relations is uncommon with cancers of any kind. First, cancer often takes a very long time to develop, and not everyone exposed to the same carcinogenic elements will develop cancer. Second, potential cancer-causing agents are so ubiquitous, yet their significance and severity of impact are so uncertain, that it is virtually impossible to create control groups not exposed to potentially cancer-causing agents.

Knowledge of breast cancer causation is also politically problematic. Claiming that something causes cancer almost always interferes with the interests of some actors (be they farmers who use pesticides, corporations who manufacture chemicals, or women's health activists who contest the claim that abortions increase breast cancer risk). What is claimed and legitimated about breast cancer causation is thus especially important because this shapes potential action by policymakers, healthcare providers, scientists, and others to prevent the disease.

While breast cancer etiology and breast cancer risk are today highly contested, the production of such knowledge has developed over time; some theories have persisted, others have been displaced after years of credibility, and still others have never gained legitimacy. Today mainstream sources typically describe breast cancer risk factors as including age, family history of breast cancer in a first-degree relative, no childbirth or late age at first childbirth, early menarche, and multiple previous benign breast biopsies (Gail et al. 1989). Women's health and breast cancer activists often add to those an emphasis on environmental factors. Against this backdrop of uncertainty, prevention efforts have been similarly problematic.

Throughout the long, arduous process of uncovering possible causes of breast cancer that has occupied scientists from multiple disciplines, with varying levels of success, for the last several hundred years, prevention strategies have been proposed and abandoned, occupying tenuous and contentious positions within the breast cancer arena. The translation of knowledge claims about breast cancer causation into specific actions in the realm of breast cancer prevention is both a problematic and an increasingly politicized process, leading to prevention choices that are underdetermined. The scientific data do not fully or clearly mandate any one choice, thus particularly opening up spaces for economic, political, and social factors to shape what are ultimately presented as scientific choices. That is, the difficulty in ascertaining what causes breast cancer in the majority of cases means that decisions about prevention cannot be based on solid scientific knowledge of causality. Instead such decisions arise from consideration of potential "risk factors." Risk factors are mutable; they vary in importance and are supported by various degrees of evidence within social, economic, cultural, and political milieus that are supportive of some decisions (e.g., public health messages should tell women to exercise and eat low-fat diets) and less so of others (e.g., public health messages should tell women to get prophylactic mastectomies).

Given what is known, or more importantly what is *unknown*, about breast cancer etiology, few options exist for primary prevention. The word "prevention" itself takes on variable meanings within the biomedical context more generally. "Primary" cancer prevention refers to the reduction of cancer mortality via a reduction in the incidence of cancer (Dunn, Kramer, and Ford 1998). This is what most of the public probably associates with cancer prevention: the ability to prohibit cancer from developing at all.

Primary cancer prevention is theoretically achieved through the identification and elimination of cancer-causing agents. However, because stopping cancer from developing in the first place has proved so difficult, when speaking of prevention today, most biomedical experts are referring to secondary cancer prevention—the reduction of mortality via the detection of cancer in its earliest and most treatable manifestations. This kind of semantic slippage has been the source of much criticism from activists and others who oppose the ubiquitous rhetoric that "early detection is your best prevention" on the basis that it misleads the public into believing that mammograms and breast self-exams are somehow "preventing" cancer, not merely detecting it.

Primary prevention is further confounded by the fact that identifying and eliminating a factor associated with increased cancer risk does not necessarily translate into disease prevention. Many women with no known risk factors still get cancer. Thus great numbers of people must change their behaviors for even a few cases of a disease to be prevented. Therefore, to prevent significant numbers of breast cancers, a risk reduction intervention must be safe enough to apply to great numbers of people, or risk calculations must be sensitive enough to pinpoint a smaller group of women for whom the intervention is likely to make a significant difference. Currently there are four main conceptualizations of what constitutes breast cancer prevention and risk reduction within biomedical and popular discourses. These include surveillance, lifestyle changes or what are often referred to as "best bets," prophylactic mastectomy, and chemoprevention.

Throughout the twentieth century, cancer prevention efforts have focused on surveillance via educating the public about the signs and symptoms of cancer and encouraging self-examination and early response. In 1913 the American Society for the Control of Cancer formed a volunteer organization emphasizing the need to overcome cancer fears and taboos and urging people to bring any concerns and early signs of cancer to their doctor (Patterson 1987, 71). By the mid-1940s, this group had changed its name to the American Cancer Society and had revitalized its mission as finding a "cure" for cancer (Patterson 1987). Constructing cancer as a "curable" disease went hand in hand with an emphasis on "early detection" (Lerner 2000).

The idea that early detection was a strategy for secondary cancer prevention (prevention of death) stemmed from the "longstanding theory that breast cancer began as a tiny focus that grew locally in a predictable and gradual manner before spreading" (Lerner 1998, 74). Within this logic, finding and removing breast tumors in their earliest stages was argued to decrease the risk of death.

By the mid-twentieth century, early detection was simultaneously being placed in the hands of women. In 1936 the American Cancer Society formed the Women's Field Army, a group of women volunteers who went door-to-door educating other women about the early signs of cancer and urging them to visit their doctors at the first indication of a problem (Patterson 1987, 122). Messages in the popular media supported this agenda, emphasizing that women, as the caregivers within the family, were the most important "detectors" of cancer and needed to be informed about early cancer signs and intervention (Fosket, Karran, and LaFia 2000). By midcentury, public health messages were encouraging both women *and* physicians to practice breast exams to detect cancer early (Lerner 1998).

However, while the theory and practice of early detection have been deeply embedded in breast cancer prevention efforts, they have also been controversial (for an extensive analysis of this controversy since the 1950s, see Lerner 1999). Not only has the concept of early detection been questioned, but so too have the techniques. Mammography and breast self-exam have maintained a stronghold while always remaining controversial. In 1997, National Cancer Institute director Richard Klausner convened a panel at the National Institutes of Health to craft a consensus on whether all women in their forties should receive regular mammograms or whether screening for all women should begin after age fifty. After two and a half reportedly divisive and hectic days, no consensus was reached: the American Cancer Society (ACS) and the National Cancer Institute (NCI) ultimately came to opposite conclusions. The ACS advocated annual screening beginning at forty; the NCI argued that there were no compelling data to suggest annual screening to all women until after the age of fifty (excepting women who fall into "high risk" categories, variously defined). The resulting official consensus was that women should discuss the issue with their doctors and come to an individualized decision about when to begin annual mammographic screening.

In 2001 mainstream doubt over mammography reemerged after an article published in the *Lancet* reported on a review of several major mammography studies and concluded that mammographic screening does not save women's lives (Olsen and Gøtzsche 2001). This set off a wave of controversy, extensively covered in popular media, and led to serious questioning of the value of mammography by various cancer groups and other experts. Some, such as the National Breast Cancer Coalition (NBCC), have decided to stop recommending that healthy women over fifty undergo regular mammographic screening due to a perceived lack of scientific evidence proving its value (Trafford 2002). In contrast, Tommy G. Thompson, then secretary of the U.S. Department of Health and Human Services, announced that a fifteen-member panel that sets federal policy guidelines on various types of medical tests concluded that all women forty years and older should undergo regular mammographic screening every one to two years (Lerner 2002).

That mammography remains a mainstay within the breast cancer prevention arena despite its many uncertainties and controversies highlights the dearth of viable strategies for breast cancer prevention. The verdict is still out as to how much mortality is actually prevented by early detection and at what age the benefits are greatest. As a result, this breast cancer prevention strategy remains inadequate. Yet it is often presented to women as the only option when they are concerned about their risk of developing breast cancer. This is due in large part to the lack of better alternatives, leading to social and political pressures to hold on to this strategy, despite its inadequacy, until something better comes along.

Other strategies recommended to women concerned about breast cancer risk include dietary practices, exercise, and general "healthy living" guidelines. At a jointly sponsored NCI and ACS conference on breast cancer risk, these general guidelines were referred to as "best bets" (Fosket 1999b). The term "best bets" was used to convey strategies that are not known to decrease risk or prevent breast cancer but are known to be good health practices that *may also* help prevent breast cancer. Best bets involve reducing or eliminating certain factors that may contribute to increased breast cancer risk. Today these factors are most commonly asserted as eating healthy diets (variously defined and understood), exercising regularly, drinking minimal amounts of alcohol, and not smoking. For instance, in 1991 the National Cancer Institute began a campaign conjointly with the food industry called the Five a Day for Better Health Program. This campaign produced and disseminated public health messages encouraging the consumption of five servings of fruits and vegetables a day to maintain good health and reduce cancer risk. The actual role of diet in breast cancer is highly uncertain, with contradictory findings being reported on a regular basis, but nonetheless the idea that eating well may reduce your risk of breast cancer persists. These factors are considered unproved, but beneficial to overall health, often described as things one "should be doing anyway" and thus are unproblematically recommended to anyone worried about breast cancer risk (Fosket 1999b).

Like early detection, best bets are prevention (or risk reduction) strategies that are considered "safe" to suggest to all women concerned about breast cancer risk. Depending on a woman's age and the frequency with which she undergoes mammographic screening, the risks associated with this technology are fairly minimal (though not nonexistent — radiation, the technology used for mammography, is a known cause of breast cancer). However, mammographic screening and breast self-exam as viable strategies for risk reduction are far from noncontroversial. First, although often depicted as such in popular media and public health campaigns, *early detection is not prevention*. One very real danger of depicting these strategies as prevention is that they may falsely suggest that breast cancer is preventable, and those women who still develop breast cancer despite undertaking such risk reduction efforts may feel guilty for not doing enough to prevent the disease. Nevertheless, because best bets and early detection pose minimal physical risks, they are considered safe strategies for all women.

However, other prevention strategies — prophylactic mastectomy and chemoprevention — carry with them significant risks and require clear designations of particular women for whom the benefits are likely to outweigh the risks. Prophylactic mastectomy, removing one or both breasts in an attempt to prevent cancer from developing in them, is the most extreme strategy of breast cancer prevention available. Prophylactic mastectomy has a long history as a method of preventing contralateral breast cancer (breast cancer in the opposite breast) among women already diagnosed with the disease in one breast. Until the practice of radical mastectomy as the assumed treatment for all breast cancer began to be questioned by women's health activists in the 1970s and 1980s (Montini and Ruzek 1989), and clinical trial data began to destabilize the medical legitimacy of the procedure (Lerner 2000), the twentieth century was a time of great enthusiasm for radical surgical interventions in breast cancer and more generally. This, coupled with a devaluing of women in general, and women's breasts in particular, led to often excessively mutilating surgical interventions, including prophylactic mastectomy. Lerner (2000, 40) reports one surgeon in 1951 remarking that women with breast cancer should have the unaffected breast removed prophylactically and that except for "possible sexual enhancement," breasts in women unlikely to breast-feed were "useless."

The theory behind prophylactic mastectomy holds that breast cancer cannot develop if all breast tissue is removed. It seems, however, that the boundaries of the body are not so clearly delineated, and this procedure does not guarantee that a woman will never develop breast cancer, which can occur in scar tissue and at other unexpected locations. Studies of the effectiveness of prophylactic mastectomy have been equivocal (Newman et al. 2000) and the procedure remains controversial in the United States because it is such a radical measure and because assessing breast cancer risk is uncertain. Thus the possibility exists that women who would never have developed breast cancer will nonetheless opt to surgically remove their breasts.

The capacity for prophylactic mastectomy to reduce breast cancer risk, coupled with the recent phenomenon of women being diagnosed as high risk via genetic testing, family histories, and other risk assessment technologies, has made mastectomy a legitimate strategy for prevention among many healthcare practitioners. However, removing one's breasts for the potential prevention of a potential disease carries with it physical, social, and psychological risks of numerous kinds, and thus the procedure remains quite controversial. Significantly, however, stories about women undergoing prophylactic mastectomies paved the way for chemoprevention to appear as a much more reasonable solution.

The Emergence of Chemoprevention

While primary prevention is chiefly thought of as the identification and removal of cancer-causing agents, another strategy aims to prevent cancer through synthetic or natural compounds that prevent, reduce, or reverse carcinogenesis (Fisher et al. 1998). Chemoprevention is based on the concept that biologically active compounds can be administered not only as tumor-destroying chemotherapy but also as tumor-preventing chemotherapy. While chemoprevention has only recently emerged in the realm of breast cancer, it has a longer history in other health realms.

The idea of administering daily doses of pharmaceuticals as a health maintenance strategy is conceptually interesting. Drugs developed as treatments for health problems given instead to healthy populations as a way to *stay healthy* highlight the intense biomedicalization of society such that technoscientific biomedical interventions are increasingly normalized as part of everyday life (Clarke et al. 2000, 2003). While chemoprevention for breast cancer is a new phenomenon, chemoprevention in general is not. In fact, the birth control pill marks the first time that women were targeted for daily doses of pharmaceuticals, not as a treatment but as a preventative (Gossell 1999; Merkin 1976).

Chemoprevention has also long been a part of heart disease prevention, where treating risk and treating disease increasingly converge. Within the realm of heart disease, drugs for lowering cholesterol and blood pressure have long been used to help prevent future coronary heart disease and stroke. These drugs are, in effect, chemoprevention, since high blood pressure and cholesterol are risk factors for cardiovascular diseases. Yet their ubiquitous, taken-for-granted use has blurred the boundaries between risk and disease, and both hypertension and hypercholesterolemia have been constructed as disease conditions in themselves, treatable with pharmaceuticals (Shim 2002a).

While the birth control pill and heart disease prevention drugs popularized and normalized the idea of otherwise healthy but at-risk people taking daily doses of prescription drugs by targeting specific health issues, the normalization of ingesting compounds for maintaining health was also accomplished through the widespread use of over-the-counter nutraceuticals (nutritional supplements like vitamins and minerals) for generalized health maintenance. In the case of cancer, the 1980s saw an increased interest in chemoprevention via nutraceuticals. In particular, the protective aspects of eating vegetables were explored, and clinical trials began to look at the impact of antioxidants such as vitamin E on cancer prevention (Fosket 1999a). Today nutraceuticals aimed at transforming specific aspects of the body—from memory to sexual libido to the immune system—are available at most mainstream markets, produced by large corporations as well as small natural-foods companies. The idea of manipulating the body to improve general health through daily doses of medications is also very much a part of the rhetoric and practices of ingesting pharmaceuticals for health maintenance.

These examples of chemoprevention situate the emergence of breast cancer chemoprevention within a broader biomedicalizing shift in the legitimate meaning of drugs from treatments to prevention. This is significant history for breast cancer chemoprevention because, as we shall see, tamoxifen's transformation from treatment to prevention has been marked by conflict and controversy. The shift toward biomedicalization and broader acceptance of pharmaceuticals for health maintenance has helped legitimate the use of tamoxifen for prevention in the face of such controversy.

Tamoxifen's Rocky Transformation from Treatment to Prevention

The only FDA-approved chemoprevention drug for breast cancer today is tamoxifen. Tamoxifen was first discovered in 1962 by Dr. Arthur L. Walpole, head of the fertility control program at the London-based Imperial Chemical Industries (ICI). This company, which formed Zeneca Pharmaceuticals in 1993, now called AstraZeneca, is the current marketer of tamoxifen (under the brand name Nolvadex [Tamoxifen Citrate]) (Jordan 1994). Originally referred to as ICI 46,474, tamoxifen's ability to interfere with estrogen action led it to first be tested (unsuccessfully) as a female contraceptive. V. Craig Jordan began studying the drug when he was a new faculty member in the Department of Pharmacology at the University of Leeds in 1972. In securing an outside examiner for his Ph.D. thesis on anti-estrogens, Jordan became connected with Dr. Arthur Walpole, opening the door to a collaboration that would result in the development of ICI 46,474 into tamoxifen. When Jordan and Walpole met in the early 1970s, ICI 46,474 had shown some effects as a treatment for advanced breast cancer in a preliminary clinical study conducted by ICI, and Jordan and others began to pursue it. In 1973, ICI 46,474, by then referred to as tamoxifen (or by its brand name, Nolvadex), was approved for the treatment of advanced breast cancer in Britain. It was not initially approved in the United States until the end of 1977 (Jordan 1994, 8).

Tamoxifen is sometimes described as an "anti-estrogen" because it competes with estrogen for binding sites in target tissues such as women's breasts. Further study revealed that tamoxifen *selectively* interferes with estrogen actions. The ability to selectively compete with estrogen explains why tamoxifen helps to reduce the incidence of breast cancer: in some organs, notably the breasts, tamoxifen occupies the estrogen-receptor site and thus blocks endogenous estrogens that would otherwise bind to the site (Overmoyer 1999). Tamoxifen blocks estrogen at the cellular level of the estrogen receptor. About 60 to 70 percent of breast cancers are classified as ER (estrogen receptor) positive (Lippman and Brown 1999). While it is clear that tamoxifen is most effective for treating ER-positive tumors, the ER changes in breast cancer are complex and poorly understood, making this an area considered uncertain and in need of further investigation (Lippman and Brown 1999).

Because estrogen is believed to encourage the growth of breast cancer, by blocking estrogen from the breasts or from a specific breast tumor or its surgical remains in the body, tamoxifen discourages cancer growth. In sites other than the breast, however, there can be significant problems with tamoxifen because it mimics estrogen. Thus, paradoxically, tamoxifen can *cause* cancer in the lining of the uterine wall because at that site it produces estrogenlike effects. The way in which tamoxifen *selectively* interferes with or mimics estrogen in different parts of the body makes it part of a class of drugs called selective estrogen receptor modulators, or SERMs.

In 1981 the U.S. National Surgical Adjuvant Breast and Bowel Project (NSABP) initiated Protocol B-14 to evaluate the efficacy of tamoxifen for women with primary breast cancer, histologically negative nodes, and estrogen-receptor-positive tumors. In 1989 the results indicated a significant prolongation of disease-free survival for the tamoxifen group (Fisher et al. 1989). Regarding tamoxifen as prevention, the results indicated a 47 percent reduction in contralateral breast cancers (new primary breast cancer in the opposite breast) after five years on tamoxifen. If tamoxifen could reduce the incidence of contralateral breast cancer, it stood to reason that it might also reduce the incidence of first-time breast cancer. These results helped prompt the initiation of the Breast Cancer Prevention Trial.

The Breast Cancer Prevention Trial NSABP-P1 (BCPT) was initiated in 1992. This study was a double-blind, randomized, placebo-controlled clinical trial with the primary objective of determining whether five years of tamoxifen treatment would reduce the incidence of invasive breast cancer in women defined as high risk for the disease, but otherwise healthy. Secondary objectives were to evaluate the incidence of ischemic heart disease, bone fractures, endometrial cancer, pulmonary embolus, deep-vein thrombosis, stroke, and cataracts. High-risk women were defined as (1) women over sixty or women who had had lobular carcinoma in situ (LCIS), and (2) women between the ages of thirty-five and sixty whose risk was calculated to be equal to that of women over sixty by the Gail model (see Fosket 2004 for an analysis of this risk calculation model). In the BCPT, 13,388 women were randomized to receive either tamoxifen or a placebo. The study was funded by the National Cancer Institute, the National Heart, Lung, and Blood Institute, and the National Institute of Arthritis and Musculoskeletal and Skin Diseases. Zeneca Pharmaceuticals (now Astra-Zeneca) provided tamoxifen and the placebo (Overmoyer 1999).

From the outset, the BCPT was plagued by ethical controversies. However, current accounts of the trial read as if none of the controversies ever occurred. Debates, contentions, and outright scandals have been smoothed over in retrospective accounts. To piece together the following account of the controversies, I spoke with researchers and activists, searched popular-media accounts and transcripts of congressional hearings, read letters to the editor in prominent scientific journals, and uncovered journal articles that both critiqued and defended the trial. Important pieces of the story — including how and why the issues were resolved enough to continue with the trial — may still be missing.

Controversies began in the design phase, where there was much concern regarding the ethics of designing a prevention trial to test a drug on *healthy* women that had until then only been used for cancer therapy. How to assess and evaluate risk in a prevention context was a critical issue in the scientific literature published at the time (e.g., Bush and Helzlsouer 1993; Love 1995). Tamoxifen was known to have some toxicity, though that toxicity was considered relatively mild in the context of cancer *treatment* — but this calculated weighing of potential risks to possible benefits would radically change upon tamoxifen's entrance into a *prevention* arena. The scientific media present an interesting credibility struggle during the 1990s, with critics voicing concerns (e.g., Williams 1995) or outright objections to the prevention trial, and supporters publishing articles that justified and explicated the design of the trial and the principle of chemoprevention (e.g., Jordan 1994). Additionally, the idea of giving a toxic drug to women who did not have breast cancer was heavily debated at several congressional subcommittees before approval for the trial was granted on April 29, 1992 (Overmoyer 1999).

The inclusion of premenopausal women in the BCPT was also a site of particular controversy during the trial's design because tamoxifen had previously been tested more thoroughly on older women. It was also controversial due to the possible increase in ovarian steroidogenesis in younger women, enhancing their later risk of ovarian cancer and potentially negating the preventive effect of tamoxifen. Finally, the possibility of pregnancies in younger women and feared teratogenesis of tamoxifen were of concern (Jordan 1995). In defense of including this younger group, Jordan (1995) argued that in animal research, preventive interventions are most effective in younger animals and that evidence appears that early intervention may also be best in humans, and thus the inclusion of premenopausal women in the BCPT was argued to be an ethically justifiable design decision.

Early in the conduct of the BCPT, informed consent became another key site of controversy when it was revealed that, when the trial began in 1992, the informed consent form failed to include information on what was then known about the risks of uterine cancer or life-threatening blood clots (Sherman 1998). Later in 1992, the National Women's Health Network and a congressional committee investigated the trial and reported that 68 percent of the consent forms either omitted or altered one or more key points from the consent form approved by the NCI committee on human subjects (Sherman 1998). Additionally, experimental evidence of tamoxifen's carcinogenicity was not included in the consent forms (Tomatis 1995). In response, a principal investigator of a tamoxifen prevention study in Italy wrote a letter to the editor in *Lancet* (Costa 1993, 8868) stating, "The study design foresees the accrual of only women who have had hysterectomy until the issue of endometrial carcinogenesis is clarified."

After these initial problems, the trial ran smoothly until 1994, when tamoxifen again became a matter of public controversy. This time, new knowledge came to light that the risk of endometrial cancer associated with tamoxifen use was far greater than reflected in the original consent forms or on the drug's warning labels (Tomatis 1995). In response, enrollment in the trial was suspended for one year, and the trial underwent a congressional review. This controversy coincided with larger problems at NSABP involving the center's director, Dr. Bernard Fisher, who had recently been asked to step down as a result of his role in falsifying data on a large-scale breast cancer treatment study conducted by NSABP. As a result, enrollment for fourteen NSABP clinical trials was halted, including the BCPT. The suspension was initiated by the National Cancer Institute, which acknowledged deficiencies in its own efforts to monitor the quality of the clinical trials and planned a thorough review (Altman 1994b). In the wake of the scandal, as questions emerged about the BCPT, some critics, such as Senator Dianne Feinstein, then co-chairwoman of the Senate Cancer Coalition, thought the BCPT should continue under the guidance of some other research organization (Altman 1994a).

Ultimately, however, after a congressional review of the BCPT in 1994, the FDA concluded that the potential benefits of tamoxifen to decrease breast cancer risk outweighed the risks of uterine cancer and recommended that the trial reopen. Zeneca sent letters to both doctors and women enrolled in the BCPT informing them that some women taking tamoxifen had developed uterine cancer (Sherman 1998), and warning materials on tamoxifen were upgraded to include this information. Accrual rates slowed after 1994 as a result of the bad publicity (Overmoyer 1999). In response, a Participant Advisory Board (PAB) was established consisting of sixteen women meant to "represent all women enrolled in the study" (Psillidis, Flach, and Padberg 1997). The inclusion of community advisory boards on clinical trials had by this time become a common strategy, initiated by AIDS activists and those responding to them (Epstein 1996). The PAB worked with recruitment sites to increase enrollments, especially targeting communities of color (ultimately only 3 percent of trial participants were women of color).

Despite these efforts, controversy continued. In 1995 tamoxifen was declared a known carcinogen in the state of California under Proposition 65, a law requiring a public list of all known carcinogens. The identification of increased endometrial cancer in women taking tamoxifen led to its investigation by the state's Carcinogen Identification Committee (CIC), which ultimately made a unanimous decision that the drug be listed as a known carcinogen (Marshal 1995). Controversy abounded in the wake of this decision, with researchers committed to tamoxifen contesting the decision and claiming it would put patients' lives at risk, as it might turn women away from tamoxifen as a breast cancer treatment. Marshall (1995, 910) cites John Glick, director of the cancer center at the University of Pennsylvania, as saying, "Many more patients would die as a result of their fear of taking tamoxifen than ever would die as a result of getting endometrial cancer." Marshall described this kind of interference with the CIC's decision as unprecedented.

In March 1998 the Breast Cancer Prevention Trial released initial estimates that indicated a 45 percent reduction in the incidence of breast cancer among high-risk women who took tamoxifen for a median duration of 3.5 years. As a result of these findings, the independent Endpoint Review, Safety Monitoring, and Advisory Committee (ERSMAC) recommended that the trial terminate to provide public access to this new information regarding tamoxifen. At this time only 57 percent of those enrolled had completed four or more years on tamoxifen. The next day, key researchers from the National Surgical Adjuvant Breast and Bowel Project working on the BCPT met with Richard Klausner, then director of the National Cancer Institute, to discuss the findings and ERSMAC recommendation. At this meeting, they agreed to end the trial and reveal which women had been taking the placebo, providing them with the opportunity to take tamoxifen. The next day, the National Cancer Institute posted the initial results of the BCPT on the Internet, before the investigators had an opportunity to complete their analysis and go through the peer-review process normally followed in the publication of clinical trial results. The NCI's unprecedented decision to circumvent the usual standard operating procedure for reporting trial findings raised suspicions about their validity; however, most (but not all) of these were laid to rest when the actual peer-reviewed publications came out.

The final results of the BCPT were published in September 1998 (Fisher et al. 1998). These results indicated that 124 women developed breast cancer (invasive and noninvasive) in the tamoxifen group, and 244 women developed breast cancer in the placebo group. However, tamoxifen was also found to increase the risk of endometrial cancer: 36 women developed uterine cancer in the tamoxifen group, and 15 in the control group (37 percent of women in each group had undergone hysterectomies before enrollment). Tamoxifen also increased the risk of pulmonary embolism (blood clot in the lung): 18 cases in the tamoxifen group, 6 in the control group. Finally, tamoxifen increased the risk of deep-vein thrombosis (blood clots in major veins): 35 in the tamoxifen group and 22 in the control group. The large numbers of these other life-threatening illnesses have led some to call the use of tamoxifen to prevent breast cancer "disease substitution" (Sherman 1998). Other side effects indicated by the trial include depression, irritability, vaginal dryness, hot flashes, visual disturbances, fatigue, and weight gain.

In a subsequent analysis of the risks and benefits of tamoxifen as indicated by the results of the BCPT, Gail and colleagues (1999) found that for white women under fifty, the benefits of tamoxifen for breast cancer risk reduction outweigh the risks associated with the drug. There were no statistically significant increases in stroke, pulmonary embolism, or endometrial cancer for these younger white women. For older white women, however, Gail and colleagues report that the benefits of tamoxifen do not always outweigh the risks for women at increased risk of pulmonary embolism, stroke, or endometrial cancer. These figures are known only for white women (because only 1.7 percent of the participants in the BCPT were African American), yet drawing from statistical literature on black women's rates of stroke, pulmonary embolism, and endometrial cancer, Gail and colleagues assert that for black women, the benefits of tamoxifen are also greatest in younger women (Gail et al. 1999).

The American Society of Clinical Oncology (ASCO) 1999 Technology Assessment of tamoxifen for breast cancer risk reduction recommended tamoxifen be offered to women whose risk is calculated as "high" vis-à-vis the Gail model, but emphasized the need for individual women and their healthcare providers to weigh their personal estimate of benefits against the risks associated with the drug (Chlebowski et al. 1999). The assessment also highlighted that we do not yet know whether tamoxifen reduces the risk of deaths from breast cancer because the BCPT did not follow women long term and the trial was terminated before its planned completion.[1] Richard Klausner, director of the National Cancer Institute, stated in a talk in San Francisco in 1999 that tamoxifen was a "first step" and compared it to the Wright brothers' plane—an important breakthrough, but "you wouldn't want to take it on your next vacation" (Fosket 1999c). This statement from a supposed advocate of chemoprevention was quite shocking for the degree to which it, perhaps inadvertently, problematized tamoxifen as an effective strategy.

While the Breast Cancer Prevention Trial was indicating favorable results in North America, clinical trials taking place in Europe were telling a different story. Both the Italian Tamoxifen Prevention Trial (Veronesi et al.

1998), which began in 1992, and the Royal Marsden Study (Powles et al. 1998) in the United Kingdom, which began in 1986, found no significant reduction in breast cancer incidence for women taking tamoxifen. Within the United States, these discrepancies are commonly explained away as the result of too few numbers in the European studies, confounding factors (such as the concurrent hormone replacement therapy permitted in the Italian trial), differential compliance, and different conceptualizations of "high risk" (Chlebowski et al. 1999; Fisher et al. 1998; Lippman and Brown 1999; Pritchard 1998).[2] In contrast, many European reports depict the North American Breast Cancer Prevention Trial, especially its early unmasking and publication of results on the Internet, as prematurely optimistic or, worse, as bad science. Ultimately, the BCPT was the only study taken completely into account in decisions regarding the U.S. FDA's approval of tamoxifen, highlighting the dominance of credibility of U.S. biomedical research to guide U.S. practice.

Despite the many controversies surrounding the BCPT, the U.S. FDA approved tamoxifen for the "reduction of risk" of breast cancer incidence. Due to the many limitations cited, tamoxifen was explicitly *not* approved for "prevention," but rather for "the short-term reduction in the risk of breast cancer in high-risk population" (Napoli 2001, 33). When reviewed by the FDA Oncologic Drugs Advisory Committee on September 2, 1998, tamoxifen was groundbreaking in being the first drug to be considered for the *prevention* of breast cancer, and the subject of risk loomed large as participants in the hearing earnestly debated the efficacy of tamoxifen compared to its risks. Ultimately, through linking the BCPT with credible tamoxifen studies of the past and highlighting the momentous events that brought tamoxifen to this point, the controversial and much-criticized aspects of the BCPT were set aside and, though not resolved, did not prove to be major obstacles to approving tamoxifen. This "setting aside" of controversy was a major strategy used by advocates of tamoxifen. However, these and other critiques did linger and contributed to the approval of tamoxifen for *risk reduction* instead of *prevention*.

Conclusions: Biomedicalizing Prevention

Despite not officially referring to tamoxifen as prevention, the events culminating in the FDA's approval of tamoxifen for otherwise healthy but high-risk women have had important consequences for shifting the way prevention is thought about and acted on in relation to breast cancer. This case is a major exemplar of biomedicalizing tendencies in the prevention arena. Headlines in the aftermath of tamoxifen's approval for risk reduction boldly made statements such as "Researchers Find the First Drug Known to Prevent Breast Cancer" (*New York Times*, April 7, 1998, A1). Researchers and funding agencies jumped on the chemoprevention bandwagon, initiating and supporting myriad large- and small-scale breast cancer chemoprevention clinical trials in the years immediately following approval of tamoxifen. Perhaps most importantly, treating risk with pharmaceuticals increasingly came to be seen as a legitimate, feasible, and worthwhile means of preventing disease both in the realm of breast cancer and more widely. Chemoprevention thus marked a significant "win" in the long "war on cancer" that biomedicine was perceived to be losing. It changed what was understood as possible in terms of preventing cancers and further focused the biomedical gaze on the interior of the human body as the most effective site of intervention for prevention.

Within processes of biomedicalization, social contexts are often obscured in understandings of health and illness as technoscientific definitions come to prevail. In the arena of breast cancer, I argue that this has meant a focus on the individual body as the source of risk, rather than on the socioenvironmental world. These trends exemplify what John McKinlay (1974/1994) calls "downstream medicine." A downstream approach to solving health problems means that interventions focus on the afflicted individual without awareness of and attention to the upstream causes and contexts of the afflictions. Such an approach potentially shifts attention away from efforts at identifying and eliminating the elements that put women at risk for breast cancer in the first place.

Chemoprevention seeks to "prevent" breast cancer through pharmaceutical intervention into selected bodies already constructed as "at risk," rather than through primary identification, control, and elimination of exposure of whole populations to carcinogens. Through biomedicalization, the shift toward downstream medicine—already well under way—is further facilitated within the realm of breast cancer prevention. The curative model of health is now being applied to prevention. Since we are already accustomed to treating symptoms of individuals with technoscientific products like pharmaceuticals, the idea of treating risk the same way is a logical extension of normative biomedicalized ideology and practice. The deep irony of approving a drug like tamoxifen with health risks that include known carcinogenesis for the prevention of cancer perhaps surpasses the illogic of downstream prevention. It creates an almost circular pattern of risk and prevention where the treatment for one risk leads to new risks that will inevitably need to be treated. The profit-making possibilities are obvious, and spiraling patient careers are likely outcomes.

In this chapter, I have traced the long and winding path that has resulted in the emergence of pharmaceutical treatments for breast cancer risk as legitimate options within the breast cancer prevention arena. Chemoprevention remains a controversial and contested practice. Conflicts and debates have followed tamoxifen since it first began to be tested as a chemoprevention agent regarding its many life-threatening health risks and the ethics of testing such a risky drug on healthy populations of women to prevent potential disease.

Yet the use of drugs to treat breast cancer risk has effectively been legitimated, either already approved by the FDA or currently being tested in major clinical trials on healthy groups of women. Despite the FDA's explicit specification of tamoxifen as a *risk-reducing*, not a preventive, strategy, the drug and the BCPT are widely seen to provide proof of the concept that chemoprevention of breast cancer is possible, feasible, and effective. Controversy is a relative and socially situated phenomenon, and the particular contexts within which chemoprevention is located contribute to its robustness in the face of active contestation.

First, chemoprevention is situated within a breast cancer prevention arena in which few other options exist. The long and arduous history of uncovering causes of breast cancer has led in the contemporary era to very few prevention strategies. As a result, chemoprevention has the distinction of seeming like a more reasonable option compared to prophylactic mastectomy and may also seem like the only real option for many women concerned about their breast cancer risk. Simultaneously, this lack of knowledge in the realm of breast cancer prevention makes for greater investment in chemoprevention by those whose reputations are staked in the cancer wars. That is, entities like the National Cancer Institute that have been waging a mostly unsuccessful war with breast cancer have presented chemoprevention as a major win in the battle to prevent breast cancer. Thus the lack of alternative prevention strategies maximizes both the focus on, and credibility of, chemoprevention despite the controversies surrounding it.

Additionally, I argued that breast cancer chemoprevention is situated within a social context in contemporary U.S. society where pharmacological agents aimed at maintaining health, as opposed to treating illness, are already well established in multiple domains. In particular, birth control pills, heart disease prevention medications, and nutraceuticals are all sites where the ingestion of drugs by healthy people to prevent illness (or pregnancy, in the case of the pill) is widely normalized. In the current U.S. climate of direct-to-consumer advertising of drugs, there are many additional examples of pharmaceuticals marketed to and taken by healthy people as strategies to *stay healthy*.

In this context, then, chemoprevention for breast cancer is part of a larger shift in the meaning of medications from something to treat illness toward something that everyone can take to avoid illness and improve quality of life. As a part of this larger shift, the controversial nature of treating healthy women with risky drugs to prevent potential breast cancer is nowhere near as contentious as it would likely have been fifty, or even fifteen, years ago. That is, while tamoxifen exemplifies biomedicalization in the many ways laid out here, it is also the case that it is very much a product of biomedicalization. Set against a backdrop of an already medicalized and increasingly biomedicalized cultural and social milieu, tamoxifen does not come across as a huge theoretical leap.

Finally, the ways in which the biomedicalization of prevention has been legitimized with the emergence of tamoxifen and the *simultaneous* persistence of controversy that surrounds that emergence reveals something further about biomedicalization. Unlike medicalization, which was often critiqued for its heavy-handed control of hapless people, biomedicalization seems to exist in continuity with self-reflexive doubts and critiques. There seems to be increased dialogue around the permeation and growing ubiquity of biomedical logics and ways of understanding in popular culture. Yet while biomedicine's reach and tendencies are increasingly transparent, this does not detract from or decrease its potency in everyday life.

Notes

1 It was possible, for instance, that while tamoxifen reduced the incidence of less-aggressive tumors, it might not have reduced the incidence of tumors that are resistant to tamoxifen treatment — tumors that may ultimately prove more life threatening. It was also possible that tamoxifen was not preventing but treating very early breast cancers, also indicating that its results could be short-lived. None of these possibilities were adequately assessed scientifically.

2 The Royal Marsden study classified high-risk status on individualized family history (Powles et al. 1998). The researchers on the U.K. trial suggest that their markedly different findings may be, in part, based on the likelihood that a larger percentage of their population (compared to the BCPT) had a genetic mutation such as BRCA1 and 2; estrogen may not be as important a factor in the genesis of cancer in women with such mutations, thus making tamoxifen a less-effective prevention for these women (Powles et al. 1998).

Jackie Orr

13 / Biopsychiatry and the Informatics of Diagnosis

GOVERNING MENTALITIES

> Governments are as responsible for the mental health as for the physical health
> of their citizens. . . . Governments, as the ultimate stewards of mental health,
> need to set policies—within the context of general health systems and financ-
> ing arrangements—that will protect and improve the mental health of the
> population.—World Health Report 2001

> The managing of a population not only concerns the collective mass of phe-
> nomena, the level of its aggregate effects, but it also implies the management
> of population in its depths and its details.—MICHEL FOUCAULT

Managing the details and potentially unruly depths of the "mentality" of populations is a central aim of contemporary techniques of governmentality.[1] The post–World War II founding of the U.S. National Institute of Mental Health (NIMH) announced a new purposefulness and promise of efficiency in the campaign to regulate the mental health and disease of citizen-subjects: "The guiding philosophy which permeates the activities of the National Institute of Mental Health," wrote its founding director in 1949, "is that prevention of mental illness, and the production of positive mental health, is an attainable goal. . . . Since this must be done as rapidly and economically as possible, *techniques for a mass approach to the problem* must be developed" (Felix 1949, 405, italics mine). As the state-sponsored ambition to produce "positive mental health" for all is generalized by 2001 into the World Health Organization's global address to governments as "the ultimate stewards of mental health," it seems reasonable to wonder—with a certain sense of historical dis-ease — what *techniques for a mass approach* to the problem of governing the mental health of entire populations have been developed in the last fifty years. How has disordered "mentality" in its aggregate effects and its symptomatic depths been made into an object of governmentality? What technologies of knowledge and power are today at work transforming the heterogeneous, elusive languages of mental suffering into the robust data useful for its rationalized regulation and administrative control?

The emergence of U.S. biopsychiatry in the last decades of the twentieth century signals the persuasive consolidation of one particular mass approach to problems of psychic distress. Situated within the multisited, postmodern processes of biomedicalization described by Clarke et al. (2003, this volume), biopsychiatry embraces a medicalized model of mental disorders while claiming a scientific status for contemporary psychiatric practices of diagnosis and treatment. At the same time, biopsychiatry has become all but inseparable from the funding and financial structures, research agendas, advertising and marketing campaigns, and globalizing ambitions of the transnational pharmaceutical industry (Breggin 1991; Healy 1996, 2002; Horwitz 2002, 211–12; Metzl 2003; Petryna, Lakoff, and Kleinman 2006). Indeed, the industrial production and consumption of psychopharmaceutical drugs appear to have delivered on the NIMH's mid-twentieth-century goal of efficient, cost-effective techniques for a mass approach to mental health.

Biopsychiatry in the United States has played a lead role in ushering in an era of mass(ive) "pharmaceutical governance" (Biehl 2004) of mental health and illness through three interrelated accomplishments. First, the standardization and operationalization of the official language of psychiatric diagnosis, established in the 1980 edition of the *Diagnostic and Statistical Manual of Mental Disorders* (*DSM-III*), introduces a new order of things in the entangled realms of psychiatry and psychotherapy, medicine, the pharmaceutical industry, the legal system, the insurance industry, social and self-identity, and popular discourse (see Kirk and Kutchins 1992; Lowe 1995; Wilson 1993). Hailed by its promoters as a "significant affirmation on the part of American psychiatry of its medical identity and its commitment to scientific medicine" (Klerman 1984, 539), *DSM-III* offers a purportedly objective, empirically oriented classification system for mental disease, where each diagnostic "thing" (each specific disorder) is defined by a systematized, standardized set of observable criteria (see American Psychiatric Association [APA] 1980). Second, bolstered by this new epistemic order of diagnostic things, biopsychiatry supports a popular commonsense and professional consensus that each discrete, mentally disorderly thing has its own discrete and targeted treatment. Each scientifically speakable disease has its own specific—and presumably scientific—cure. Third, biopsychiatry naturalizes this coupling of standardized psychiatric diagnostic entities with scientifically sanctioned treatments by modeling mental disorder as an expression of an underlying brain or biological dysfunction. Like the medical model of organismic disease that it tries to mirror and extend, the biopsychiatric imaginary of brain dysfunction as the fundamental context for understanding psychic distress legitimates biological intervention as an appropriate, even exemplary, technique for regulating disorderly mentalities. Psychopharmaceutical drugs as effective technoscientific treatments for a wide range of psychic and emotional disturbances have become, by the early twenty-first century, a normalized practice in biopsychiatry, general medicine, and everyday life.

The contemporary routinization of the psychopharmaceutical governance of mental suffering and psychic difference converges historically not only with the ascendance of U.S. biopsychiatry and the related technoscientific, economic, and social contexts associated with processes of biomedicalization but also, I argue, with broader post–World War II transformations in the very techniques of governmentality itself. Governmentality, as Foucault theorized its eighteenth-century emergence in European nation-states, marks the "birth of a new art" for exercising power and constituting knowledges (Foucault 1978/2000, 217). Organized explicitly around the problem of the health and welfare of the population and the possibilities of knowledge materialized by social statistics (rates of birth, disease, and death; the patterned movements of people, epidemics, and capital), the art of modern governing builds on a biopolitical regime of power/knowledge in which the life—or "bios"—of a population is perceived as a new *subject* of needs and desires, and a new *object* of perpetual administration and government.

If, as Clarke et al. (2003, this volume) suggest, today's institutional and technoscientific networks of computerized information constitute one key feature of "biomedical governmentality," then it may be useful to consider how computer and information technologies more broadly are shaping contemporary practices of governing the life and the lived "mentalities" of populations. How to think about biomedical and biopsychiatric governmentality as a symptom of something we might call "cybernetic governmentality," or the birth of a new art of regulating populations through the post–World War II communication and information sciences—christened "cybernetics" by their founding figure, Norbert Wiener (1948, 1950)? Derived from the Greek word *kybernetics* (referring to automated mechanisms of governing or steering, as in a steamship), cybernetics—which influenced the design of the first computers and helped launch the informatic infrastructures of power in which we move today—may have shifted late-twentieth- and early-twenty-first-century techniques of governmentality in directions that Foucault never adequately theorized. Cybernetic governmentality, founded on new material and imaginary circuits of communicative feedback and information exchange, extended through techniques of electronic automation and computer simulation, enfleshed in flexible networks of modulated and continuous control, may touch on what the sociologist C. Wright Mills (1959, 166) struggled to theorize in 1959 as an emergent "postmodern" era. Electrified by new forms of social power, cold war U.S. society was, Mills feared, drifting toward a political economy and a social psychology dominated by electronic communications technologies. Even as Arnold Schwarzenegger, telematic cyborg, today governs one of the more powerful economic and cultural regions on the planet, cybernetics as a technology of social governance remains a curiously undertheorized concern.

So while my limited aim in this chapter is to tell a story of the rise of computer-simulated diagnosis of mental disorders—or the "informatics of diagnosis"[2]—as a key element of U.S. biopsychiatry and the biomedicalization of psychic distress, I want to situate that story within a broader, speculative history of cybernetic techniques for governing mentalities. Simulation, automation, and informatic modeling today are becoming pervasive, everyday techno-social practices, assembling humans, computers, information, "mentalities," and imaginaries of control into a flexible array of social-cybernetic forms and technoscientific formats. How to reassemble our own theoretical imaginations and empirical investigations into an adequate analytics for mapping such shifts, even as those shifts partially and perceptually remap us? At the empirical heart of my story is the rewriting in 1980 of the *DSM-III*, the APA's official psychiatric diagnostic classification system, in an informatic language of codable symptoms and computerizable criteria; I trace this transformation to over a decade of experimental efforts to develop computer simulations of psychiatric diagnosis and to automate couplings of specific diagnoses with specific drug treatments. The *DSM-III*'s informatics of diagnosis makes possible the first national epidemiological studies of the prevalence of discrete, differentiated mental disorders in the U.S. population, generating "mentally disorderly" statistical subpopulations that can become new objects of biomedical governance and new subjects of psychopharmaceutical desires. My theoretical aim is to locate this transformation of the *DSM* and the ascendance of U.S. biopsychiatry within a historically specific techno-social dream of automated, informatic control of human mentalities—a (social) science fiction dream of cybernetic power that begins at the uncanny crossroads of the U.S. military and a public mental institution. Here psychiatry, cybernetics, and the cold war U.S. campaign to inhabit outer space together start to imagine the technologics necessary for the cybernetic regulation of psychophysiological processes—in institutionalized mental patients and human astronauts—and the informatic infrastructures of their everyday governance.

Psycho-cyborgs at Rockland State

> After the implosion of informatics and biologics, simulation is not derivative and inferior but primary and constitutive. "All life is an experiment."
> —DONNA HARAWAY

> Not only simulation of central nervous system activity, but a direct tie-in—sort of a psycho-cyborg—is at least conceivable.
> —KLINE and LASKA

The ruins of Rockland State mental institution are today available for online viewing. On my computer screen, a series of digital images displays the abandoned, dilapidated buildings that once housed nearly eight thousand mental patients; an empty dentist chair in a vacant room, barred window in the background; in the children's ward (the caption tells me), a small steel bed frame lying on its side, and a note inside an open drawer —"GET OUT"—scrawled in crayon.[3] Rockland State Hospital in Orangeburg, New York, is still open for business, and the website includes a digitalized photo of the recently built main hospital complex that now towers over the ghostly sprawl of decaying buildings. Here, in the ruins of informatic images and electronic traces of absence, a history of cybernetic governmentality flickers through the very framings of cyber-technology itself. This is a history of the present, then, and of techno-social forms of perception and power that inform the very subjects we are becoming.

In 1960, two researchers from Rockland State travel to Brooks Air Force Base, Texas, to present their paper "Drugs, Space, and Cybernetics: Evolution to Cyborgs" (Kline and Clynes 1961). They attend a symposium, organized by the U.S. Air Force, addressing the unprecedented "psychic stressors" and behavioral challenges of sending humans into outer space. An array of "informed physicians and scientists" is invited by the U.S. military to identify and solve the interdisciplinary, biopsychosocial problems of manned space flight: "Man's behavior is a joint function of many kinds of influences arising both inside and outside the body, and . . . the understanding and control of this behavior can be achieved only through the joint efforts of all the life sciences" (Benson 1961, vi). Paper topics at the symposium range from the neurophysiology of stress to experiments with monkeys inside U.S. missiles, hypnosis as a space tool, and the psychosocial problems of small groups.

In their contribution to the Air Force conference, Nathan Kline and Manfred Clynes of Rockland State Hospital introduce for the first time the figure of the "cyborg"—an imagined assemblage of *cyb*ernetic and *or*ganismic processes necessary for the successful control of human behavior and psychophysiological functioning in extraterrestrial space (Kline and Clynes 1961, 347–48). While the cyborg is a kind of hybrid techno-human creation, it is also Kline and Clynes's name for a method—"the Cyborg technique"—which supplements or simulates bodily processes via the "biochemical, physiological, and electronic modification of man's existing modus vivendi" (346). Pharmacology is the most significant cyborg technique imagined by Kline and Clynes for the automated, self-regulating control of the cybernetic man-machine system in outer space. Modeling the central nervous system as an automated cybernetic mechanism, Kline and Clynes envision a science fiction series of pharmaceutical devices for simulating biological and psychological equilibrium (or homeostasis) through a control loop of constant informational feedback. For example, should the unusual stress of long-term space flight precipitate a psychotic episode in the cyborg, the authors suggest the possible administration of psychopharmaceutical drugs via remote control from earth (369–70).

These early fantasies of cyborg control techniques are not the ravings of mad scientists but the cheerful astral projections of two scientists of "madness" whose research activities on spaceship Earth are profoundly influential and richly funded. Nathan Kline, a psychiatrist and director of the research department at Rockland State Hospital, plays a major role in establishing the new field of psychopharmacology for the treatment of mental illness in the 1950s and 1960s (see Kline 1956). As director of the Dynamic Simulation Lab at Rockland, Manfred Clynes conducts award-winning research on "the organization of the body's nervous system and its cybernetic control" and the basic "brain algorithms" underlying the "biocybernetics" of emotions (Clynes 1995, 45; also Clynes 1973). While the cyborg is originally conceived as a solution to problems in outer space, cyborg research more generally, Kline and Clynes argue, can contribute to "a clearer understanding of man's needs in his home environment" (Kline and Clynes 1961, 346). Their sentiments are echoed in the concluding remarks at the air force symposium: "What was discussed here is not only space oriented. It has implications throughout the field of mental health.... The possible *use of special devices to . . . automate man . . .* offers interesting opportunities for research that will benefit not only space travel but general medicine as well" (Flaherty 1961, 375–76; italics mine).

To understand, then, the full scope of the techno-social dream of continuous, modulated, automated control embedded in cybernetic governmentality, we need to look more broadly at the ambitious terrestrial deployment of "cyborg techniques" at the institutional and organizational levels of Rockland State Hospital. "Part of what makes the world real," notes Donna Haraway, cyborg theorist and historian, is the worldly "dreamwork" materializing in the shifting, shimmering borderlands between social reality and science fictions. "Clynes and Kline are a great example. They were actually involved in real projects, in an institutional environment of multiple real projects. Social reality was being made to happen there, and it was fantastically dreamworked" (2006, 153). If the cyborg is a kind of militarized dream of the automatic regulation of extraterrestrial "man," cybernetic governmentality is the institutionalized dreamwork trying to make real the techno-social (science) fictions of automation and informatic management made more materially possible through the cold war spread of computer and information technologies.

Rockland State Hospital becomes well known (famous or infamous, depending on the audience) as a center for psychopharmacological experiments with mental patients, starting in the early 1950s. By the mid-1960s, Nathan Kline has built the Research Department at Rockland into an internationally renowned center for the testing of dozens of potential new psychopharmaceuticals, with a new "model" research ward, a staff of two hundred people, and a $1 million budget (over half of the budget comes from pharmaceutical industry funding, and private and federal government sources).[4] Less well known is the centrality of computers and the dream of electronic automation to Rockland State's research, clinical, and administrative activities in the 1960s and 1970s. In *Computers and Electronic Devices in Psychiatry* (1968), Nathan Kline and Eugene Laska situate their notion of the psycho-cyborg within the technoscientific and organizational context of "psychoelectronics," that is, the embeddedness of electronic technologies at all levels of the psychiatric institution: from "treatment" techniques like electroshock therapy to the computer evaluation of clinical and epidemiological data; the computer automation of psychiatric diagnosis and patient case histories; and the electronic tracking of "the whereabouts of each patient at all times" (Kline and Laska 1968, v–vi).

During the same years that the pharmaceutically enhanced, extraterrestrial cyborg materializes out of the experimentally drugged mental patient at Rockland State Hospital, Kline and his colleagues are modeling an early exercise in cybernetic governmentality within the experimental spaces of the computer-enhanced mental institution.[5] In 1963, with grant money from the NIH, Kline opens a computer laboratory at Rockland equipped with a state-of-the-art IBM computing machine (to supplement the existing psychoelectronics laboratory and Manfred Clynes's Dynamic Simulation Lab). The computer lab is a precursor of what will become the Multi-State Information System for Psychiatric Patients (MSIS), a computer-based clinical and administrative management information system operational by 1975 in public and private mental health institutions in seven states, with a central computer facility at Rockland State (Spitzer and Endicott 1975). Emerging from the information management capacities of the new computers, and the observation that there exist "enormous areas of overlap" in the information needed for psychiatric treatment decisions, hospital administration, and clinical and epidemiological research, the MSIS attempts to automate the circulation, storage, and analysis of data in a "total hospital system" (Laska and Kline 1968, 7–8).

"At the center of everything is . . . a centralized, integrated computer

file," writes Kline in 1967 as he tries to merge a speculative future of cybernetic information systems with the pragmatic present of the modern mental institution (Kline and Logemann 1967, 544, 547). By using standardized
forms that can easily be coded for computerized data input, "uniform information can be collected from hospitals and psychiatrists the world over"
(Laska et al. 1967, 120). SCRIBE, a computer program developed by Kline
and his colleagues at Rockland for the "Systematic, Complete, Interlingual,
Brief, Expandable" production of automated patient case histories, promises to translate a sequence of numerically coded questionnaire responses
into a structured, computer-produced clinical narrative available in multiple languages. Not only will the automated case history aid the "over-
worked psychiatrist," but it will allow the efficient, standardized exchange
of patient data between clinical and research activities within the hospital,
and between different hospitals using the same forms (Laska et al. 1967,
120). In the longer term, such data can be stored in computerized archives,
enabling the hospital administrator to "discover trends in his patient population" or to validate the efficacy of drugs for the treatment of specific disorders (Kline and Logemann 1967, 547–48).

The Multi-State Information System for Psychiatric Patients (MSIS) —
designed and housed in Rockland State's Information Sciences Division,
in collaboration with New York State's Department of Mental Hygiene —
begins to implement these automated, electrified fantasies in the early
1970s. An "integrated" set of standardized forms is developed for use in
participating psychiatric hospitals, and the MSIS can produce narrative case
histories "written" by computer (although real-world problems quickly set
in as clinicians report that the automated histories are too boring to read)
(Spitzer and Endicott 1971).

But the most dramatic value of the MSIS to the modern mental hospital appears to lie in the systematic, integrated reach of new forms of electronic measurement, surveillance, and control enabled — sometimes automatically performed — across an expanding range of objects and practices.
Designed initially as a record-keeping and administrative system to help
rationally manage large patient populations, the MSIS produces "automated data" that allow hospital administrators to "know" the populations
they manage in new ways, through computer-generated statistics that reveal institution-specific patterns previously invisible to the administrative
gaze (Spitzer and Endicott 1975, 816). And if once upon a time the target

object of surveillance and control in mental hospitals was the institutionalized patient, the MSIS begins to organize — even encourage — an integrated
surveillance of both patient population *and* doctors' performance, both
mental illness *and* its professionalized treatment. While in one hospital,
nurses use the MSIS to produce daily individual and aggregate reports on
"patient behavior" on each ward, in another, psychiatrists state their goals
at the beginning of treatment and then receive periodic computer queries
about their progress in reaching those goals (Spitzer and Endicott 1975,
816).

Most significantly, mental illness itself is monitored and managed in
new ways through the computer simulation of psychiatric diagnosis and
the automation of drug treatment decisions — two additional features built
into the informatic infrastructure of the MSIS. While still experimental and
controversial, these features of the MSIS promote the value of empirically
based rules — capable of being incorporated as computerized algorithms —
in directing both diagnostic and drug treatment practices in the mental
hospital. "Computer-assisted drug prescription could be more effective
than 'doctor's choice' of medication," report Spitzer and Endicott in their
review of the MSIS and other automated psychiatric information systems
(1975, 824). And evidence of the computer's strengths in simulating the
diagnostic practices of "expert" clinicians suggests that computer-based
diagnosis of mental illness can be more reliable, more explicit, and more
"scientific" than the notoriously unreliable and nonstandardized diagnoses
of doctors (ibid., 822; Spitzer and Endicott 1974). "Future work" on computers in psychiatry promises to bring together these two fields — diagnosis and drug treatment — by "programming the logic of current therapeutic
knowledge regarding interaction between patient characteristics and drug
response," that is, by maximizing informational feedback between specific
symptoms of a mental disorder and the prediction of how those symptoms
will respond to a specific drug (Spitzer and Endicott 1975, 825).

If the cyborg is first envisioned in 1960 as a technique for regulating
and controlling humans' psychophysiological functioning in outer space,
cybernetic governmentality, as experimentally launched at Rockland State,
stakes out related but far more ambitious methods for the informatic management of mental distress and of everyday practices of diagnosis, treatment, measuring, and monitoring in the mental institution. While the
cyborg marks an emergent form of techno-social *subjectivity*, cybernetic

governmentality marks an emergent form of techno-social *power*. Based on the real and imagined capacities of computerized information systems to surveil and simulate a range of complex human-machine-institutional processes, cybernetic governmentality in the mental hospital attempts to network populations of patients and doctors, symptoms and technoscientific treatments, electronic information and institutional procedures, future goals and present performance, into a systems architecture of integrated management and modulated controls. Just as U.S. psychiatry in the 1970s grows disenchanted with the psychoanalytic power of dreams, a (resolutely unanalyzed) dream of informatic control begins to motivate and materialize biopsychiatry's new language of diagnosis and its embrace of psychopharmacology as a mass approach to psychic suffering. The publication in 1980 of the *DSM-III* signals not only the ascendance of U.S. biopsychiatry but the incorporation of an informatics of diagnosis at the very heart of biopsychiatric practice, and the possibility of cybernetic techniques of governing "mentalities" diffusing beyond the institutional and imaginary infrastructures of the mental asylum.

Algorithmic Imaginaries: Computing the *DSM-III*

> Any concept that can be operationally defined . . . can be coded for computer analysis. . . . As psychiatry becomes increasingly based on actual knowledge, rather than on theoretical speculation, so will the value of computers to psychiatry increase.
> —SPITZER and ENDICOTT

> The "algorithmization" of mental illness constitutes perhaps the most remarkable new world view within the heart of psychiatry.
> —ALASDAIR DONALD

Since its first edition in 1952, the *Diagnostic and Statistical Manual of Mental Disorders* (DSM) has been published by the American Psychiatric Association as a professional guidebook for the classification and diagnosis of psychic and emotional distress. Adapted from the psychiatric classification system developed by the U.S. Army during World War II, the *DSM-I* identifies just over one hundred disorders, described primarily in a psychodynamic and psychoanalytic language; a majority of the disorders are defined as psychological (and not biologically based) disturbances, with anxiety and unconscious defenses playing a central role (APA 1952). With the third

edition of the *DSM* published in 1980, the APA introduces a radical break
with the previous language of psychiatric classification: the more than 260
diagnostic entities identified in the *DSM-III* are operationalized in an avow-
edly empirical and "atheoretical" language of explicit, specified, observable
diagnostic criteria (APA 1980, 7). From a psychoanalytically inflected vocabulary of neurotic personality and unconscious conflict, the language of
psychiatric disease takes a symptomatic turn toward the styleless style of
scientific objectivity and medicalized precision. Robert Spitzer, chair of the
APA task force that produces the *DSM-III*, calls the extraordinary changes
in the manual's form and content a "signal achievement for psychiatry" and
an important advance toward the "fulfillment of the scientific aspirations of
the profession" (Bayer and Spitzer 1985, 187).

The transformation in psychiatric diagnostic language inaugurated by
the *DSM-III* also marks a foundational moment in the establishment and
hegemony of U.S. biopsychiatry. "Jurisdiction over categorical diseases
was a prerequisite for the entry of psychiatry into the new prestige system
of biomedicine" (Horwitz 2002, 61). Defining discrete categories of mental disorders through carefully enumerated criteria, the *DSM-III* creates a
standardized diagnostic currency that can be valued, circulated, and ex-
changed in ways that parallel the circulation of "diseases" in the biomedical
realm. Critics warn that the construction of the *DSM-III* marks a "remedicalization of American psychiatry" and a brazen demotion of a clinically
oriented biopsychosocial approach that had been dominant for decades,
promoting instead a new "research-based medical model" of mental disease (Wilson 1993, 399–400). With biopsychiatry's "diseases of the brain"
displacing psychosocial notions of unconscious conflict or environmental stress, the shifting language of psychiatric diagnosis both expresses
and makes possible fundamental transformations in psychiatric practice.
While purportedly remaining neutral vis-à-vis questions of causation, the
new language of the *DSM-III* lays the epistemic groundwork for a brain-
focused, biologically based etiology of mental disorder and its contemporary corollary—pharmaceutically based treatment techniques (Horwitz
2002, 64–67; Healy 1997, 231–37).

A number of economic, cultural, political, and professional factors are
regularly cited when explaining the revolution in U.S. psychiatric diagnostic language accomplished by the *DSM-III*: psychiatry's desire to mimic
medical science in the face of political critiques of mental illness as a myth

and psychiatry as a form of social control; the demand for specificity in both mental illness categories and their treatments created by the insurance industry and new FDA regulations regarding drug efficacy; a bid for greater professional power by research psychiatrists invested in scientific method, biological approaches to mental disease, and the growing business of clinical drug trials; and state and federal governments' push for more rationalized mental health policymaking and more reliable psychiatric statistics on disease epidemiology and treatment outcomes.[6]

But such accounts pay virtually no attention to an emerging informatics of mental health management and related techno-social trajectories of computer and information systems that also animate the history of U.S. biopsychiatry and the scientized tongues with which psychiatry learns to speak. In the archive of passionate controversy that accompanied the publication of the DSM-III, the issue of computerization rarely appears.[7] Once, in a widely publicized debate with Robert Spitzer over the new DSM at the 1982 annual meetings of the American Psychiatric Association, George Vaillant proclaims that psychiatry "has more in common with the inevitable ambiguity of great drama than with *the DSM-III's quest for algorithms compatible with the cold binary logic of computer science*" (Vaillant 1984, 544; italics mine). Is it possible that behind the DSM-III's move to operationalize each mental disorder with a systematic, rule-bound set of explicit criteria lies another desire — perhaps more powerful even than the pursuit of scientific procedure? Another drama at least as compelling and contagious as the reign of rational method in psychiatry?

In the years just before his appointment as chair of the APA task force overseeing the publication of the DSM-III, Robert Spitzer works extensively on computer simulation of psychiatric diagnosis, trying to replicate in the realm of electronic code what clinical psychiatrists perform in their daily practice. In the late 1960s and early 1970s, Spitzer and his colleague Jean Endicott design a series of computerized diagnostic programs — DIAGNO I, II, and III — for use with an IBM computer system (see Spitzer and Endicott 1968, 1969; Fleiss et al. 1972; Spitzer et al. 1974). Against all claims that the complexities of clinical judgment can never be translated into computerized information, Spitzer and Endicott state that, on the contrary, any concept that can be defined operationally in a limited number of words can be coded for computer analysis (Spitzer and Endicott 1975, 835). Formulating diagnostic categories in such a way that they can be translated into computerized data, Spitzer and Endicott pursue the transformation of mental disorders into patterns of information that can be manipulated by computer algorithms.

By the mid-1970s, DIAGNO III is installed for use at Rockland State Hospital as part of the Multi State Information System for Psychiatric Patients (MSIS) (Spitzer and Endicott 1975, 823). Spitzer and Endicott report enthusiastically on the potential not only for computerized diagnosis of mental illness but also for linking automated diagnoses to automated drug treatment recommendations. They cite a computerized system, developed by the psychiatrist Donald Klein and his colleagues, that classifies patients into diagnostic categories and predicts how the diagnosed patient will react to different drug treatments. "The computerized diagnoses were in substantial agreement with carefully made clinical diagnoses and predicted response to drugs as well as did the clinical diagnoses," they note (824–25).

In 1974, summarizing their computer work to date, Spitzer and his colleagues argue that the ultimate obstacle to "simulat[ing] the diagnostic practices of expert diagnosticians" lies not with the technological constraints of the computer but "in the traditional diagnostic system itself" (Spitzer et al. 1974, 198, 202). In particular, the low reliability of the psychiatric classification system hinders the development of valid diagnostic procedures for computers as well as clinicians.[8] And so the effort to computerize diagnosis turns into the drive to improve the reliability of psychiatric diagnosis. "Our current effort," Spitzer reports, "is in the direction of a change in the diagnostic system itself with emphasis on simplification, *explicit criteria*, and limiting the categories to those conditions for which validity evidence exists" (203; italics mine). With support from the NIMH, Spitzer and a team of colleagues begin working on the Research Diagnostic Criteria (RDC)—a set of specified criteria operationalizing each of twenty-five selected mental disorders—as the next, necessary step in improving diagnostic reliability (Spitzer, Endicott, and Robins 1978).

When in 1974 Spitzer is appointed chair of the APA committee overseeing the development of the third edition of the DSM, he invites Jean Endicott, Donald Klein, and two psychiatrists from Washington University at St. Louis (where early experiments in computerized diagnosis are also conducted) to join the original nine-person DSM-III task force. This critical nucleus of the task force's members shares not only a biologically based approach to mental disorder but a background in computer-simulated approaches to psychiatric diagnosis. The pursuit of greater reliability in a psychiatric classification system targeted for computer simulation morphed into the development of the 1978 Research Diagnostic Criteria, which served as the template for the explicit, specified criteria that became the hallmark of the new diagnostic language ushered in by the DSM-III in 1980. Jean Endicott recollects how, in moving from the design of automated diagnostics to the development of the Research Diagnostic Criteria, "we used to laugh and kind of say, okay, we'll stop trying to teach the computer to act like a clinician. And we're trying to teach the clinician to apply logical rules, kind of more like a computer."[9] (Endicott interview). By the time the logical rules of diagnosis are embedded in a set of specific inclusion and exclusion criteria for each DSM-III diagnostic category, the automation of psychiatric diagnosis has been partially achieved—not by successful computer simulation but through a shift in diagnostic language and the diagnostic performance it demands. Here, in the DSM-III's informatic reformatting of each diagnostic entity, *the clinician simulates the computer simulation of psychiatric diagnosis.* The heterogeneous interpretive practices and variations in clinical experience and judgment that had previously informed psychiatric diagnosis are replaced by an *informatics of diagnosis* institutionalized by the DSM-III, in which the performance of diagnosis mimics the operations of an automated computer system—moving through a discrete set of logical, standardized steps that can ultimately be simulated by a computer.

And soon they are. In 1981 researchers at Washington University at St. Louis develop the Diagnostic Interview Schedule (DIS), a structured interview for use by nonclinicians to diagnose mental disorders in the general population (Robins et al. 1981). The DIS produces data that are then coded as computer input and analyzed by a computer program that uses one of three different sets of diagnostic criteria (including the DSM-III and the Research Diagnostic Criteria) to make an automated psychiatric diagnosis. The DIS makes psychiatric diagnoses, Robert Spitzer explains, "on the basis of algorithms that translate the DSM-III diagnostic criteria into inflexible rules which are then applied to the coded data after the interview has been completed" (Spitzer 1984, 281). In 1983, Spitzer concludes his presidential address to the American Psychopathological Association with the following provocation: "The DIS has put the proverbial ball in the clinician's court and the score is 40-love in favor of the DIS. The burden of proof is now on the clinician to show that advances in technology have not made the clinician superfluous in the task of diagnostic assessment" (287).

But the real challenge of the DIS and the informatics of diagnosis that it deploys and extends may lie elsewhere than the threat (or sci-fi fantasy) of eliminating human clinicians from diagnostic decisions. The human clinician can instead simply become an extension of the informatic management of mental health that the DSM-III and several of its most influential designers were historically pursuing. As long as clinicians' diagnoses *simulate the computer simulation* of psychiatric diagnosis, there is little need to replace clinicians with electronic machines—the substitution has already been made elsewhere, in the performance imperatives and procedural standardization of the diagnosis itself. As a technique or technology for governing the "mentality" of those who govern mental disorders, the DSM-III initiates a cognitive automation of diagnostic behavior, regardless of whether that behavior is performed by computers or humans. If the cyborg assembles human organisms and cybernetic machines by modeling both as information and communication processing devices, cybernetic governmentality splices together human organisms, cybernetic machines, professional discourses, institutional procedures, cultural perceptions, and social practices by assembling and intensifying the automated, informatic control features across each of these fields. Computer and information systems play a key techno-social role in constructing such assemblages, though the exercise of control—here the control of diagnostic behavior—remains distributed throughout a multitextured, multilevel architecture.

What then are the emergent "control features" of an informatics of diagnosis incorporated in the Diagnostic Interview Schedule? Developed under contract with the Division of Biometry and Epidemiology at the NIMH, the DIS and its automated diagnostics are used to carry out the first large-scale U.S. epidemiological survey of prevalence rates for specific mental disorders in the general population, the Epidemiologic Catchment Area (ECA) study (Regier et al. 1984; Robins and Regier 1991). Celebrated as "a landmark in . . . American contributions to the psychiatric knowledge base" (Freedman 1984, 931), the ECA survey delivers up national statistical populations composed of the estimated number of cases for different mental disorders defined by the DSM-III. Redressing the previous lack of any credible national, state, or local statistics for incidence of mental disorders,

ECA researchers claim to offer "the most comprehensive report on psychiatric disorders in America ever assembled" (Regier, Myers, Kramer, et al. 1984, 939). Based on DIS interviews and the computer-automated diagnoses of twenty thousand people in five U.S. cities, the ECA stands — until its findings are disputed by the next major epidemiological study of mental disorders conducted with a second-generation computer-automated diagnostic system, the CIDI (Kessler et al. 1994) — as the first effort to aggregate populations via differentiated mental disorders.

The kindred fields of epidemiology and social statistics both originate as key features of nineteenth-century "sciences of the state and its populations" (Krieger 2000, 155). Like other social statistics, the population estimates for specific mental disorders made possible in the 1980s by automated psychiatric diagnostics are, in part, a response to state-generated demands for information useful to the management of resources and the development of policy (see Kramer 1975; President's Commission on Mental Health 1978). The informatics of diagnosis that gives rise to the DIS, which in turn enables the ECA's cost-effective (interviews are conducted by laypeople, and diagnoses are rendered by computer) and standardized (DSM-III diagnostic criteria are used throughout) national epidemiological estimates, can usefully be understood within a genealogy of statistical techniques of governmentality employed by the state. But situated within the brief genealogy of cybernetic governmentality I have sketched here, something even more dis-eased and more disturbing is also taking place with this new visibility, this unprecedented and literal accounting of mental disorders via a computerized and cost-effective epidemiology.

In the computer laboratories of Rockland State Hospital and the social-science-fiction fantasies of Nathan Kline and his colleagues in the 1960s, the cybernetic governing of mentalities involved the automation not only of diagnosis but of an informatic link between diagnosis of a particular mental disorder and selection of a particular drug treatment. Injecting psychopharmacology directly into the technoscientific and administrative circuits of the mental health management of cyborgs — and institutionalized mental patients — was an early animating feature of what might be called "cyber-psychiatry" (Orr 2006). Today a direct injection of drugs into the technoscientific and administrative circuits of the mental health management of entire populations is a (social) science fiction being actively dreamed and massively inhabited. By the early twenty-first century, cybernetic governmentality is effectively deinstitutionalized, discharged and diffused beyond the institutional enclosures of the mental hospital into the informatic regulation of the mentalities of a vastly expanded general population.[10] With psychopharmacology as the control technology of choice, the statistical aggregates of specific mental disorders constructed through psychiatric epidemiological surveys like the ECA study serve as target (sub)populations for the administration of drugs; the emergent diagnostic group "is alternatively an epidemiological population, a market segment, and a community of self-identity" (Lakoff 2006, 22). And how is this link made between a population of disordered mentalities and its market-based pharmacological cure? Is it as simple as they once dreamed, inside the experimental spaces of Rockland State, each disease informatically coded to a desired drug regime along the electrified routes of computer-automated communications?

Compass Information Services Inc. seems to think so. Part of a burgeoning field of digital diagnostics and medical informatics, the company has developed a software system for mental health screening, COMPASS-PC, designed primarily for use on a hand-held computer by patients in primary care settings. Conceived as one element in an overall "behavioral health care disease management system," COMPASS-PC prints out a kind of lab report for physicians regarding patients' mental health status, based on symptom scales for seven common mental disorders defined in the DSM-III-R (1987) (Grissom and Howard 2000). In addition, COMPASS-PC "helps the physician to determine the choice of medication" and provides a longer-term electronic archive for monitoring ongoing psychological symptoms and the effects of drug treatment (258).

But if the deinstitutionalization of cybernetic governmentality has led to an uncanny generalization of the techno-structures of a 1960s mental hospital into a widening assemblage of medical and social spaces, there is also much that the cyber-visionaries at Rockland State did not foresee, including the networking of electronic information technologies into the everyday, individualized perceptual systems of the general population. Today the most influential communications systems that informatically couple specific mental disorders with psychopharmaceutical drugs may not be professionally designed computerized diagnostic systems. The informatic management of mental health today may be working most powerfully and persuasively through popular electronic communication and information systems built into the technoscientific infrastructures of everyday life: television and the Internet. Over the last decade, as the FDA averts its regulatory gaze, pharmaceutical companies have begun direct-to-consumer drug advertising, including drugs for DSM-defined mental disorders. Whereas previously the pharmaceutical industry focused its advertising and marketing efforts on psychiatrists and prescribing physicians, by 2005 drug companies are spending an estimated $4 billion annually on consumer advertising (Saul 2005), including pharmaceutical-company-sponsored web pages with computerized self-tests to screen for the presence of specific mental disorders. Advertisements and Web-based information on psychopharmacological drugs are now a ubiquitous part of everyday electronic media, repetitiously linking specific mental disorders with specific drug treatments. Today "the pharmaceuticalization of certain disorders is sometimes more a product of popular culture . . . than of professional interests" (Petryna and Kleinman 2006, 9). With patients asking directly for brand-name pharmaceuticals from their doctors, and research suggesting that mass-media marketing significantly boosts prescription drug sales, the Pharmaceutical Research and Manufacturers of America quickly affirmed that the new drug ads communicate important "information" to consumers (quoted in Mirken 1996). Creating electronically communicated and enormously profitable information feedback between bipolar disorder and Zyprexa, depression and Zoloft, panic attacks and Paxil, pharmaceutical industry advertising takes the wildest dreams of informatic control of mentalities at Rockland State and recircuits them through the normalized, deinstitutionalized spaces of popular culture, commodity markets, and a (symptomatically) expanding population of disordered mentalities.

Governing Markets: Global Ambitions

Schizophrenia, at least theoretically, now means the same thing in Beijing
as it does in Birmingham, agoraphobia the same in Athens as it does in Atlanta.
—ANDREWS and WITTTCHEN

S
The salting of sores
The sacking of altars
The sanctifying of evils
. . . The shattering of nerves

The shifting of blame
The snatching of lightbulbs . . .
—BERN PORTER

As the informatic management of mental health begins to structure new market-based assemblages of symptoms and information systems, pills and populations, the globalizing power of an informatics of diagnosis helps to situate such assemblages in the universalizing trajectories of an era of global pharmaceuticals (Petryna, Lakoff, and Kleinman 2006). From its inception, the DSM-III was seen as an instrumental tool for promoting a worldwide standardization of psychiatric diagnostic language. As early as 1982, Gerald Klerman—arguably the most powerful psychiatrist in the United States at the time—preemptively declared "the triumph of DSM-III on the international scene" (Klerman 1984, 540). Klerman was hired that same year by the Upjohn Company to oversee one of the first industry-sponsored multinational drug trials of a global psychopharmaceutical: the Cross-National Collaborative Panic Study, which attempted to establish the new DSM-III diagnosis of panic disorder as a universally homogeneous mental disease, and Upjohn's Xanax as its universally effective cure (see Healy 1996, 194–99; Klerman, Coleman, and Purpura 1986; Orr 2006, 251–62). Myrna Weissman, a nationally renowned epidemiologist who worked on the study, reports how the project, which was both designed and funded by Upjohn, "created a generation of young investigators throughout the world who had used DSM-III and who could talk to each other in the same language" (quoted in Healy 1998, 532).

The "diagnostic liquidity" (Lakoff 2006, 18–22) performed by the new language of the DSM-III generates the promise of a global tide of psychiatric communication—and corporate pharmaceutical sales and profits—across an unprecedented number of cultural, linguistic, national, and psychic borders. But delivering on that promise rests in part on the development of persuasive, cost-effective epidemiological techniques for constructing globalized populations of mental disorders and a transnational geography of "unmet needs" (for global mental health policy) or "market potentials" (for a globalizing drug industry). As in the making of U.S. epidemiological populations after the publication of the DSM-III, computer-simulated diagnostics play a foundational role in the transnational psychiatric epidemiology emerging in the 1990s. Myrna Weissman's study in 1997 on the epidemiology of panic disorder in ten countries—in East Asia, Europe, the

Middle East, North America, and the Caribbean — uses the computerized Diagnostic Interview Schedule to produce psychiatric diagnoses for over forty thousand research subjects (Weissman et al. 1997). Financed initially by individual governments, Weissman's epidemiological studies are later funded by the Upjohn Company as a valuable form of transnational market research (Weissman interview).

By the mid-1990s, the DIS mutates into a more globally comprehensive computer-automated system, the Composite International Diagnostic Interview (CIDI), designed specifically for use in multinational, cross-cultural epidemiological studies of mental disorders (Robins et al. 1988; Essau and Wittchen 1993). Sponsored by the World Health Organization in collaboration with the U.S. Alcohol, Drug Abuse, and Mental Health Administration (ADAMHA), the CIDI uses computerized algorithms to score interview data into discrete diagnoses based on DSM and ICD (International Classification of Diseases) criteria. Available by 2007 in twenty-five languages and in a fully computerized format (CIDI-Auto) — which now automates the diagnostic interview itself and can be self-administered in front of a computer screen — the CIDI is designed "to guide the collection and interpretation of data for diagnosing large numbers of subjects who speak different languages and vary in the degree to which they are ... accepting of the idea of responding frankly to personal questions put to them by strangers in the service of science" (Robins et al. 1988, 1070). Transcoding potentially unspeakable stories of local psychic distress into informatically mediated global epidemiological data, the CIDI also guides, with coded control, the language of the scientific strangers who come calling, now, with transnationally automated inquiries and algorithmic anticipations of mental disease. And its cure?

Again the informatics of psychiatric diagnosis appears to move in dynamic techno-social feedback relations with an informatics of drug treatment — now on an increasingly globalized scale. The International Psychopharmacology Algorithm Project (IPAP) brings together psychiatrists, pharmacologists, and informaticians to develop drug treatment algorithms — logical flowcharts for the use and sequencing of different drug medications — in the treatment of major DSM-defined mental disorders. Funded by pharmaceutical companies (including Pfizer, Eli Lilly, and Johnson and Johnson) and private donors, IPAP is currently in consultation with the World Health Organization "to explore ways in which psychopharmacology algorithms and guidelines crafted by an expert international faculty might be useful in WHO's efforts to promote mental health" (Jobson 2004, 26; see also http://www.ipap.org); the WHO has already adopted IPAP's Schizophrenia Algorithm as a recommended drug treatment regime for schizophrenia.[11] The algorithms, which exist in both computerized and nonelectronic form, create the same kind of automated decision making within a bounded set of procedural rules in the field of drug treatment that the DSM-III created in the field of psychiatric diagnosis.

This informaticization of drug treatment is not only, or even most urgently, about psychopharmacology becoming networked into global circuits of computerized decision making funded by the pharmaceutical industry (although this is plenty scary). It is about a specific materialization of informatic power, the administration of psychopharmaceutical drugs, becoming a routinized, in-built feature of human-machine-institutional-market-mental relations worldwide. It is about an increasing transnational hegemony of biopsychiatric knowledge systems — embedded in computerized *and* human practices (and their human-machine assemblages) of diagnosis, treatment, epidemiological accounting, population management — guiding sociopolitical responses to symptoms of mental distress and psychic suffering. In the WHO's *World Health Report 2001* on the global geography of mental health and illness, national governments are instructed to address the burden of psychic disease borne by an estimated 450 million people who suffer from mental and behavioral disorders, most of which go untreated. The WHO calls for more comprehensive "epidemiological information" to provide "quantitative information on the extent and type of problems" in the population (World Health Organization 2001, chap. 4). As the top priority in a series of biopolitical strategies for managing such diseased populations, the WHO recommends the universal availability of psychotropic drugs (chap. 5).

Locating the emergence of a contemporary informatics of psychiatric diagnosis and epidemiology within a genealogy of the informatic management of the mentality of populations allows us to at least pose and pursue the following questions. First, how do we take seriously the material histories and technoscientific imaginaries of a cybernetic governing of mentality while not forgetting the heterogeneous, demanding realities of psychic suffering? What difference does it make today to notice the culturally specific dream of automated control animating the socio-psycho-pharmaco-logics of Rockland State Hospital and the cold war founders of U.S. biopsychiatry, when faced with the WHO's evocation of a massive, socioeconomically stratified, and largely unmet "global disease burden" of mental disorder and disability (World Health Organization 2001; Bebbington 2001)? I position the WHO's *World Health Report 2001* as a frank statement of the unprecedented pursuit of biopolitical governance of global mental health; it is, at the same time, an affirmation of the seriousness and ongoing stigmatization of forms of psychic distress, a world consensus on the urgency of addressing mental distress within a framework of human and civil rights, and public health discourse.[12] The production of specific disease populations and consumer markets through an informatics of transnational psychiatric epidemiology can look, from within this framework, like a promising technoscientific and social advance. Better an informatic management of mentalities than the benign cruelties of negligence or the systematic sadisms of the mental asylum.[13]

Which brings us to a second set of questions, related and equally urgent. Why such an emphasis on the imaginaries of control embedded in the informatics of mental health administration, when Clarke et al. (2003, this volume) are careful to note that there "are no one-way arrows of causation, no unchallenged asymmetries of power, no simple good versus bad," in the complex terrain of biomedicalization? Both the force and the limit of my critique here rest on an attempt to make visible a contemporary, historically specific form—or format—of power, not on a nuanced investigation of power asymmetries and resistances to them. Cybernetic governmentality names a form of power made possible through twentieth-century communication and information systems and the techno-social architectures of automated, modulated, continuous control they perform and proliferate.[14] If these new forms of informatic control are circuited through *affective* and *psychic* networks, if material mentalities as well as material bodies are more and more governed through what Gilles Deleuze (1990) calls "societies of control" in which power becomes "ever more immanent to the social field, distributed throughout the brains and bodies of the citizens" (Hardt and Negri 2000, 23), then is even the *desire for resistance* opened to modulations by the control technologies we might desire to resist? Are biomedicalization and biopsychiatry centrally situated techno-social sites of innovation for "ultrarapid forms of free-floating control," including "the extraordinary pharmaceutical productions, the molecular engineering, the genetic manipulations," that Deleuze sees displacing a disciplinary society based on hierarchical ranking and institutional enclosures (1990, 4)? Is it possible that there are new *forms of power* operating today that need to be diagnosed and challenged, perhaps in new ways? What other forms or formats of psychic difference and resistance can emerge from "inside out" of a transformative biopolitical economy of health and illness, life and death?

Third, if the pharmaceutically enhanced cyborg dreamed of by Kline and Clynes was, as they enthused in 1960, about adapting a man-machine system to new environments (i.e., "outer space") for which evolution had not prepared it, then is the cybernetic governance of mentalities today also deeply implicated in a biopolitical reengineering of psychic "life itself" for survival in environments that would be, for at least some of us, uninhabitable *without* psychopharmaceutical supplementation? Is the "spiritual challenge to take an active part in [man's] own biological evolution," which Kline and Clynes (1961, 345) link to the cybernetic regulation of human psychophysiology, an early enunciation of what Sarah Franklin identifies as the dual imperative of a "vital" governmentality, that is, to "take evolution in one hand and to govern it with the other" (2000, 188)? Kline and Clynes foresaw the power of automatic or "autonomic" nonconscious processes of informatic feedback and exchange to transform psychophysiological functions "so as to biologically optimize them for the particular environment chosen. Such a step, previously carried out by evolution through selective survival, would now eventuate through the purposeful construction of Cyborgs" (1961, 361). This original cyber-science fiction of "participant evolution" (345) is today a social reality, evident in Franklin's analysis of a new "genomic governmentality" exercised via "technologically assisted genealogy" (quoted in Clarke et al., this volume). Is participant evolution also a social reality in the making in the biopsychiatric and informatic dreamwork of cybernetic governmentality, in which individuals and entire (sub)populations are psychopharmaceutically adapted to function in environments—social, economic, corporeal, cultural—that would otherwise be psychically unsustainable? If madness and its medicalized kin, "mental disorder," are always partially sociohistorical symptoms of stress and distress, then is the informatic management of mental health and disease today an automated and increasingly population-based strategy for technoscientifically transforming the environmental limits of psychic functioning? Is psychopharmaceutical governance a kind of informatically induced vitality, creating conditions for psychic habitation where none really, socially, environmentally, exist?

Finally, running through each of these questions is the challenge of thinking through the specificities of psychic or mental life as one speculative currency in a biopolitical economy trading in techno-social transformations of "life itself." How to articulate the ethical and political specificities of issues raised by transformations in the psychic materialities of mental processes like memory, affect, perception, thought, desire? If the molecular level of matter is one key, technoscientifically expanding surface on which biomedicalization operates its alchemies from "the inside out," then *how to think through the implications of thought itself* (and its complex bio-electro-neuronchemistries) *becoming pharmaceutical*? Without reinstantiating old dualisms of mind versus body rejected by both biopsychiatry and four decades of feminist politics, how to approach the governing of mentalities as an especially fraught scene of material consequence and imaginary and political possibilities? This was an acute insight of Marge Piercy's as she worked through the archives of Rockland State Mental Hospital on her way to writing her science fiction fantasy *Woman on the Edge of Time* (1976). If today entire populations living outside Rockland State also live strangely within its techno-social dream of cybernetic governance, then what would it mean to "GET OUT"? If thought itself today is becoming pharmaceutical — if this essay is written within the neurochemical embrace of a nightly triazolobenzodiazepine dose — then what form of theory, what practices of historical imagination beyond the haunting archive of a Cold War mental asylum, might be adequate to the task of thinking toward other futures?

Notes

1 See Foucault 1978/2000 for his conceptualization of governmentality, and Burchell et al. 1991 for an influential elaboration. In the English language, the work of Nikolas Rose has consistently deployed Foucault to study the social management of mentalities through what Rose calls the "psycho-sciences": see especially Rose 1992, 1996. For related discussions, see also Healy 2002, 334–89; and Lakoff 2004, 256–58.

2 I am playing seriously here with Donna Haraway's notion of the "informatics of domination" (1985). Much of Haraway's published work in the years before publication of her influential "Cyborg Manifesto" attempts to theorize cybernetics in relation to questions of power and social governance. See especially Haraway 1981, 1983.

3 Digital images were downloaded on March 13, 2002, from http://www.mid nightsociety.com/web/abandoned/rockland.

4 Annual Report of the Rockland State Hospital, vols. 30–35, 1960–1965.

5 Drug research at Rockland State in the 1950s and 1960s—like drug research at many U.S. mental institutions during that era—was profoundly experimental for at least two reasons: (1) drugs were administered to patient populations regardless of their diagnostic status, and (2) the effects of the psychoactive drugs were for the most part unknown. See Nathan Kline's description of his early research with reserpine at Rockland State in Kline 1956. For a critical science fiction account of drug and surgical experimentation on psychiatric patients institutionalized at "Rockover State Hospital," see Piercy 1976. Piercy conducted archival research at Rockland State Hospital in preparation for writing the novel.

6 For a careful sociological account of the multiple forces at work in the rise of the DSM-III, see Horwitz 2002, 66–79. For other recent analyses of the origins of DSM-III, see Healy 1997, 231–38; Kirk and Kutchins 1992; Lakoff 2006, 10–14; Lewis 2006, 97–120.

7 For a sampling of the literature that critiques the DSM-III, see Faust and Miner 1986; Kaplan 1983; Kutchins and Kirk 1988; and Schacht 1985.

8 To address the reliability problem in psychiatric diagnosis through computer- ized techniques, the Biometrics Department at the New York State Psychiatric Institute—headed by Robert Spitzer—introduces in 1972 a computerized statis- tical procedure, "the kappa coefficient," as a quantified measure of diagnostic re- liability. See Fleiss et al. 1972. For a critique of the kappa coefficient, see Kirk and Kutchins 1992, 37–45.

9 Interview by the author with Jean Endicott, New York, March 25, 1996.

10 While the rest of the chapter focuses on noninstitutionalized mentalities and their informatic management, I want to note Alasdair Donald's ethnographic analysis of the "new asylum," in which Total Quality Management techniques of managed care introduce an "algorithm of care" for institutionalized patients, including "the production of 'treatment protocols,' which are forms which consist of a list of each possible specific diagnosis and have an *already printed out treatment plan* felt by the hospital committees to represent the optimal treatment for that particular diagnosis. Clinicians need only to tick the boxes. . . . The authentic representation of clinical reality has been lost, replaced by algorithmically efficient treatment planning. . . . [A] simulacrum of 'real clinical reality' has arrived at the floor of the psychiatric unit" (Donald 2001, 433, 435). The treatment protocols, of course, cen- trally involve the administration of psychopharmaceuticals.

11 See http://www.who.int/mental_health/management/schizophrenia/en.

12 The WHO's *World Mental Health Report 2001* was based on nation-specific data gathered through Project Atlas, an ongoing attempt to map globally the differ- ential distribution of resources (including national mental health policies and budgets, government-sponsored services, mental health facilities, availability of treatments including drugs, professional expertise, information and data collec-

tion systems, organizational resources, etc.) for addressing mental and neurological disorders. For the most recent (2005) WHO Project Atlas reports, see http:// www.who.int/globalatlas/default.asp.

13 The social and political issues at stake here are starkly evident in the controversial transnational spread in the last decade of post-traumatic stress disorder (PTSD) diagnoses and trauma-specific mental health interventions within the complex psychogeographies of war, population displacement and refugee experiences, state and nonstate terrorism, and other forms of political violence including mass rape. See Van Ommeren, Saxena, and Saraceno 2005 for a WHO-sponsored effort to review the controversy and promote the ongoing usefulness of trauma-intervention programs, particularly in low-income countries. For influential critiques of the transnational export of PTSD diagnoses and treatments, see Bracken, Giller, and Summerfield 1995; and Summerfield 1999. My own analysis would foreground how globalizing PTSD diagnostics produce new, proliferating populations for surveillance and regulation at the level of mentalities, and I would certainly note the recent publication (in English, Spanish, Bahasa [Indonesia], and Chinese) of an algorithm for the drug management of PTSD by the International Psychopharmacology Algorithm Project.

14 My theoretical fiction of cybernetic governmentality is resonant with other recent theorizations of power and control, including Scott Lash's notion of "power through the algorithm" (2007, 70–71); Patricia Clough's biopolitics of "affect itself," in which the "target of control is not the production of subjects whose behaviors express internalized social norms; rather, control aims at a never-ending modulation of moods, capacities, affects, and potentialities, assembled in . . . bodies of data and information (including the human body as information and data)" (2007, 19); and Brian Massumi's concept of "command power" as the political execution of self-effecting, automatic, machinelike controls operating in the name of security (2006). All three of these authors are, in part, indebted to the theoretical provocations of Deleuze (1990) and Hardt and Negri (2000) briefly discussed earlier.

Adele E. Clarke

EPILOGUE / Thoughts on Biomedicalization
in Its Transnational Travels

In 1968 Owsei Temkin, director of the Institute of the History of Medicine at Johns Hopkins, warned that "the tradition of the Western approach to the history of medicine is too narrow a basis for the historical comprehension of medicine as it is developing in our era." The Western or scientific medicine to which he had devoted so much study was, in his view, changing rapidly into global medicine. "If world health is a common concern," he concluded, "history [and social science] focused upon global medicine is a legitimate aim" (Temkin 1968, 362, 365, 368).[1] Sustaining Temkin's concerns forty years later, this brief epilogue worries about and articulates some aspects of how the framework of biomedicalization theory might contribute to such projects.[2]

One of our goals for this volume is to provoke continued empirical studies of the rise of medicine, medicalization, and biomedicalization, both as phenomena in and of themselves and in relation to particular drugs, technologies, procedures, practices, and biocapital. Our focus has been almost exclusively on the U.S. case, holding the nation-state relatively constant as the case studies examined different instantiations of biomedicalization (see discussion in the introduction). But both "things medical" (my generic term) and the dynamics of (bio)medicalization travel widely, and we urge scholars to research empirical cases in and across other sites.

This epilogue reflects first on some complications that might be anticipated in such projects. Second, it offers some key concepts — from globalization and transnationalization to stratified biomedicalization. Last it presents an array of possible "pathways in" for such projects that have successfully been used by other scholars to date. It is aimed especially at scholars new to transnational research, with the exception of the final sections on strategies for pursuing local sociohistories of the rise of medicine, medicalization, and biomedicalization and on healthscapes as traveling assemblages. These sections outline innovative pathways into such projects through researching healthscapes (discussed in chapter 3 of this volume).

Sites for Concern

There are no direct or easy translations into other geopolitical sites for understanding the rise of medicine, medicalization, and biomedicalization as the historicized concepts we have presented in this volume. Several complications immediately crop up in trying to think beyond the U.S. case — and within the United States too there are multiple configurations and stratifications. Here I briefly discuss the complicating issues of (1) which medicines; (2) colonialisms, postcolonialisms, and medicines; (3) the implications of the absence of "direct translations" for the rise of medicine, medicalization, and biomedicalization; (4) the commonly blurred and contested boundaries between (bio)medicine and public health; and (5) linkages between (bio)medicine and modernity.

The first complication concerns the historically and still vexed issue of what kind(s) of medicine(s) we are talking about. There are many well-organized and successful forms of medicine in practice today on this planet. Key here is that most forms of medicine are traveling and have been doing so for centuries (Lock and Nichter 2002). For example, traditional Korean medicine is now also located in several sites in Australia where Koreans have immigrated, and elsewhere (e.g., Han 2000). Many forms of traditional Asian medicines are well established as alternative modes in the United States (e.g., Alter 2005), including being the focus of what has been called biopiracy for corporate appropriation (Adams 2002). Many forms of African medicines were introduced into "the New World" along vectors of the slave trade and have endured and transformed in the Caribbean and the United States (Johnson 2007). American imperial medicine in the Philippines at the beginning of the twentieth century influenced the shape of urban American public health programs into the twenty-first (Anderson 2006). And of course Western scientific medicines have been traveling from the earliest moments of Western imperialism centuries ago (Kiple 1993) through international health and development programs of the twentieth century to the now rapidly elaborating biopolitical economic formation called "global health" (Anderson 2004; Arnold 1994; Brown, Cueto, and Fee 2006; Novotny and Adams 2007).

Thus the issue of which medicine(s) are the focus of biomedicalization studies is significant, and their sociohistories are especially important in site-specific studies. I recommend reserving the term "biomedicine" for the technoscientific forms that have emerged from what have historically been termed "Western" or "scientific" medicines, most especially drugs, but also a wide range of surgical and dental procedures, technologies, and infrastructures through which they are distributed. Accordingly, then, the rise of medicine, medicalization, and biomedicalization would focus on (bio)medicine wherever these were being examined.

However, it is crucial to simultaneously assert the importance of other forms of medicines. These cannot be underestimated—both on their own terms and in relation to biomedicine historically and today. For example, actively pursuing health (as well as illness and disease) by biomedical means is a hallmark of biomedicalization, but in the United States and elsewhere, health is also pursued by both westerners and others using non-Western medicines (e.g., Johnson 2004b). There are a number of major recent sociohistorical and contemporary studies of other forms of medicine (in English as well as in other languages) helpful in situating research on biomedicalization elsewhere. These include works centered on medicines in their indigenous manifestations,[3] studies of non-Western medicines traveling to the West and elsewhere,[4] and works retheorizing the travel of things medical per se.[5]

The key point is that biomedicalization studies need to be fully situated in the histories of the full range of medicines and approaches to health and healing of their geopolitical sites of analysis. For example, in our original work on the U.S. case (Clarke et al. 2003), we carefully discussed the political economy of alternative medicine in relation to biomedicine. In our introduction to this volume, we follow through on these concerns manifest in various countertrends against processes of biomedicalization, especially in more rhizomatic alternative health practices and movements.

Moreover, in much of the world, biomedicine is not the primary form of medicine, healthcare or healing available or accessible. Biomedicine may be available only in urban areas, only through international aid programs (such as those for maternal and child health or HIV/AIDS), only episodically, and so on. Research that pursues aspects of biomedicine in such situations needs to clearly and deeply understand what else is going on and the distinctive local meanings and symbologies of the various approaches to health, healing, illness, and medicine that circulate in a particular site.

The second major complication concerns the complexities of the histories of imperialisms, colonialisms, and postcolonialisms, as these not only intersected with but also were and are constitutive of the histories of all kinds of medicines and approaches to healing. In cutting-edge scholarship today, conventional assumptions about late-nineteenth- and early-twentieth-century European colonialisms as adequately representative of forms of imperialism are being actively destabilized (e.g., Stoler, McGranahan, and Perdue 2008; Fu 2007). For example, the significance of Japanese imperialism for much of Asia and the Pacific is beginning to be elaborated (e.g., Chin 1998; Low 2005). Such work will complicate studies of (bio)-medicalization by demanding studies of what constituted the rise of medicine in non-Western colonial situations (such as Korea or Taiwan under the Japanese between ca. 1900 and 1945). Such studies will augment the already ambitious and sophisticated literature on medicines and colonialisms in many places (e.g., Anderson 2006; Andrews and Sutphen 2003; Arnold 1994; Hunt 1999; Manderson 1996; Packard 1998; Vaughan 1991).

The postcolonial poses yet other problems, including the term itself. Hall's (1996b) question "When was the postcolonial?" implied that colonialisms have continued long past dates of independence. Today we more often see the terms "neocolonial" and "neoliberal expansions" used to refer to varied forms of colonial practices. Until recently, divides of emphasis among sociology (the West), anthropology (the colonized), and history (the past) concentrated study of the postcolonial in anthropology. Today we see convergences as sociologists increasingly study transnationally, anthropologists study "experts," and historians take up the postcolonial period. Traffic among these disciplines is exciting, and geography is entering the mix (e.g., Inhorn 2007). In the extant literature, there is much less on the postcolonial/neocolonial era (for reviews, see Anderson 2002, 2004; Anderson and Adams 2008; and Packard 2003). But explicit work on the postcolonial/neocolonial and things medical is emerging (e.g., Anderson, forthcoming; Chin 1998; Leung and Furth, forthcoming; Ma 2007; Towghi 2007).

The third major complication concerns the fact that there are no direct translations for the rise of medicine, medicalization, and biomedicalization as the historicized concepts we present here in the U.S. case. Each will differ in different places and across time. Even when we hold tightly to our definitions, appropriate given that we are asserting that these processes are focused on the "Western/scientific/global medicine" as Temkin advised, problems remain. The three definitions are:

1. The rise of medicine: the serious growth of scientific (bio)medicine, especially new medical educational and service delivery organizations and institutions.
2. Medicalization: the expansion of both what gets defined as (bio)medical and the jurisdictional extension of medicine into new domains of life.
3. Biomedicalization: the elaboration of capacities for technoscientific interventions in biomedical diagnostics, treatments, practices, and health to exert more and faster transformations of bodies, selves, and lives.

In the United States, these were each and all relatively distributed geographically (if unequally so), at least enough for a national public discourse about the oxymoron "the American health system." As my examination of American healthscapes in chapter 3 demonstrates, the definitional and jurisdictional innovations of the medicalization era built on those of the rise of medicine. Institutional infrastructure was established and distributed. Technoscientific innovations implemented during the biomedicalization era are built on the infrastructures established during the medicalization era. And all three processes (rise, medicalization, and biomedicalization) continue and are cumulative across time. But *none* of these patterns may hold elsewhere. Even when they do, there will be notable if not fundamental differences that empirical research needs to address.

The fourth complication concerns the commonly blurred and contested boundaries between what today would be called biomedicine and public health. Historically, the assumptions, value, and significance of Western medicine (Gordon 1988) have long been contested vis-à-vis their effectiveness as compared to classic public-health measures such as clean water, adequate sanitation, et cetera (e.g., Bashford 2006; Cunningham and Andrews 1997; Illich 1976/1982; McKeown 1979). Development and modernization projects have been devoted to such measures, often still sorely needed. These projects date from colonial and imperial-era practices (e.g., Anderson 2006; Andrews and Sutphen 2003) to early foundation interventions (e.g., Farley 2004; Schneider 2002) to post–World War II development programs (e.g., Cooper and Packard 2005; Edelman and Haugerud 2005).

More recently, however, a new phenomenon troubles the vexed boundary between biomedicine and public health — the biomedicalization of public health. This involves the transformation of such practices into much more technoscience-dependent forms — rather than locally sustainable or "appropriate" technological forms (e.g., Castro and Singer 2004; Rodriguez-Ocana 2002). The lively question then becomes: under what conditions *is* public health a form of biomedicine? Specifically, it is not only biomedicine that is transnationalizing/globalizing today but also many forms and practices of public health themselves that are increasingly reliant on high-tech biomedicine. Certainly HIV/AIDS has been the major impetus and vector (e.g., Barnes 2000; Patton 2002). But "prepackaged" forms of public health policy such as tobacco control (e.g., Brandt 2007; Reid 2005) and harm reduction (Stimson 2007; Chen 2009) can be highly biomedicalized and biomedicalizing as well.

Today, echoing Temkin, the emergent rubric replacing the old international (public) health is "global health," with new and deeply biomedically oriented master's and doctoral programs established in the West for application "elsewhere" (e.g., Buse and Walt 2002; King 2002; O'Neil 2006; Reich 2002).[6] One programmatic emphasis is a

> new approach to global health grounded in sensible, ethical, and informed health diplomacy . . . [to] provide interdisciplinary training of health professionals to improve delivery of global health services, development assistance, and scientific investigation. Such training will support the U.S. Department of Health and Human Services' call for global public health preparedness, security, and responsiveness, as well

as the larger global health community's efforts to grapple with the new resources available in international health philanthropy. (Novotny and Adams 2007, 1)

New philanthropic resources are largely provided by the Bill and Melinda Gates Foundation. Now the wealthiest on the planet, it succeeds the Rockefeller Foundation in this domain and runs both similar and different risks in terms of biomedical imperialism (e.g., Birn 2006; Brown 1979; Brown, Cueto, and Fee 2006; Farley 2004; Schneider 2002). The Gates Foundation has assets of about $37.6 billion; after focusing on specific diseases, it recently committed $24 million to the Center for Global Development in Washington to "have a larger impact by influencing policy from the ground up" (Bumiller 2008, A12).

There are calls for such philanthropic organizations to be more directly accountable to those they actually serve by acknowledging the diversity in models of medicine and what constitutes public health in those sites (e.g., Sensenig 2007). More commonly, accountability is to Western/Northern standards rather than indigenous/Southern frameworks (e.g., Jackson and Scambler 2007; Kim 2007). An example of biomedicalizing public health to Western/Northern standards is the current trial in northeastern Thailand of a high-tech method of cervical cancer screening that both requires samples be sent elsewhere for expensive molecular testing (feeding biocapital) and displaces local nurses trained in delivering low-tech but labor-intensive forms of effective screening and treatment (Gaffikin et al. 2008).

As we think more transnationally, boundaries between biomedicine/ healthcare and public health become even more problematic and needful of analysis. Differences may be stark. An HIV/AIDS clinic run by a public health department in the United States operates within a complex system of biomedical care provision available to a significant proportion of the population. In contrast, such a clinic in other geographical locations may be the sole site of availability of any biomedicine for the local or even regional population. Petryna's (2006, 2007) studies of clinical trials in eastern Europe found that services offered by the clinical trial organization (CTO) were sometimes the only biomedical services available—and only available to participants. The CTO *was* the system. There was no public health.

Generating dependency on biocapitalized forms of biomedicine rather than locally sustainable and more appropriate, labor-intensive forms deserves attention. The fiscal power of global health philanthropy assuredly has the capacity to overwhelm other approaches, displacing and/or appropriating them.

The fifth complication in thinking about biomedicalization in its transnational travels concerns linkages between (bio)medicine and modernity—or modernities, or modernity not as a condition but as a figure of a condition—another unstable site. Simmel argued that modernities are complex and unpredictable mixes of possibilities of greater freedom and prosperity as well as exploitation and alienation (Miller 1997). Consumption of things medical *and* its absence often link to issues of personal, collective, and even national identity and modernity (Clarke et al. 2003; chapter 1 this volume). Early advertising executives "proudly proclaimed themselves missionaries of modernity" (Laird 1998, 364). Advertisements for things medical in the mass media and increasingly today on the Internet do such work across the planet. "The universalizing tendencies of modernity have always called upon the service of the translator to spread its gospel elsewhere" (Liu 1999, 1). In some ways, biomedicine is modernity "in translation."

These religious metaphors of missionaries and gospels are particularly apt if, as often argued, biomedicine and science are becoming increasingly common (if contested) sites of faith and action. Certainly what could be called "the international development machine" (*pace* Spivak 1993) that has been operating since at least the late nineteenth century has been deeply organized by and through things medical. Some of the most interesting work today takes up precisely the moral and spiritual issues emergent at intersections of development projects, things medical, and faith (e.g., Adams and Pigg 2005). According to Liu (1999, 2), "The task of the theorist then is to analyze the intellectual and material conditions under which the reciprocity of meaning-value or the denial thereof occurs at significant moments of cultural encounter." Things medical, signs, and meanings do not travel unproblematically. It is not enough to know that the television series *ER* was first aired in Thailand in 1992, for example; we also need to know how it was and is "read" there and by whom (Mensah 2000, 139).

Partaking of modern medicine may define what Rosaldo (2003) calls "the borders of belonging" in modernity. Pigg's (2005) analysis of "globalizing the facts of life" in development projects concerned with reproduction and sexuality vividly demonstrates that such logics actively displace other ways of conceptualizing sex and reproduction and are seductive in doing so through their "modernity." Taking pharmaceuticals, participating in clinical trials, and receiving shots each and all allow people "to be modern" in just such ways (e.g., Towghi 2007).

There are no answers for these complications I have raised. All must be worried about empirically. Moreover, as the term "biomedicalization" travels transnationally, it too will need to be handled as what Williams (1976/1985) called a "keyword"—a concept whose meaning changes over time, circumstance, and location—worthy of exploration.

Key Concepts

As researchers take up (bio)medicalization beyond the United States, concepts are needed that allow and encourage articulations of differences, specificities, and distributions of things medical. Researchers need to address the range of variation in the situations—from "bare life" and states of exception (e.g., Agamben 2005; Mbembe 2003; Ong 2008; Rabinow and Rose 2003/2006) to "boutique biomedicalization" for optimization (e.g., Clarke et al. 2003; Rose 2007). Some extant concepts are useful: transnationalization, medical pluralisms, medical partialisms, stratifications of the rise of medicine, medicalization and biomedicalization, and health-scapes.

TRANSNATIONALIZATION

Serious debates surround the use of the terms "globalization" and "transnationalization." The literature on globalization debates its extent, depth, forms, and so on (e.g., Coronil 2001; Escobar 2003; Featherstone 2006; Randeria 2007; Tsing 2005). Feminist scholars have been differentiating between transnational and global assertions for some time. Grewel and Kaplan (1994, 17) noted, for example, that the term "global feminism" "elided the diversity of women's agency in favor of a universalized Western model . . . that celebrates individuality and modernity." Scholars in the United States tend to use "global" to refer to everywhere but the United States (e.g., Kim-Puri 2005). There is considerable debate about whether there is "really" globalization (e.g., Guillen 2001) or more accurately networks of global cities (Sassen 2002). Vis-à-vis biomedicine, this question is deeply significant, as much of the planet remains profoundly lacking in biomedical infrastructure. This makes the term "transnationalization" more accurate. Yet the intent and directionality of the movement of biocapital is certainly global (Sunder Rajan 2006), from big pharma (e.g., Lakoff 2006; Petryna et al. 2006) to global health (King 2002; Navarro 2007; Nichter 2008).

At the same time, what Appadurai (1996, 32) calls "the dynamics of indigenization" counter and complicate the homogenizing aspects of globalization. The active processes of localization of things medical imported from elsewhere are profoundly heterogeneous (e.g., Whyte, Van der Geest, and Hardon 2002). The term "glocalization" has emerged to address these sticky processes (Eoyang 2007; Kraidy 2002; Raz 1999). For example, in Taiwan, consensus conferences brought experts and laypeople together to debate whether surrogate motherhood should be legalized (Chen and Deng 2007).

In studying biomedicalization in its transnational travels, the research goal is to empirically grasp what Reid (2005, 244; italics mine), in a comparative study of tobacco control, describes as "global singularities," "neither the simple expression of national or regional histories nor the product of global circumstance, but *somewhere in between*." That is, processes of transnationalization must be studied as localized, grounded, and situated temporally, spatially, historically, economically, politically, culturally, socially and so on.

MEDICAL PLURALISMS, PARTIALISMS, AND STRATIFICATIONS

The localization of biomedical innovations is everywhere tempered and complicated by medical pluralisms, partialisms, and multiple forms and loci of stratification. The concept of medical pluralism, its roots deep in anthropology, addresses the simultaneous availability of multiple kinds of medicines in the same places (e.g., Ernst 2002). This concept is among the most robust in medical anthropology, paralleling medicalization theory in medical sociology. It raises the vexed issue of what exactly to call different approaches more categorically (rather than specifically, such as Ayurvedism, Tantrism, etc.). In 1976 the pioneering medical anthropologist Fred Dunn coined the term "cosmopolitan" medicine, attempting to step outside the traditional-versus-modern medicine binary, and its use was sustained by Charles Leslie (Lock and Nichter 2002, 2). Yet the term "cosmopolitan" medicine linked more to the cultural and geographic location of practices than to their locus or mode of understanding of health and illness.

The term "biomedicine" was then taken up in medical anthropology and segments of sociology for those forms that located health, illness, and diseases through biological disciplines (e.g., Lock and Gordon 1988).

Attending to the epistemological legitimacy of other approaches, Leslie also questioned the restriction of the term "scientific" to medicine with roots in Western laboratories and clinics; he argued for broad "acknowledgement of the scientific and rational principles present" in other systems as well (Lock and Nichter 2002, 2). Leslie thus anticipated what today is discussed as "epistemologically privileging" one or another approach to healing, usually those of "the West" over "the rest" (Hall 1996a; Connell 2007).[7] Asserting the scientific status of forms other than Western medicine today remains a radical act (Sensenig 2007). Doing so is also growing rather than receding in importance, as the relevance and authority of different forms of knowledge are increasingly contested (Featherstone 2006). In short, Western-centric knowledges are insufficient for understanding global health in the senses that Temkin raised.

While medical pluralism characterizes most places in the world, the local, regional, and national statuses of different approaches vary as legitimate, to whom, and especially vis-à-vis state or private insurance coverage of the costs of obtaining such care. Lock and Nichter (2002, 2) note that in many places, older medical traditions have "retained a firm foothold in the face of modernization and biomedicine . . . [and] the politics of nationalism was in many instances giving the indigenous literate traditions a powerful boost, assuring continued legitimation and even expansion." Studies of how such situations developed in different sites are emerging and valuable (e.g., Leslie 1978; Langford 2002; Ma 2007; Wahlberg 2006).

Yet in many sites, different approaches are in profound competition, with the more powerful hegemonically attempting to set the terms of legitimacy according to their own standards of adequacy and effectiveness and attempting to enroll the state in their efforts, including educational funding and coverage of care (e.g., Adams 2002). This pattern is commonly understood today through Foucauldian lenses of governmentality (Foucault 1991a; Petersen and Bunton 1991). Such efforts can provoke resistance, which can take the form of "medical revivalisms . . . looking backward to find something recoverable — something felt to be lost in the present . . . for the sake of a better future" (Lock and Nichter 2002, 7).

Medical pluralisms are cross-cut by what I term *medical partialisms—*

the only partial and often contingent actual availabilities of various kinds of medicines in many sites and at different times. This is often ignored. The usually tacit assumption is of some consistency of availability of the different forms in practice (provided one has access). Yet things medical appear and disappear, clinics and pharmacies open and shut, providers die or move away, and so on. The concept of medical partialisms alerts us to issues of availability along with those of access — of all kinds of medicines.

There is also stratified access to medical goods and services for individuals and social groups also vis-à-vis all kinds of medicines. The class- and race-based inequalities that characterize healthscapes in the United States are often both different and deeper elsewhere. Moreover, inequalities among "the rest" are exacerbated by imperialisms, neocolonialisms, global expansions of neoliberalism, fundamentalisms, and so on. Gatekeeping may be enforced by one's ability to pay out-of-pocket, via insurance coverage, by citizenship status, or by the duration of a development program in the local area and its focal population(s). Lack of access may constitute social suffering (Kleinman, Das, and Lock 1997).

A transnational gaze also reveals broad patterns of institutional stratification. Foremost, these include stratification of the rise of Western/scientific/global biomedicine — the uneven development and distribution of institutions of biomedicine — from medical schools to hospitals and clinics. Without permanent local and regional institutional infrastructure, biomedical interventions are temporary rather than sustainable over time and often instigated wholly from without. Recognition of this problematic has recently pervaded the field of global health concerned with HIV/AIDS and with the migration of medical professionals from the rest to the West in a classic postcolonial pattern (Hall 1996a).[8] For example, my own university, UCSF, is participating in an "Academic Twinning Partnership" program with Muhimbili University of Health and Allied Sciences (MUHAS) in Tanzania as a U.S.-Africa collaborative institutional model for addressing health workforce problems (e.g., education) and healthcare needs (e.g., access to basic surgery), anticipated as a decades-long series of commitments.[9] Such projects are elaborating, clear, and present examples of biomedicalization in its transnational travels — and well worthy of study.

In addition to institutional stratification, there exists stratified medicalization — the targeting of different categories of persons, groups, and populations for expanded treatment of conditions defined as medical (Clarke et al. 2000, 2003).[10] For example, women, especially vis-à-vis reproduction, have long been targets of medicalization in the United States and trans-nationally (Ruzek 1978; Ginsburg and Rapp 1995; Adams and Pigg 2005). We also discussed stratified biomedicalization as unequal distribution of, and access to, technoscientific biomedicine (e.g., Clarke et al. 2003, this volume; see also Whyte, Van der Geest, and Hardon 2002). Patterns of stratification vary geopolitically with citizenship as well (see the discussion of biological citizenship in the introduction). Stratifications of all kinds often intersect with things medical in ways worthy of empirical attention.

HEALTHSCAPES

The healthscape is another useful concept for thinking about biomedical-ization transnationally. As I elaborated elsewhere in this volume, health-scapes are ways of grasping the patterns that occur where health and medi-cine are performed, who is involved, the sciences and technologies in use, its material cultures, media coverage, political and economic elements, and changing ideological and cultural framings of health, illness, healthcare, and medicine per se, including the full range of human actors. In short, healthscapes include whatever is imagined and utilized as things medical.

The healthscapes concept can be useful for research in at least two ways. First, healthscapes can be conceptualized as traveling assemblages. Second, one can read local, regional, national, and transnational sociohistories of the rise of medicine, medicalization, and biomedicalization as sociohisto-ries of healthscapes (discussed at the end of this chapter).

The assemblage concept is especially useful where biomedicine is *not* distributed across a locale (region, nation-state). In such sites the processes of the rise of medicine, medicalization, and biomedicalization may be col-lapsed into one another, producing pockets of intense biomedicalization. Two examples of such assemblages that have flowed into one another over time are first the essentially freestanding fully biomedical(izing) medical center that largely serves local elite and transnational patients, and second, more recent sites of medical tourism.[11] The marketing of essentially free-standing fully biomedical(izing) medical centers to international patients has a long history in the United States and Europe (Hutchins 1998). More recently, such centers have been developed in "the rest," usually the only such entities in a wide geographic region, also drawing interna-tional patients. In Thailand, Bumrungrad Hospital and Bangkok International Hospital are famous examples (Turner 2007), actually reviewed in the magazine *Travel and Leisure* in an article titled "The Medical Vacation" (Kamps 2006). The second example of healthscape as traveling assemblage is the growing transnational medical tourism industry (Ramirez de Arellano 2007), described as "first world health care at third world prices" (Turner 2007, 303). Today this transnational biocapital-intensive economic sector is "driven by government agencies, public-private partnerships, private hospital associations, airlines, hotel chains, investors and private equity funds, and medical brokerages. . . . Destination nations regard medical tourism as a resource for economic development" (Turner 2007, 303). For example, Bumrungrad Hospital, now specializing in medical tourism, is internationally accredited, offers heart surgery, hip resurfacing, knee replacement, complete physical checkups, prostate cancer treatment, and plastic surgery.[12] The medical records and main language of care is English, and most of the healthcare providers are trained in the United States. In New Delhi, Escorts Heart Institute and Research Center specializes in heart care, has an elaborate website, and offers concierge-like services tailored for international patients.[13] There are also current efforts to franchise such forms (Turner 2007). The concepts of assemblage and healthscape facilitate understanding such packaged commodification and distribution of things medical.

Pathways In

A number of pathways into empirical work on biomedicine in its transnational travels exist. Here I briefly note some recent projects that take up such work through medical anthropological, sociological, and historical lenses combined with those of science, technology, and medicine studies (ST&MS).[14] Last, I discuss studying the sociohistories of specific healthscapes as traveling assemblages.

FOLLOW THAT "X"

ST&MS lenses often methodologically suggest that researchers "follow" this or that phenomenon to understand its networks and broader situation or arena of action. In biomedicine this involves following a particular molecule, drug, device, procedure, innovative program, industry sector, health social movement, and so on. Projects that have done so are inspirational for future research that explicitly takes up biomedicalization in its transnational travels.

The strategy of "following that molecule" has been used especially in studies focusing on how pharmaceuticals in general or particular drugs are produced, distributed, and consumed transnationally. The historian Nancy Tomes (2001, 531, 534) noted that "health in a bottle and a book" is much more widely accessible than other forms of care. "In 2004, global pharmaceutical sales surpassed the $500 billion threshold for the first time. With fewer than 5% of the world's population, the United States accounts for almost 50% of total sales" (Tone and Watkins 2007, 4). Self-medication with both over-the-counter and prescription pharmaceuticals (regardless of whether actually prescribed) is likely the most frequent form of health-care intervention on the planet beyond routinized dietary approaches such as "comiendo bien," or eating well for its health effects (Martinez, in prep.).

Some years ago, Nichter and Vuckovik (1994) offered an agenda for social science research on pharmaceuticals, recently updated by Petryna, Lakoff, and Kleinman (2006) in their theoretically sophisticated *Global Pharmaceuticals: Ethics, Markets, Practices*. They offer an array of case studies, as do Whyte, Van der Geest, and Hardon (2002), whose work focuses with a strong comparative eye on "the social lives of medicines" from relaxants to antibiotics and from the Netherlands to the Philippines. Lakoff (2006) examines the globalization of the new biomedical psychiatry and its diagnostics and psychotropic drugs, focusing on the Argentine case. Frazzetto, Keenan, and Singh (2007) have focused on Italian debates about children and drugs framed there as "the right to Ritalin vs. the right to childhood." And Hayden (2007) examines emergent politics of corporate and state benefit sharing with local areas rather than biopiracy of natural resources used in pharmaceuticals.

AIDS treatments have been studied widely in their transnational trials and tribulations. Nguyen (2005) calls this "antiretroviral globalism," raising issues of "therapeutic citizenship." Biehl (2007) tracked treatment distribution in Brazil and that nation's development of generic AIDS drugs to provide affordable treatment for the poor, rather than rely on big pharma in its transnational expansions. Barnes (2002) examined collaborations between United States and Mexican HIV/AIDS sectors as creating a binational political-organizational field. Differential distribution of AIDS interventions has also been a focus (Patton 2002), along with gender relations, economic contexts, migration, and violence in implementing interventions (e.g., Dworkin and Ehrhardt 2007).

Contraceptive drugs, devices, and other means of birth control including abortion and sterilization and their transnational distribution have been of particular interest to feminists and others interested in women's health and rights for decades (Adams and Pigg 2005; Petchesky 2003; Saetnan, Oudshoorn, and Kirejczyk 2000; Sen, George, and Ostlin 2002). Recent work focuses on Haiti (Maternowska 2006), China (Greenhalgh 2008), East Asia (Clarke 2009; Wu et al. 2009), and Balochistan (Towghi 2007), and on topics such as testing new forms like the male pill transnationally (Oudshoorn 2003; Van Kammen and Oudshoorn 2002).

Significantly, it is not only drugs themselves in increasingly transnational circulation, but also clinical trials of drugs and other biomedical interventions.[15] Pioneering studies have focused on globalizing human-subjects research (Adams 2002; Adams et al. 2005; Petryna 2006; Whitmarsh 2008). Drawing on ST&MS perspectives, Lakoff (2008, 741) argues that pharmaceutical "effects are not embedded in the drug itself; rather they arise at the nexus of chemical substance with governmental regulation, biochemical expertise, commercial interest, and patient experience" as pharmaceuticals circulate globally. One of the challenges thus involves finding "the right patients for the drug" via (re)classifying illnesses and using rating scales and genomics research to identify patients (see also Fishman 2004; Fosket 2004).

Finding the "right patients" can also involve "bioethnic conscription" of particular populations for participation in clinical trials and research (e.g., Montoya 2007). Here Whitmarsh (2008) studied how the medical leadership of distinctively African-heritage Barbados offered the nation's population for genetics trials as a means of "becoming modern" and upgrading medical services. To staff such efforts, new master's degree programs organized to recruit people from a variety of disciplines and backgrounds to do clinical research "elsewhere" are also appearing in the United States.[16] These are potent vectors for the transnationalization of biomedicine.

Research can also follow healthcare providers. An excellent study in urban Trinidad by Reznik, Murphy, and Belgrave (2007) focused on providers' perceptions of the globalization of mainstream medicine and how this manifests in their everyday practices. These providers articulated (1) a notion of history as an autonomous force with globalization as the current stage, (2) assumptions of a neoliberal commodified market exchange as the driving social force, and (3) the fragmentation of society into atomistic, self-interested, and competitive individuals; and (4) they advocated the adoption of standardized ideals for social order regardless of geopolitical location. Medical globalization was linked with medicalization, and most vivid was the "mutual reinforcement of mainstream practices and the abstract social process of globalization" (Reznik, Murphy, and Belgrave 2007, 539).

The "follow that x" strategy also works for researching the transnational uptake of new surgeries, procedures and technologies. Organ transplantation surgeries are the most studied to date, as there are established and growing transnational markets in organs, both legal and black markets, including kidney-theft rings (e.g., Gentleman 2008; Scheper-Hughes 2005) — sites of the transnational social lives of human organs (Cohen 2005; Lock 2008). MRIS (Joyce, this volume; see also Prasad 2006) and other visualization technologies also travel widely and are used differentially in ways important to understanding the particularities of biomedicalization (Burri and Dumit 2007). Infamous here is the use of fetal ultrasound for sexing, selectively followed by abortion of fetuses of the "wrong" sex (usually female) (e.g., Van Balen and Inhorn 2003). Redfield (2008) offers a fascinating study of the development of emergency response "kits" — essentially biomedical "packages" — for use in emergencies by the nongovernmental organization Doctors without Borders, vividly illustrating the packageability and portability of biomedicalization. Other possibilities include following surgeries for obesity (Boero, this volume) or lesbian use of ARTS (Mamo, this volume) in their transnational travels.

"Following that program" has long been a strategy of medical anthropologists and scholars in international health and development studies for evaluation and critique (e.g., Justice 1986; Garrett 2007). In recent work, Adams and Pigg (2005) examined the moral and political aspects of sexuality implicated in development programs. Murphy (2003, forthcoming) follows the technology of menstrual regulation or extraction from its feminist self-help activist sitings to its integration in USAID programs in Bangladesh and elsewhere as part of foreign-aid policy in decolonizing nations.

The potency of increasingly scientized patient groups has become a major site of scholarly interest (e.g., Epstein 2007, 2008), especially their increasing influence on treatment policies (Callon 2003; Gibbon and

Novas 2008). A number of studies of transnational groups or formations have focused on HIV/AIDS (e.g., Barnes 2002; Biehl 2007; Epstein 1996; Patton 2002). And a wide array of projects examine transnational health social movements around women's health issues (Adams and Pigg 2005; Davis 2007; Halfon 2007; Petchesky 2003). These groups' different advocacies of more or less (bio)medicalization are a fascinating topic, as are questions of their appropriation by the state (e.g., the British National Health Service's National Centre for Involvement).[17]

International organizations remain relatively understudied vis-à-vis biomedicine, despite their relative autonomy and importance to (bio)medicalization (Boli and Thomas 1999; Buse and Walt 2002; Cueto 2007; Iriye 2002). The International Classification of Diseases (ICD), for example, is a major site of transnational standardization and rationalization, establishing core infrastructures along which things medical travel (Bowker and Star 1999). As pharmaceuticals begin to target people of particular "races" (e.g., see Kahn in this volume starting on p. 263), racial and ethnic differences have come to the fore in ICD classification debates and in the FDA, making both worthy of further study (Epstein 2007). Kuo (2007) examines another significant but largely invisible transnational regulatory organization: the International Conference on Harmonization of Technical Requirements for Registration of Pharmaceuticals for Human Use.

Following biocapital at this historical moment of the intensive transnational expansion of biomedicine is perhaps one of the most important strategies for studying biomedicalization in action. For example, in 2006 the Organization for Economic Co-operation and Development (OECD) initiated an "international futures program" called "The Bioeconomy to 2030" to "assess the economic potential of the biosciences and map out the regulatory and business responses that could maximize the potential of biotechnological innovation" (Lezaun 2007, 381). ST&MS scholars are following it in real time. Hilgartner (2007) is studying the mechanisms through which planners are making the elements of the bioeconomy measurable. And Parry (2007) is analyzing how the OECD is attempting to corner the "futures market in bio-epistemologies" by rhetorical means, constructing biofutures as different and needful of "specialist political husbandry" because of heightened risk (see also King 2002, 2005; Nichter 2008). Following biocapital can also be accomplished through studying a biomedical industry or sector. Possibilities include clinical trials organizations (CTOS) (e.g., Petryna 2007; Fisher 2009), big pharma (see Mirowski 2007), biotechnology (Cooper 2008; Jasanoff 2007), and others. Novas (2006) is studying biosociality vis-à-vis biotechnology companies in relation to patients. The medical software industry is another focus (Cartwright 2000b; Oudshoorn, Brouns, and van Oost 2005). Prasad and Prasad (2007) examine the new Indian "imaginary" of "human capital" via transnational medical transcription services.

On the flip side of biocapital, one can also "follow that philanthropy" in its transnational endeavors. As noted, both Rockefeller and Gates Foundations have been exceptionally active (Brown 1979; Farley 2004; Page and Valone 2007; Schneider 2002). New "social investors" are bringing corporate strategies and expectations of accountability to bear—called "philanthrocapitalism"—that focus intently on biomedical domains (Bishop and Green 2008). The Gates Foundation has also moved into broader global health policymaking, supporting a think tank and making movies (Bumiller 2008; Eiseman and Fossum 2005). Looking at philanthropy regionally, such as Hewa and Hove's (1997) work on Western philanthropy in parts of Asia, is another fruitful research path.

An innovative direction is to "follow those scientists" as they move transnationally to pursue multisite projects. For example, Anderson (2008a) examines the scientists, labs, funding sources, and natives involved in "turning whitemen into kuru scientists" since around 1957. He follows the specimens (brains and other body parts) collected from natives in New Guinea and their global travels along with those of the scientists and the natives as Kuru and mad cow became prion diseases. Transnational collaborations between reproductive scientists and feminists focused on microbicides—a technology simultaneously preventing AIDS and serving as a new mode of contraception (Bell 2003).

Explicitly comparative work on (bio)medicalization remains relatively rare but exciting. Lloyd (2006) analyzed the clinical clash over the medicalization of shyness in France as raising issues of the Americanization of French experience. Parthasarathy (2007) compared the geneticization of breast cancer in the United States and the United Kingdom, revealing the profound consequences of nationalized versus private healthcare. Löwy and Gaudillière (2008) have studied testing practices and ways of managing breast cancer risk in France and compare them with those in the United States and the U.K. They demonstrate how complex interactions among global and local factors shape the multiple meanings assumed by the phrase "cancer risk." Leach and Fairhead (2007) compare the kinds of "vaccine anxieties" that circulate in the U.K. with those in parts of Africa, especially around child health. Van Heteren, Gijswijt-Horstra, and Tansey (2002) offer "biographies of remedies"—comparative studies of drugs, medicines, and contraceptives in Dutch and Anglo-American healing cultures, vividly demonstrating "local biologies" (Lock 1993). A unique comparative offering is a volume combining photographs and narratives of hospitals across four continents (Feldman and Douplitzky 2002).

Governmentalities have also been the focus of research and are of increasing comparative interest. For example, Daemmrich (2004) compares drug regulation in the United States and Germany. Several works ambitiously compare European nations and the United States in terms of governing genomics, biobanks, and other biotech innovations (Gottweis 2005; Gottweis and Petersen 2008; Jasanoff 2007). Reid (2005) compares antismoking policies and campaigns in California, France, and Japan. Baggott and Forster (2008) compare health consumer and patients' organizations in Europe vis-à-vis differing health expectations. Comparative scholarship on disability policy is emerging (e.g., Davidson 2008; Ingstad and Whyte 2007). Recent biopolitical governmentality work also compares the colonial medicines of settler states and consequences for the histories of indigenous health (Anderson 2007, forthcoming).

This sampler of the rich array of recent research following quite different aspects of biomedicine in different geopolitical sites provides both inspiration and context for future projects pursing biomedicalization in its transnational travels. They also alert us to possible complexities. Let us now turn to a different approach.

DOING LOCAL SOCIOHISTORIES OF THE RISE OF
MEDICINE, MEDICALIZATION, AND BIOMEDICALIZATION

My argument here is that the concepts of healthscapes and assemblages and my study of the iconography of things medical in the United States (chapter 3 in this volume), along with our chart across time (chapter 2), can together be used as a provocative model or template by scholars wishing to pursue studies of the rise of medicine, medicalization, and biomedicalization elsewhere. The deeply empirical processes of making such a historical chart and gathering a wide array of images from cultural and geopolitical

areas of interest that form healthscapes should be analytically rich. The results would offer an empirical base from which to theorize the transnationalization of things medical in a more grounded, in-actual-local-practice
fashion. Comparative work would also be facilitated.

In pursuing a study of the rise of medicine, medicalization, and/or biomedicalization elsewhere, then, all the following domains deserve analytic
consideration: political economic and cultural histories of health, healing,
medicines, and public health (including consumption and user studies);
histories of medical education (for all medicines in practice); histories of
healthcare infrastructural development, including public health, in terms
of clinics, hospitals, and/or other sites of service delivery; economic histories of production and importation of medical goods and services from
pharmaceuticals to technologies; histories of clinical research and trials;
the history of clean water, public sanitation, and transportation; and last
but far from least the historical patterns of fiscal sponsorship (private, governmental, philanthropic, international organizational, NGO, etc.) of the
development and distribution of all kinds of health and illness care, education, and training, and the interventions used (from pharmaceuticals to
lab tests to traditional healers and spiritual activities, etc.).

Key events in these histories belong in the chart which should be wildly
inclusive at least at the outset, as the salience of something may well not be
initially transparent. Such charts essentially become historical databases.
They may be (re)organized like ours around the five basic processes of
biomedicalization or around some other empirically generated principles.
Their analytic potency may be augmented by collecting and studying local
iconographies of things medical. In my work on the U.S. case, where things
medical are widely distributed, I decided I wanted representative icons—
images that millions of people would have seen at about the same time.
Criteria for collecting images elsewhere might well differ. For some locales,
gathering images that represent all the different medicines in circulation
across time might be a good initial strategy. Media and visual studies can
be a most useful resource (Clarke 2005, chap. 6; Harrison 2002; Sturken
and Cartwright 2001).

If—and only if—there appear to be eras of the rise of medicine, the beginnings of medicalization, and biomedicalization in the locale under investigation, the chart should allow analyses of their periodization. It should
also allow theorizing of their particular local nature and characteristics.
Social life around health, illness, medicine, and bodies is continually (re)-

organizing itself, but there may be tipping points where the overall emphasis seems to shift, potentially moving from one era into another, that deserve serious analytic attention. The specific forms this takes—the "global singularities" as Reid (2005) calls them—will vary from place to place, from nation to nation, perhaps region to region, over time. That is, the history and nature of the rise of medicine, the emergence and coalescence of medicalization, and the emergence of biomedicalization need to be pursued locally, regionally, nationally, and transnationally. Attention to differences is of greatest import.

There are also sites where there has been no rise of medicine, hence no widely distributed medicalization or biomedicalization. Even where there has been a rise of medicine, flows are not unidirectional and can be blocked, rerouted, appropriated, turned back. The "global situation" is notoriously unstable, rife with resistances and recalcitrancies (Tsing 2005). The sociologist Everett Hughes (1971) urged his students to routinely consider "how things could have been otherwise," to attend to contingencies. I echo him here. For example, parallel to some of the countertrends against biomedicalization in the United States detailed in our introduction, Lew-Ting's (2005) study of the complicated relations between biomedicine and other forms of medicine in Taiwan revealed an "antibiomedicine" movement in a society renowned for its medical pluralism. Wahlberg (2006, 123) discusses how the revival of traditional medicine in Vietnam under Ho Chi Minh is sustained in the present by a postcolonial biopolitics "that aims to promote the 'appropriate' use of traditional herbal medicines." After decades of being deemed quackery, herbal medicine has been deeply integrated into the national public health system, as also occurred in China and Korea. In Korea, traditional practitioners succeeded at professionalizing their knowledge through new forms of standardization and also linked new institutional forms to state medical systems (Ma 2007).

In sum, the healthscapes approach to studying biomedicalization in its transnational travels focuses on one geopolitical region and examines flows of "things medical" over time. The final approach, discussed next, focuses instead on what is traveling.

STUDYING HEALTHSCAPES AS TRAVELING ASSEMBLAGES

The spatiality of change is obviously an issue here. A now very dated, implicitly functionalist, and even imperial concept asserted that "cultural lag" occurred when two or more parts of a culture that are "correlated" actually change at different times. Such a framing normalizes cultural integration, "progress" and "failure to progress," and patrols a deep linearity (Ogburn 1922/1964). Such assumptions undergirded modernization and development theories and programs initiated after World War II, and have been profoundly critiqued for decades (e.g., Cooper and Packard 2005; Edelman and Haugerud 2005). Instead, we talk today of "gaps" or "disjunctures" between what is (bio)medically and (bio)technologically possible and what is actually available, socially understood, and conventionalized—and where. That is, the distributions of things medical are in fact uneven all over the planet in varying ways that, under close scrutiny, increasingly challenge not only the framework of "the West and the rest" (Hall 1996a) but also the nation-state as an adequate, meaningful category in terms of grasping (bio)-medicalization. Sites or healthscapes of super-high-tech biomedicalization sit cheek by jowl with sites where the rise of medicine has not occurred (or has even disappeared, along with welfare or public health). And they do so in the West *and* the rest. The implications of such spatial disjunctures for understanding transnationalizing biomedicine are huge. Local and regional patterns may well be more important than national ones, and variations need to be carefully specified through empirical and richly ethnographic studies (see, e.g., Tsing 2005; Adams and Pigg 2005).

Another empirical alternative, then, is to study biomedicalization in its transnational travels by studying particular healthscapes as traveling assemblages. The concept of healthscapes is a kind of assemblage, an infrastructure of assumptions as well as people, things, places, images (e.g., Appadurai 1996; Deleuze and Guattari 1987; Marcus and Saka 2006). The concept allows for the simultaneous *potential* importance of structural elements, political economies, concrete practices, persons, and discourses—with their *actual* import in a particular case—to be analyzed empirically.

Assemblages often have distinctive capacities for travel. How things (like sciences, medicines, and technologies) travel has been a concern of ST&MS for decades (Hackett et al. 2008). In much of the world, the organization of things medical more closely resembles assemblages—temporary arrangements of things, people, and practices—rather than sustained integrated systems. For example, Ong and Collier (2005, 3–5) talk of "global assemblages" as "spatial forms that are nonisomorphic with standard units of analysis . . . abstractable, mobile and dynamic. As global forms are articulated in specific situations—territorialized in assemblages—they define new material, collective and discursive relationships." In some sites, explicit studies of (bio)medicalization processes and practices may usefully be framed as "global assemblages" reworked into "local healthscapes" parallel to Lock's (1993) "local biologies."

Specifically, as noted earlier, where biomedicine in its fully organized and institutionalized forms is *not* widely distributed, the rise of medicine, medicalization, and biomedicalization may fold together. This may produce sites or pockets of intense biomedicalization—highly localized but fully biomedicalized and biomedicalizing healthscapes intended to serve particular individuals or groups. Extreme forms today are biomedical tourism sites in second- and third-world countries (discussed earlier). Other forms remain unstudied.

In short, a transnational or "traffic between" framing of research issues, with attention to inequities in that traffic, would be useful. Sassen's (2002) global networks of cities may have parallels in terms of networks of biomedical assemblages and institutions. Key questions about such sites, pockets, and networks will include but not be limited to issues of biocapital and stratification—whether wider distribution of access to biomedicine to local populations is part of the project or not.

Conclusions

I began this epilogue with Temkin's (1968) advocacy of doing history and social science focused on global medicine, and worrying about how the framework of biomedicalization theory might contribute to such projects. We especially hope to provoke research in which "sensitivity is evident to both globalized political and economic issues as well as to a situated meaning-centered approach to ethnography" (Lock and Nichter 2002, 1). Explicit studies of biomedicalization in its transnational travels should attend to "the rupture and relocation of material, social and national boundary demarcations" effected by biomedical developments (Lock 2008, 875). The time has come for an intensified focus on situated and comparative studies or at least studies with a more comparative consciousness and awareness. Biomedicalization theory and its processes as analytics, the concepts of healthscapes and assemblages, and the methods of charting and iconography should all be of use in such efforts.

Notes

I am grateful to the Rockefeller Foundation Bellagio Conference and Study Center for a residency on "Medicines and Globalization: Historical and Cultural Analysis" with Warwick Anderson in 2005, which made writing this epilogue possible. I presented these ideas at the East Asian Science and Technology Studies Journal (EASTS) Conference, National Taiwan University (August 2007), and at the Life Sciences and Society Conference, Centro de Estudos Sociais and Centro de Neurosciencias e Biologia Celular, University of Coimbra, Portugal (October 2007). I especially thank Diawie Fu, Chia-Ling Wu, Joao Nunes, and Vololona Rabeharisoa for discussion. Vincanne Adams, Monica Casper, Michelle Murphy, Janet Shim, Jenny Fosket, and the anonymous Duke University Press reviewers commented generously on earlier versions. I thank Jia-Shin Chen and Peter Davidson for citational assistance.

1 Thanks to Warwick Anderson (forthcoming) for this quote.

2 I cite lightly here, focusing on recent works that can serve as resources to access the vast extant literatures.

3 On medicines in their indigenous manifestations and contestations, see, e.g., I. Anderson et al. 2006; W. Anderson 2007, 2008b; Low 2005; and Watson-Verran and Turnbull 1995.

4 For studies of non-Western medicines traveling to the West and elsewhere, see, e.g., Alter 2005; Hog and Hsu 2002; Low 2005; and Sheehan and Brenton 2002.

5 On retheorizing traveling, especially but not only of things medical, see, e.g., Connell 2007; Escobar 2003; Ferguson and Gupta 2002; Fu 2007; Olesen 2002; Stoler and Cooper 1997; and Stoler, McGranahan, and Perdue 2008.

6 See, e.g., http://www.usaid.gov/our_work/global_health; and http://www.cdc.gov/cogh. Universities offering such programs include Washington, Pennsylvania, Wisconsin, and UC San Francisco. The University of California is considering a systemwide program (http://globalhealthsciences.ucsf.edu).

7 Ludwik Fleck, a physician and philosopher of science and medicine, asserted the legitimacy of multiple approaches in the 1930s, influenced by the sociology of knowledge and contemporary anthropology.

8 The GAVI Alliance has recently committed $500 million to building infrastructure. *Los Angeles Times*, January 20, 2007; http://www.gavialliance.org.

9 See http://www.globalhealthsciences.ucsf.edu/programs/AcademicTwinning.aspx. Leslie and Kargon (2006) have written about exporting MIT engineering in ways that parallel this exporting UCSF by "twinning."

10 We drew here on Colen's (1995) concept of stratified reproduction—the varied legitimacies of reproducing for different categories of women. See also Collins 1999.

11 The establishment of (outpatient) medical services for (bio)medicalized tourists has long been considered requisite local infrastructure for the tourist indus-

try (e.g., Wilder-Smith, Schwartz, and Shaw 2007). Critical studies of such sites and their relations with local medical and health needs and infrastructures seem absent from the literature to date.

12 On accreditation, see http://www.bumrungrad.com/Overseas-Medical-Care/ about-us/hospital-accreditation.aspx. On the hospital generally, see http://www .bumrungrad.com.

13 See http://www.ehirc.com and http://www.ehirc.com/international_patients/ index.asp. See also Mudur 2004.

14 See Hackett et al. (2008) for reviews. Saillant and Genest (2007) assess medical anthropological work country by country in North America, western Europe, and Brazil.

15 See also "The Construction and Governance of Randomized Controlled Trials," special issue, *BioSocieties* 2, no. 1 (2007).

16 See, e.g., http://www.epibiostat.ucsf.edu/courses/post.html.

17 See http://www.nhscentreforinvolvement.nhs.uk and the journal *Health Expectations: An International Journal of Public Participation in Health Care and Health Policy.*

REFERENCES

Abate, Tom. 2000a. "Gene Research Project Sets New Standard for Supercomputer Power." *San Francisco Chronicle*, December 18, D1.

——. 2000b. "World Health Leaders Discuss Ways to Bridge the Medical Divide." *San Francisco Chronicle*, February 21, B1, B5.

Abbott, Andrew. 1988. *The System of Professions: An Essay on the Division of Expert Labor*. Chicago: University of Chicago Press.

Abir-Am, Pnina. 1985. "Themes, Genres, and Orders of Legitimation in the Consolidation of New Disciplines: Deconstructing the Historiography of Molecular Biology." *History of Science* 23:73–117.

Ackerman, Michael. 2004. "Science and the Shadow of Ideology in the American Health Foods Movement, 1930s–1960s." In *The Politics of Healing: Histories of Alternative Medicine in Twentieth Century North America*, ed. Robert D. Johnston, 55–70. New York: Routledge.

Adams, Vincanne. 2002. "Randomized Controlled Crime: Indirect Criminalization of Alternative Medicine in the United States." *Social Studies of Science* 32 (5–6): 659–90.

Adams, Vincanne, Suellen Miller, Sienna Craig, Arlene Samen, Lhakpen Nyima, Droyoung Sonam, and Michael Varner. 2005. "The Challenge of Cross Cultural Clinical Trials Research: Case Report from the Tibetan Autonomous Region, People's Republic of China." *Medical Anthropology Quarterly* 19 (3): 267–89.

Adams, Vincanne, Michelle Murphy, and Adele Clarke. 2009. "Anticipation: Technoscience, Affect, Temporality, Speculation." *Subjectivity* 28:246–65.

Adams, Vincanne, and Stacy Pigg, eds. 2005. *Sex in Development: Science, Sexuality, and Morality in Global Perspective.* Durham, N.C.: Duke University Press.

Adkins, Lisa. 2001. "Risk Culture, Self-Reflexivity, and the Making of Sexual Hierarchies." *Body and Society* 7 (1): 5–55.

Adler, Robert. 2003. "Inside the Damaged Brain: New Dynamic Techniques Provide a Deeper Look at Alzheimer's and Schizophrenia." *Boston Globe,* May 6, B14.

Agamben, Giorgio. 2005. *State of Exception.* Chicago: University of Chicago Press.

AHeFT.org. 2005. "A-HeFT and Genomic Medicine." http://www.aheft.org/genomic.asp (accessed March 22, 2005).

Ahuja, Anjana. 2004. "We Can Treat Your Heart Disease . . . If You're Black." *Times* (London), October 29, 2004.

Aizley, Harlyn. 2003. *Buying Dad: One Woman's Search for the Perfect Sperm Donor.* Los Angeles: Alyson.

Akrich, Madeleine, João Nunes, F. Paterson, and Vololona Rabeharisoa, with the participation of Marisa Matias and Angela Marques Filipe. 2008. *The Dynamics of Patient Organizations in Europe.* Paris: Presses de l'École des Mines.

Almeling, Rene. 2007. "Selling Genes, Selling Genders: Egg Agencies, Sperm Banks, and the Medical Market in Genetic Material." *American Sociological Review* 72 (2): 319–40.

Alpers, Svetlana. 1983. *The Art of Describing: Dutch Art in the Seventeenth Century.* London: John Murray.

Altenstetter, Christa, and Reinhard Busse. 2005. "Health Care Reform in Germany: Patchwork Change within Established Governance Structures." *Journal of Health Politics, Policy, and Law* 30 (1–2): 121–42.

Alter, Joseph S., ed. 2005. *Asian Medicine and Globalization: Writings on the Transmission of Chinese, Indian, and Other Asian Medical Practices to Western Societies.* Philadelphia: University of Pennsylvania Press.

Altheide, David. 2002. *Creating Fear: News and the Construction of Crisis.* Piscataway, N.J.: Aldine Transaction.

Althof, Stanley E. 2002. "Quality of Life and Erectile Dysfunction." *Urology* 59:803–10.

Altman, Lawrence K. 1994a. "Cancer Study Overseers Are Assailed at Hearing." *New York Times,* May 12, A17.

———. 1994b. "U.S. Halts Recruitment of Cancer Patients for Studies, Pointing to Flaws in Oversight." *New York Times,* March 30.

American Heart Association. 2001. *2002 Heart and Stroke Statistical Update.* Dallas: American Heart Association.

American Psychiatric Association. 1952. *Diagnostic and Statistical Manual of Mental Disorders.* 1st ed. Washington: American Psychiatric Association.

———. 1980. *Diagnostic and Statistical Manual of Mental Disorders.* 3rd ed. Washington: American Psychiatric Association.

Ames, Bruce N., Ilan Elson-Schwab, and Eli A. Silver. 2002. "High-Dose Vitamin Therapy Stimulates Variant Enzymes with Decreased Coenzyme Binding Affinity (Increased K[m]): Relevance to Genetic Disease and Polymorphisms." *American Journal of Clinical Nutrition* 75:616–58.

Amsterdamska, Olga, and Anja Hiddinga. 2004. "Trading Zones or Citadels? Professionalization and Intellectual Change in the History of Medicine." In *Locating Medical History: The Stories and Their Meanings,* ed. Frank Huisman and John Harley Warner, 237–61. Baltimore: Johns Hopkins University Press.

Anderson, Benedict. 1983. *Imagined Communities.* London: Verso.

Anderson, I., S. Crengle, M. Kamaka, T. H. Chen, N. Palafox, and L. Jackson-Pulver. 2006. "Indigenous Health in Australia, New Zealand, and the Pacific." *Lancet* 367:1775–85.

Anderson, Warwick. 2002. "Postcolonial Technoscience Studies: An Introduction." *Social Studies of Science* 32 (5–6): 643–58.

———. 2004. "Postcolonial Histories of Medicine." In *Locating Medical History: The Stories and Their Meanings,* ed. Frank Huisman and John Harley Warner, 285–306. Baltimore: Johns Hopkins University Press.

———. 2006. *Colonial Pathologies: Hygiene, Race, and Science in the Philippines.* Durham, N.C.: Duke University Press.

———. 2007. "The Colonial Medicine of Settler States: Comparing Histories of Indigenous Health." *Health and History* 9 (2): 1–11.

———. 2008a. *The Collectors of Lost Souls: Turning Whitemen into Kuru Scientists.* Baltimore: Johns Hopkins University Press.

———. 2008b. "Indigenous Health in a Global Frame: From Community Development to Human Rights." *Health and History* 10:94–108.

———. Forthcoming. "Biomedicine in Chinese East Asia: From Semicolonial to Postcolonial?" In *Health and Hygiene in Modern Chinese East Asia,* ed. Angela Ki Che Leung and Charlotte Furth. Durham, N.C.: Duke University Press.

Anderson, Warwick, and Vincanne Adams. 2008. "Pramoedya's Chickens: Postcolonial Studies of Technoscience." In *The Handbook of Science and Technology Studies,* ed. Edward J. Hackett, Olga Amsterdamska, Michael Lynch, and Judy Wajcman, 181–204. 3rd ed. Cambridge: MIT Press.

Andrews, Bridie, and Mary P. Sutphen, eds. 2003. *Medicine and Colonial Identity.* London: Routledge.

Andrews, Gavin, and Hans-Ulrich Wittchen. 1995. "Editorial: Clinical Practice, Measurement, and Information Technology." *Psychological Medicine* 25 (3): 443–46.

Anker, Suzanne, and Dorothy Nelkin. 2004. *The Molecular Gaze: Art in the Genetic Age.* Cold Spring Harbor, N.Y.: Cold Spring Harbor.

Appadurai, Arjun. 1986. *The Social Life of Things: Commodities in Cultural Perspective.* Cambridge: Cambridge University Press.

———. 1996. *Modernity at Large: Cultural Dimensions of Globalization.* Minneapolis: University of Minnesota Press.

Applbaum, Kalman. 2004. *The Marketing Era: From Professional Practice to Global Provisioning.* New York: Routledge.

Apple, Rima D. 1996. *Vitamania: Vitamins in American Culture.* New Brunswick, N.J.: Rutgers University Press.

Arksey, Hilary. 1994. "Expert and Lay Participation in the Construction of Medical Knowledge." *Sociology of Health and Illness* 16 (4): 448–68.

Armstrong, David. 1995. "The Rise of Surveillance Medicine." *Sociology of Health and Illness* 17 (3): 393–404.

———. 2000. "Social Theorizing about Health and Illness." In *Handbook of Social Studies in Health and Medicine,* ed. G. L. Albrecht, R. Fitzpatrick, and S. C. Scrimshaw, 24–35. Thousand Oaks, Calif.: Sage.

———. 2005. "The Myth of Concordance: Response to Stevenson and Scambler." *Health* 9 (1): 23–27.

———. 2006. "Embodiment and Ethics: Constructing Medicine's Two Bodies." *Sociology of Health and Illness* 28 (6): 866–81.

Armstrong, David, and Deborah Caldwell. 2004. "Origins of the Concept of Quality of Life in Health Care: A Rhetorical Solution to a Political Problem." *Social Theory and Health* 2 (4): 361–71.

Armstrong, Elizabeth. 2003. *Conceiving Risk, Bearing Responsibility: Fetal Alcohol Syndrome and the Diagnosis of Moral Disorder.* Baltimore: Johns Hopkins University Press.

Armstrong, Elizabeth, and Barbara Katz Rothman, eds. 2008. *Advances in Medical Sociology: Sociological Perspectives on Bioethical Issues.* Oxford: Elsevier.

Arnason, V., and S. Hjörleifsson. 2007. "Geneticization and Bioethics: Advancing Debate and Research." *Medicine, Health Care, and Philosophy* 10 (4): 417–31.

Arney, William Ray, and Bernard J. Bergen. 1984. *Medicine and the Management of Living.* Chicago: University of Chicago Press.

Arnold, David, ed. 1988. *Imperial Medicine and Indigenous Societies.* Manchester: Manchester University Press.

———. 1994. "Medicine and Colonialism." In *Companion Encyclopedia of the History of Medicine,* ed. W. F. Bynum and Roy Porter, 1393–1416. New York: Routledge.

Ashenburg, Katherine. 2007. *The Dirt on Clean: An Unsanitized History.* New York: Farrar, Straus and Giroux.

Associated Press. 2004. "The Big Buzz Is Magnets: Once Rare, MRI Machines Have Become More Profitable." *San Antonio Express-News,* June 19, D1.

Association of American Medical Colleges [AAMC]. 2004. "Careers in Medicine: Specialty Information Radiology." http://www.aamc.org/students/cim/pub_radiology.htm (accessed March 2005).

AstraZeneca. 2004. "AstraZeneca Presents New Data for CRESTOR in African-American Patients with High Cholesterol at American Heart Association Annual Meeting." November 9. http://sev.prnewswire.com/health-care-hospitals/20041110/nytU07009112004–1.html (accessed February 23, 2005).

Atkinson, Paul, Peter Glasner, and Helen Greenslade. 2007. *New Genetics, New Identities.* London: Routledge.

Baggott, Rob, and Rudolph Forster. 2008. "Health Consumer and Patients' Organizations in Europe: Towards a Comparative Analysis." *Health Expectations* 11:85–94.

Baglia, Jay. 2005. *The Viagra Ad Venture: Masculinity, Media, and the Performance of Sexual Health.* New York: Peter Lang.

Bagnall, Janet. 2001. "Test Genes Should Be Public." *Montreal Gazette*, November 8, B3.

Ballard, Karen, and Mary Ann Elston. 2005. "Medicalization: A Multi-dimensional Concept." *Social Theory and Health* 3 (3): 228–41.

Bancroft, John. 1982. "Erectile Impotence: Psyche or Soma?" *International Journal of Andrology* 5:353–55.

Barbot, Janine. 2006. "How to Build an 'Active' Patient? The Work of AIDS Associations in France." *Social Science and Medicine* 62 (3): 538–51.

Barboza, David. 2006. "Body Business Booms to Meet Exhibitions' Needs." *San Francisco Chronicle*, August 8, 2.

Barker, Kristin. 2003. "Birthing and Bureaucratic Women: Needs Talk and the Definitional Legacy of the Sheppard-Towner Act." *Feminist Studies* 29 (2): 333–55.

Barnard, Anne. 2000. "Clinics Market Scans for the Symptom-Free." *Boston Globe*, August 26, 1, A1.

Barnes, Linda L., and Susan S. Sered, eds. 2005. *Religion and Healing in America.* New York: Oxford University Press.

Barnes, N. 2002. "Collaboration between the U.S. and Mexican HIV/AIDS Sectors: The Role of Community-Based Organizations and Federal HIV/AIDS Funding Policies in Creating a Binational Political-Organizational Field." *International Journal of Sociology and Social Policy* 22 (4–6): 21–46.

Basala, George. 1976. "Pop Science: The Depiction of Science in Popular Culture." In *Science and Its Public: The Changed Relationship*, ed. Gerald Holton and William A. Blanpied, 261–78. Boston: Riedel.

Bashford, Alison. 2006. "Global Biopolitics and the History of World Health." *History of the Human Sciences* 19 (1): 67–88.

Bastian, Hilda. 2002. "Promoting Drugs through Hairdressers? Is Nothing Sacred?" *British Medical Journal* 325 (November): 1180.

Battista, Judy. 2002. "MRI Reveals Partial Tear in Abraham's Left Knee." *New York Times*, August 6, D1.

Baudrillard, Jean. 1996. *The System of Objects.* Trans. James Benedict. New York: Verso.
———. 1998. *The Consumer Society: Myths and Structures.* Thousand Oaks, Calif.: Sage.

Bauer, Martin. 1998. "The Medicalization of Science News—from the 'Rocket-Scalpel' to the 'Gene-Meteorite' Complex." *Social Science Information/Information sur les Sciences Sociales* 37:731–51.

Bayer, Ronald, and Robert L. Spitzer. 1985. "Neurosis, Psychodynamics, and DSM-III." *Archives of General Psychiatry* 42 (2): 187–96.

Beaulieu, Anne. 2000. "The Brain at the End of the Rainbow: The Promises of Brain Scans in the Research Field and in the Media." In *Wild Science: Reading Feminism,*

Medicine, and the Media, ed. J. Marchessault and K. Sawchuk, 39–52. London: Routledge.

———. 2002. "Images Are Not the (Only) Truth: Brain Mapping, Visual Knowledge, and Iconoclasm." *Science, Technology, and Human Values* 271:53–86.

Bebbington, Paul. 2001. "The World Health Report 2001." *Social Psychiatry and Psychiatric Epidemiology* 36:473–74.

Beck, Ulrich. 1992. *Risk Society: Towards a New Modernity.* Thousand Oaks, Calif.: Sage.

Becker, Gay. 2000a. *The Elusive Embryo: How Men and Women Approach New Reproductive Technologies.* Berkeley: University of California Press.

———. 2000b. "Selling Hope: Marketing and Consuming the New Reproductive Technologies in the United States." *Sciences Sociales et Santé* 18:105–25.

Becker, Gay, and Robert D. Nachtigall. 1992. "Eager for Medicalization: The Social Production of Infertility as a Disease." *Sociology of Health and Illness* 14 (4): 456–71.

Becker, Howard. 1963. *Outsiders: Studies in the Sociology of Deviance.* New York: Free Press.

Becker, Howard S., Blanche Geer, Everett C. Hughes, and Anselm L. Strauss. 1961. *Boys in White: Student Culture in Medical School.* Chicago: University of Chicago Press.

Bekelman, J. E., Y. Li, and C. P. Gross. 2003. "Scope and Impact of Financial Conflicts of Interest in Biomedical Research: A Systematic Review." *Journal of the American Medical Association* 289 (4): 454–65.

Belkin, Lisa. 1996. "Charity Begins at . . . the Marketing Meeting, the Gala Event, the Product Tie-In." *New York Times Magazine*, December 22, 40–58.

Bell, Robert. 2001. "MRI Utilization Mystery." *Decisions in Imaging Economics: The Journal of Imaging Technology Management*, May–June. http://www.imagingeconomics.com/library (accessed March 2005).

———. 2004. "Magnetic Resonance in Medicine in 2020." *Decisions in Imaging Economics: The Journal of Imaging Technology Management*, December. http://www.imagingeconomics.com/library (accessed March 2005).

Bell, Susan E. 2002. "Photo Images: Jo Spence's Narratives of Living with Illness." *Health* 6 (1): 5–30.

———. 2003. "Sexual Synthetics: Women, Science, and Microbicides." In *Synthetic Planet: Chemical Politics and the Hazards of Modern Life*, ed. Monica Casper, 197–212. New York: Routledge.

Bellafante, Ginia. 2007. "Doctor, Give Me the News, and Sew Me Up Pretty." *New York Times*, June 18, B4.

Bendelow, Gillian, Mick Carpenter, Caroline Vautier, and Simon Williams, eds. 2001. *Gender, Health, and Healing: The Public/Private Divide.* London: Routledge.

Bender, George A., and Robert A. Thom. 1966. *Great Moments in Medicine: The Stories and Paintings in the Series "A History of Medicine in Pictures" by Parke, Davis and Company.* Stories by George A. Bender. Paintings by Robert A. Thom. Detroit: Northwood Institute.

Benford, Robert D., and David A. Snow. 2000. "Framing Processes and Social Movements: An Overview and Assessment." *Annual Review of Sociology* 26:611–39.

Benson, Heidi. 2008. "Nanking's Unsung Heroes." *San Francisco Chronicle*, January 9, E1–E2.

Benson, Otis, Jr. 1961. Preface to *Psychopharmacological Aspects of Space Flight*, ed. Bernard E. Flaherty. New York: Columbia University Press.

Berenson, Alex. 2005. "Sales of Drugs for Impotence Are Declining." *New York Times*, December 4, 34.

Berenson, Alex, and Reed Abelson. 2008. "The Evidence Gap: Weighing the Costs of a CT Scan's Look inside the Heart." *New York Times*, June 29, B1.

Berg, Marc. 1997. *Rationalizing Medical Work: Decision-Support Techniques and Medical Practices*. Cambridge: MIT Press.

———. 2000. "Orders and Their Others: On the Constitution of Universalities in Medical Work." *Configurations* 8:31–61.

Berkman, Lisa F., and Ichiro Kawachi. 2000a. "A Historical Framework for Social Epidemiology." In *Social Epidemiology*, ed. Lisa F. Berkman and Ichiro Kawachi, 3–12. New York: Oxford University Press.

———, eds. 2000b. *Social Epidemiology*. New York: Oxford University Press.

Berlant, Lauren G. 1997. *The Queen of America Goes to Washington City: Essays on Sex and Citizenship*. Durham, N.C.: Duke University Press.

Bernstein, Douglas, and Peggy Nash. 2001. *Essentials of Psychology*. 2nd ed. Boston: Houghton Mifflin.

Best, Joel. 1989. *Images of Issues: Typifying Contemporary Social Problems*. New York: Aldine de Gruyter.

Biehl, João. 2004. "The Activist State: Global Pharmaceuticals, AIDS, and Citizenship in Brazil." *Social Text* 22 (3): 105–32.

———. 2006. "Pharmaceutical Governance." In *Global Pharmaceuticals: Ethics, Markets, Practices*, ed. Adriana Petryna, Andrew Lakoff, and Arthur Kleinman, 206–39. Durham, N.C.: Duke University Press.

———. 2007. *The Will to Live: AIDS Therapies and the Politics of Survival*. Princeton, N.J.: Princeton University Press.

Binstock, Robert H., and Jennifer R. Fishman. 2010. "Social Dimensions of Anti-aging Science and Medicine." In *The International Handbook of Social Gerontology*, ed. Dale Dannefer and Chris Phillipson. Chicago: University of Chicago Press.

Bird, Chloe E., Peter Conrad, and Allan M. Fremont, eds. 2000. *Handbook of Medical Sociology*, 5th ed. Upper Saddle River, N.J.: Prentice Hall.

Birn, Anne-Emanuelle. 2006. *A Marriage of Convenience: Rockefeller International Health and Revolutionary Mexico*. Rochester: University of Rochester Press.

Bishop, J. W., C. J. Marshall, and J. S. Bentz. 2000. "New Technologies in Gynecologic Cytology." *Journal of Reproductive Medicine* 45:701–19.

Bishop, Matthew, and Michael Green. 2008. *Philanthrocapitalism: How the Rich Can Save the World*. London: Bloomsbury.

Blizzard, Deborah. 2007. *Looking Within: A Sociocultural Examination of Fetoscopy*. Cambridge: MIT Press.

Bloche, M. Gregg. 2004. "Health Care Disparities: Science, Politics, and Race." *New England Journal of Medicine* 350:1574.

Block, Jennifer. 2007. *Pushed: The Painful Truth about Childbirth and Modern Maternity Care*. Cambridge, Mass.: Da Capo.

Blois, Marsden S. 1984. *Information and Medicine: The Nature of Medical Descriptions*. Berkeley: University of California Press.

Blum, Linda M., and Nena F. Stracuzzi. 2004. "Gender in the Prozac Nation: Popular Discourse and Productive Femininity." *Gender and Society* 18 (3): 269–86.

Blume, Stuart. 1992. *Insight and Industry: On the Dynamics of Technological Change in Medicine*. Cambridge: MIT Press.

———. 2003. "Medicine, Technology, and Industry." In *Companion to Medicine in the Twentieth Century*, ed. Roger Cooter and John Pickstone, 171–86. London: Routledge.

Boero, Natalie. 2007. "All the News That's Fat to Print: The American 'Obesity Epidemic' and the Media." *Qualitative Sociology* 30 (1): 41–61

———. 2009. "Fat Kids, Working Moms, and the 'Epidemic of Obesity': Race, Class, and Mother Blame." In *The Fat Studies Reader*, ed. Esther Rothblum and Sondra Solovay, 113–20. New York: New York University Press.

Boli, John, and George M. Thomas, eds. 1999. *Constructing World Culture: International Nongovernmental Organizations Since 1875*. Stanford: Stanford University Press.

Bolter, J. David, and Richard Grusin. 1999. *Remediation: Understanding New Media*. Cambridge: MIT Press.

Bonaccorso, Monica M. Forthcoming. "Narrating Conception, the Baby, Kinship: Heterosexual, Lesbian, and Gay Couples' Accounts of Gamete Donation." *Journal of the Royal Anthropological Institute*.

Bond, Patricia, and Robert Weissman. 1997. "The Costs of Mergers and Acquisitions." *International Journal of Health Services* 27:88–97.

Bordo, Susan. 1993. *Unbearable Weight: Feminism, Western Culture, and the Body*. Berkeley: University of California Press.

———. 1999. *The Male Body*. New York: Farrar, Straus and Giroux.

Boston Women's Health Book Collective. 1971. *Our Bodies, Ourselves*. Boston: South End.

Bowker, Geoffrey C. 2006. *Memory Practices in the Sciences*. Cambridge: MIT Press.

Bowker, Geoffrey C., and Susan Leigh Star. 1999. *Sorting Things Out: Classification and Its Consequences*. Cambridge: MIT Press.

Bracken, Patrick, Joan E. Giller, and Derek Summerfield. 1995. "Psychological Responses to War and Atrocity: The Limitations of Current Concepts." *Social Science and Medicine* 40 (8): 1073–82.

Brand, Rachel. 2004. "Souvenir Sonograms: Firms Draw Criticism for Selling Ultrasound Memories to Parents." *Rocky Mountain News*, August 9, B1.

Brandt, Allan M. 2007. *The Cigarette Century*. New York: Basic.

Brandt, Allan M., and Martha Gardener. 2003. "The Golden Age of Medicine?" In *Companion to Medicine in the Twentieth Century*, ed. Roger Cooter and John Pickstone, 21–37. London: Routledge.

Brass, Larisa. 2005. "New PET Policy Boosts CTI Shares: Reimbursement Plan to Allow Payment for Scans to Detect, Monitor Cancer." *Knoxville News-Sentinel*, February 1, C1.

Breast Cancer Fund of America. 1998. *Art.Rage.Us.: The Art and Outrage of Breast Cancer*. San Francisco: Chronicle.

Breggin, Peter. 1991. *Toxic Psychiatry*. New York: St. Martin's.

Briggs, Charles, and Daniel C. Hallin. 2007. "Biocommunicability: The Neoliberal Subject and Its Contradictions in News Coverage of Health Issues." *Social Text* 25 (4): 43–66.

Brodie, Mollyann, Ursula Foehr, Vicky Rideout, Neal Baer, Carolyn Miller, Rebecca Flournoy, and Drew Altman. 2001. "Communicating Health Information through the Entertainment Media." *Health Affairs* 20 (1): 192–99.

Broom, Alex. 2005. "Medical Specialists' Accounts of the Impact of the Internet on the Doctor/Patient Relationship." *Health* 9 (3): 319–38.

Brown, E. Richard. 1979. *Rockefeller Medicine Men: Medicine and Capitalism in America*. Berkeley: University of California Press.

Brown, Elspeth H. 2005. *The Corporate Eye: Photography and the Rationalization of American Commercial Culture, 1884–1929*. Baltimore: Johns Hopkins University Press.

Brown, Nik, and Andrew Webster. 2004. *New Medical Technologies and Society: Reordering Life*. Cambridge: Polity.

Brown, Phil. 1987. "Popular Epidemiology: Community Response to Toxic Waste Induced Disease in Woburn, Massachusetts, and Other Sites." *Science, Technology, and Human Values* 12 (3–4): 76–85.

———. 1992. "Toxic Waste Contamination and Popular Epidemiology: Lay and Professional Ways of Knowing." *Journal of Health and Social Behavior* 33 (3): 267–81.

———. 1995. "Popular Epidemiology, Toxic Waste, and Social Movements." In *Medicine, Health, and Risk: Sociological Approaches*, ed. J. Gabe, 91–112. Oxford: Blackwell.

———. 1997. "Popular Epidemiology Revisited." *Current Sociology* 45 (3): 137–56.

———. 2007. *Toxic Exposures: Contested Illnesses and the Environmental Health Movement*. New York: Columbia University Press.

Brown, Phil, and Steve Zavetoski, eds. 2005. *Social Movements in Health*. Malden, Mass.: Blackwell.

Brown, Phil, Stephen Zavetoski, Sabrina McCormick, Brian Mayer, Rachel Morello-Frosch, and Rebecca Gasior Altman. 2004. "Embodied Health Movements: New Approaches to Social Movements in Health." *Sociology of Health and Illness* 26 (1): 50–80.

Brown, Phil, Steve Zavestoski, Sabrina McCormick, Joshua Mandelbaum, Theo Luebke, and Meadow Linder. 2001. "A Gulf War Difference: Disputes over Gulf War–Related Illnesses." *Journal of Health and Social Behavior* 42:235–57.

Brown, Theodore M., Marcus Cueto, and Elizabeth Fee. 2006. "The World Health Organization and the Transition from 'International' to 'Global' Public Health." *American Journal of Public Health* 96 (1): 62–72.

Brownlee, Shannon. 2007. *Overtreated: Why Too Much Medicine Is Making Us Sicker and Poorer.* New York: Bloomsbury.

Brulle, Robert J., and David N. Pellow. 2006. "Environmental Justice: Human Health and Environmental Inequalities." *Annual Review of Public Health* 27:103–24.

Bryson, Norman, ed. 1994. *Visual Culture: Images and Interpretation.* Hanover, N.H.: University Press of New England.

Buchanan, Allen, Dan W. Brock, Norman Daniels, and Daniel Winkler. 2000. *From Chance to Choice: Genetics and Justice.* Cambridge: Cambridge University Press.

Bucher, Rue. 1962. "Pathology: A Study of Social Movements within a Profession." *Social Problems* 10 (1): 40–51.

Bucher, Rue, and Anselm L. Strauss. 1961. "Professions in Process." *American Journal of Sociology* 66:325–34.

Bud, Robert, Bernard Finn, and Helmuth Trischler, eds. 1999. *Manifesting Medicine: Bodies and Machines.* Amsterdam: Harwood Academic.

Bumiller, Elisabeth. 2008. "Research Groups Boom in Washington." *New York Times,* January 30, A12.

Bunton, Robin, and Alan Peterson. 2002. *New Genetics and the New Public Health.* New York: Routledge.

Bunton, Robin, Sara Nettleton, and Roger Burrows. 1995a. *The Sociology of Health Promotion: Critical Analyses of Consumption, Lifestyle, and Risk.* New York: Routledge.

———, eds. 1995b. *The Sociology of Health Promotion: Health, Risk, and Consumption under Late Modernity.* London: Routledge.

Burchell, Graham, Colin Gordon, and Peter Miller, eds. 1991. *The Foucault Effect: Studies in Governmentality.* Chicago: University of Chicago Press.

Burri, Regula Valerie, and Joseph Dumit, eds. 2007. *Biomedicine as Culture: Instrumental Practices, Technoscientific Knowledge, and New Modes of Life.* New York: Routledge.

———. 2008. "Social Studies of Scientific Imaging and Visualization." In *The Handbook of Science and Technology Studies,* ed. Edward J. Hackett, Olga Amsterdamska, Michael Lynch, and Judy Wajcman, 297–318. Cambridge: MIT Press.

Bury, Michael R. 1986. "Social Constructionism and the Development of Medical Sociology." *Sociology of Health and Illness* 8 (2): 137–69.

———. 2000. "Health, Ageing, and the Lifecourse." In *Health, Medicine, and Society: Key Theories, Future Agendas,* ed. S. Williams, J. Gabe, and M. Calnan, 87–105. New York: Routledge.

———. 2006. "Dominance from Above and Below." *Society* 43 (6): 37–40.

Buse, Kent, and Gill Walt. 2002. "The WHO and Global Public-Private Health Partnerships: In Search of 'Good' Global Health Governance." In *Public-Private Health Partnerships for Public Health,* ed. Michael R. Reich, 169–95. Harvard Series on Population and International Health. Cambridge: Harvard University Press.

Bush, T. L., and K. J. Helzlsouer. 1993. "Tamoxifen for the Primary Prevention of Breast Cancer: A Review and Critique of the Concept and Trial." *Epidemiological Review* 15:233–43.

Butler, Judith. 1993. *Bodies That Matter: On the Discursive Limits of "Sex."* New York: Routledge.

———. 2002. "Is Kinship Always Already Heterosexual?" *Differences: A Journal of Feminist Cultural Studies* 13 (1): 14–44.

Butler, Robert N., and Mitzi L. Davis. 1988. *Love and Sex after Sixty*. New York: Harper and Row.

Byrd, W. Michael, and Linda A. Clayton. 2000–2002. *An American Health Dilemma*. 2 vols. New York: Routledge.

Cahan, Vicky, and Doug Dollemore. 2004. "National Institute of Aging, Industry Launch Partnership, $60 Million Alzheimer's Disease NeuroImaging Initiative." U.S. Department of Health and Human Services National Institutes of Health Press Release, October 13.

Callahan, Daniel. 1998. *False Hopes: Overcoming the Obstacles to a Sustainable, Affordable Medicine*. New Brunswick, N.J.: Rutgers University Press.

Callon, Michel. 2003. "The Increasing Involvement of Concerned Groups in R&D Policies: What Lessons for Public Powers?" In *Science and Innovation: Rethinking the Rationales for Funding and Governance*, ed. A. Geuna, A. J. Salter, and W. E. Steinmueller, 30–68. Cheltenham, England: Edward Elgar.

Campos, Paul. 2004. *The Obesity Myth: Why America's Obsession with Weight Is Hazardous to Your Health*. New York: Gotham.

Canadian Institute for Heath Information [CIHI]. 2005. *Medical Imaging in Canada, 2004*. Ottawa, Ontario: CIHI.

Canguilhem, Georges. 1966/1994. "The Concept of Life." In *A Vital Rationalist: Writings from Georges Canguilhem*, ed. François Delaporte, trans. Arthur Goldhammer, 302–20. New York: Zone.

———. 1978. *On the Normal and the Pathological*. Trans. C. R. Fawcett. Dordrecht: Reidel.

———. 1992. *Machine and Organism*. New York: Zone.

Cantor, David. 2007. "The Cartoon Medicine Show." *Newsletter of the History of Science Society* 36 (4) :1–5.

Carmichael, Ann G., and Richard M. Ratzan, eds. 1991. *Medicine: A Treasury of Art and Literature*. Beaux Arts Editions. New York: Hugh Lauter Levin.

Carson, Peter, Susan Ziesche, Gary Johnson, and Jay Cohn. 1999. "Racial Differences in Response to Therapy for Heart Failure: Analysis of the Vasodilator–Heart Failure Trials." *Journal of Cardiac Failure* 5:178–87.

Cartwright, Lisa. 1995. *Screening the Body: Tracing Medicine's Visual Culture*. Minneapolis: University of Minnesota Press.

———. 1998. "Gender Artifacts: Technologies of Bodily Display in Medical Culture." In *Visual Display: Culture beyond Appearances*, ed. Lynne Cooke and Peter Wollen, 218–35. New York: New Press.

————. 2000a. "Community and the Public Body in Breast Cancer Media Activism." In *Wild Science: Reading Feminism, Medicine, and the Media*, ed. Janine Marchessault and Kim Sawchuk, 120–38. London: Routledge.

————. 2000b. "Reach Out and Heal Someone: Telemedicine and the Globalization of Health Care." *Health* 4 (3): 347–77.

Casper, Monica J. 1998. *The Making of the Unborn Patient: A Social Anatomy of Fetal Surgery*. New Brunswick, N.J.: Rutgers University Press.

Casper, Monica J. and Marc Berg. 1995. "Introduction to Special Issue on Constructivist Perspectives on Medical Work: Medical Practices in Science and Technology Studies." *Science, Technology and Human Values* 20 (4): 395–407.

Casper, Monica J., and Barbara Koenig. 1996. "Introduction: Reconfiguring Nature and Culture; Intersections of Medical Anthropology and Technoscience Studies." *Medical Anthropology Quarterly* 10 (4): 523–36.

Casper, Monica J., and Courtney Muse. 2006. "Genital Fixations: What's Wrong with Treating Intersex in the Womb?" *American Sexuality Magazine* (NSRC). http://www.nsrc.sfsu.edu/MagArticle.cfm?Article=595andReturnURL=1.

Castel, Robert. 1991. "From Dangerousness to Risk." In *The Foucault Effect: Studies in Governmentality*, ed. Graham Burchell, Colin Gordon, and Peter Miller, 281–98. Chicago: University of Chicago Press.

Castro, Arachu, and Merrill Singer, eds. 2004. *Unhealthy Health Policy: A Critical Anthropological Examination*. Walnut Creek, Calif.: AltaMira.

Catalyst. 2008. *Catalyst: Newsletter of the UCSF Foundation*, no. 17.

Centers for Disease Control. 2000. *Assisted Reproductive Technology Success Rates*. Atlanta: CDC.

Centers for Medicare and Medicaid Services. 2006. Medicare Coverage Database. http://www.cms.hhs.gov/mcd/results.asp?show=all&t=2006622175417 (accessed June 22, 2006).

Chadarevian, Soraya de, and Harmke Kamminga. 1998. *Molecularizing Biology and Medicine: New Practices and Alliances, 1910s–1970s*. Amsterdam, The Netherlands: Harwood Academic.

Chapman, Audrey R., and Mark S. Frankel, eds. 2003. *Designing Our Descendants: The Promise and Perils of Genetic Modifications*. Baltimore: Johns Hopkins University Press.

Charatan, F. 2000. "U.S. Prescription Drug Sales Boosted by Advertising." *British Medical Journal* 321:783.

Charmaz, Kathy. 1992. *Good Days, Bad Days: The Self in Chronic Illness and Time*. New Brunswick, N.J.: Rutgers University Press.

————. 2006. *Constructing Grounded Theory: A Practical Guide through Qualitative Analysis*. Thousand Oaks, Calif.: Sage.

Chen, Dung-Sheng, and Chung-Yeh Deng. 2007. "Interactions between Citizens and Experts in Public Deliberation: A Case Study of Consensus Conferences in Taiwan." *East Asian Science, Technology, and Society: An International Journal* 1:77–97.

Chen, Jia-shin. 2009. "The Science-Policy Complex of Harm Reduction in Taiwan." Ph.D. diss., University of California, San Francisco.

Chervenak, F. A., and L. B. McCullough. 2005. "An Ethical Critique of Boutique Fetal Imaging: A Case for the Medicalization of Fetal Imaging." *American Journal of Obstetrics and Gynecology* 192 (1): 31–33.

Chin, Hsein-yu. 1998. "Colonial Medical Police and Postcolonial Medical Surveillance Systems in Taiwan, 1895–1950s." *Osiris* 13:326–38.

China Daily. 2006. "Health Ministry Tightens Control on Sperm Banks." All China Women's Federation, April 13. http://www.womenofchina.cn (accessed December 12, 2007).

Chlebowski, Rowan T., Deborah E. Collyar, Mark R. Somerfield, and David G. Pfister. 1999. "American Society for Clinical Oncology Technology Assessment on Breast Cancer Risk Reduction Strategies: Tamoxifen and Raloxifene." *Journal of Clinical Oncology* 17:1939–55.

Chobanian, A. V., G. L. Bakris, H. R. Black, W. C. Cushman, L. A. Green, J. L. Izzo Jr., D. W. Jones, B. J. Materson, S. Oparil, J. T. Wright Jr., E. J. Roccella, National Heart, Lung, and Blood Institute Joint National Committee on Prevention, Detection, Evaluation, and Treatment of High Blood Pressure, and National High Blood Pressure Education Program Coordinating Committee. 2003. "The Seventh Report of the Joint National Committee on Prevention, Detection, Evaluation, and Treatment of High Blood Pressure: The JNC 7 Report." *Journal of the American Medical Association* 289 (19): 2560–72.

Christiani, David C. 1996. "Utilization of Biomarker Data for Clinical and Environmental Intervention." *Environmental Health Perspectives* 104:921–25.

Clark, Stephanie Brown. 2004. "Frankenflicks: Medical Monsters in Classic Horror Films." In *Cultural Sutures: Medicine and Media*, ed. Lester D. Friedman, 129–48. Durham, N.C.: Duke University Press.

Clarke, Adele E. 1987. "Research Materials and Reproductive Science in the United States, 1910–1940." In *Physiology in the American Context, 1850–1940*, ed. G. L. Geison, 323–50. Bethesda: American Physiological Society.

———. 1988. "Getting Down to Business: The American Life Sciences as Start-Up and Consolidating Industries, c. 1890–1945." Presented at the Stanford Conference on Organizations, Asilomar.

———. 1990. "Controversy and the Development of Reproductive Sciences." *Social Problems* 37 (1): 18–37.

———. 1991. "Social Worlds/Arenas Theory as Organizational Theory." In *Social Organization and Social Process: Essays in Honor of Anselm Strauss*, ed. R. Maines, 128–35. Hawthorne, N.Y.: Aldine de Gruyter.

———. 1995. "Modernity, Postmodernity, and Reproductive Processes ca. 1890–1990 or, 'Mommy, Where Do Cyborgs Come from Anyway?'" In *The Cyborg Handbook*, ed. C. H. Gray, H. J. Figueroa-Sarriera, and S. Mentor, 139–55. New York: Routledge.

———. 1998. *Disciplining Reproduction: Modernity, American Life Sciences, and "the Problems of Sex."* Berkeley: University of California Press.

———. 2005. *Situational Analysis: Grounded Theory after the Postmodern Turn.* Thousand Oaks, Calif.: Sage.

———. 2007. "Reflections on the Reproductive Sciences in Agriculture in the U.K. and U.S., c. 1900–2000." *History and Philosophy of the Biological and Biomedical Sciences* 38 (2): 316–339.

———. 2009. "Introduction: Reproductive Technologies and Gender in East Asia." EASTS: *East Asian Science and Technology Studies: An International Journal* 3 (1).

Clarke, Adele E., Jennifer R. Fishman, Jennifer R. Fosket, Laura Mamo, and Janet K. Shim. 2000. "Technoscience and the New Biomedicalization: Western Roots, Global Rhizomes." *Sciences Sociales et Santé* 18 (2): 11–42.

———. 2004. "Biomedicalization: Technoscientific Transformation of Health, Illness, and U.S. Biomedicine." In *The Sociology of Health and Illness*, 7th ed., ed. Peter Conrad, 442–55. New York: St. Martin's Press.

Clarke, Adele E., and Joan H. Fujimura. 1992. *The Right Tools for the Job: At Work in Twentieth-Century Life Sciences.* Princeton, N.J.: Princeton University Press.

Clarke, Adele E., and Theresa Montini. 1993. "The Many Faces of RU486: Tales of Situated Knowledges and Technological Contestations." *Science, Technology, and Human Values* 18: 42–78.

Clarke, Adele E., and Virginia L. Olesen. 1999. "Revising, Diffracting, Acting." In *Revisioning Women, Health, and Healing: Feminist, Cultural, and Technoscience Perspectives*, ed. A. E. Clarke and V. L. Olesen, 3–48. New York: Routledge.

Clarke, Adele E., Janet K. Shim, Laura Mamo, Jennifer Ruth Fosket, and Jennifer R. Fishman. 2003. "Biomedicalization: Technoscientific Transformations of Health, Illness, and U.S. Biomedicine." *American Sociological Review* 68 (2): 161–94.

Clarke, Adele. E., Janet K. Shim, Sara Shostak, and Alondra Nelson. 2009. "Biomedicalising Health, Diseases and Identities." In *Handbook of Genetics and Society: Mapping the New Genomic Era*, ed. Paul Akinson, Peter Glasner, and Margaret Lock, 21–40. London: Routledge.

Clarke, Adele E., and Susan Leigh Star. 2008. "The Social Worlds/Arenas Framework as a Theory-Methods Package." In *The Handbook of Science and Technology Studies*, ed. Edward Hackett, Olga Amsterdamska, Michael Lynch, and Judy Wajcman, 113–37. Cambridge: MIT Press.

Clarke, Juanne N. 1983. "Sexism, Feminism, and Medicalism: A Decade Review of Literature on Gender and Illness." *Sociology of Health and Illness* 5 (1): 62–82.

Clough, Patricia. 2007. Introduction to *The Affective Turn: Theorizing the Social*, ed. Patricia Ticineto Clough and Jean Halley, 1–33. Durham, N.C.: Duke University Press.

Clynes, Manfred. 1973. "Sentics: Biocybernetics of Emotion Communication." *Annals of the New York Academy of Sciences* 220: 55–131.

———. 1995. "An Interview with Manfred Clynes." In *The Cyborg Handbook*, ed. Chris Hables Gray, 43–53. New York: Routledge.

Cockburn, Cynthia, and R. Furst-Dilic. 1994. *Bringing Technology Home: Gender and Technology in a Changing Europe*. Philadelphia: Open University Press.

Cockerham, William C. 2001. *The Blackwell Companion to Medical Sociology*. Malden, Mass.: Blackwell Publishing.

———, ed. 2005a. *The Blackwell Companion to Medical Sociology*. Malden, Mass.: Blackwell.

———. 2005b. "Health Lifestyle Theory and the Convergence of Agency and Structure." *Journal of Health and Social Behavior* 46 (1): 51–67.

———. 2007. *Social Causes of Health and Disease*. Cambridge, England: Polity.

Cohen, Carl I. 1991. "Old Age, Gender, and Physical Activity: The Biomedicalization of Aging." *Journal of Sport History* 18:64–80.

Cohen, David, Gwynedd Lloyd, and Joan Stead, eds. 2007. *Critical New Perspectives on ADHD*. London: Routledge.

Cohen, J. 2005. "Public Health: Gates Foundation Picks Winners in Grand Challenges in Global Health." *Science* 309 (5731): 33.

Cohen, Joel W., and Nancy A. Krauss. 2003. "Spending and Service Use among People with the Fifteen Most Costly Medical Conditions, 1997." *Health Affairs* 22 (2): 129–38.

Cohen, Jon. 2003a. "AIDS Vaccine Trial Produces Disappointment and Confusion." *Science* 299:1290–91.

———. 2003b. "VaxGen's Sketchy Statistics." *Science Now*, February 27. http://sciencenow.sciencemag.org/cgi/content/full/2003/227/1 (accessed March 17, 2003).

Cohen, Lawrence. 1993. "The Biomedicalization of Psychiatry: A Critical Overview." *Community Mental Health Journal* 29:509–21.

———. 1999. "Where It Hurts: Indian Material for an Ethics of Organ Transplantation." *Daedalus* 128:135–65.

———. 2005. "Operability, Bioavailability, and Exception." In *Global Assemblages: Technology, Politics, and Ethics as Anthropological Problems*, ed. Aihwa Ong and Stephen J. Collier, 79–90. Malden, Mass.: Blackwell.

Cohen, Marc R., and Audrey Shafer. 2004. "Images and Healers: A Visual History of Scientific Medicine." In *Cultural Sutures: Medicine and Media*, ed. Lester D. Friedman, 197–214. Durham, N.C.: Duke University Press.

Cohn, Jay. 1991. "Lessons from V-HeFT: Questions for V-HeFT II and the Future Therapy of Heart Failure." *Hertz* 16:267–71.

———. 1993. "Introduction." *Circulation, Supplement VI* 87 (6): 1–4.

———. 1994. "Invited Editorial: Treatment of Infarct Related Heart Failure: Vasodilators Other than ACE Inhibitors." *Cardiovascular Drugs and Therapy 1994* 8:119–22.

Cohn, Jay, David Archibald, Susan Ziesche, et al. 1991. "A Comparison of Enalapril with Hydralazine-Isosorbide Dinitrate in the Treatment of Chronic Congestive Heart Failure." *New England Journal of Medicine* 325:303–10.

Cohn, Simon. 2004. "Increasing Resolution, Intensifying Ambiguity: An Ethnographic Account of Seeing Life in Brain Scans." *Economy and Society* 33 (1): 52–76.

Cole, Joanna, with illustration by Bruce Degen. 1990. *The Magic School Bus inside the Human Body*. New York: Scholastic.

Colen, Shellee. 1995. "'Like a Mother to Them': Stratified Reproduction and West Indian Childcare Workers and Employers in New York." In *Conceiving the New World Order: The Global Politics of Reproduction*, ed. F. Ginsberg and R. Rapp. Berkeley: University of California Press.

Collier, Stephen, and Andrew Lakoff. 2005. "On Regimes of Living." In *Global Assemblages: Technology, Politics, and Ethics as Anthropological Problems*, ed. Aihwa Ong and Stephen J. Collier, 22–39. Malden, Mass.: Blackwell.

Collins, Patricia Hill. 1986. "Learning from the Outsider Within: The Sociological Significance of Black Feminist Thought." *Social Problems* 33 (6): S14–S32.

———. 1990/2000. *Black Feminist Thought: Knowledge, Consciousness, and the Politics of Empowerment*. 2nd rev. ed. New York: Routledge.

———. 1999. "Will the 'Real' Mother Please Stand Up? The Logic of Eugenics and American National Family Planning." In *Revisioning Women, Health, and Healing: Feminist, Cultural, and Technoscience Perspectives*, ed. Adele E. Clarke and Virginia L. Olesen, 266–82. New York: Routledge.

Computer Retrieval of Information on Scientific Projects [CRISP]. 2005. "Current and Historical Awards (1972–2005)." http://crisp.cit.nih.gov (accessed February 2005).

Connell, R. W. 1995. *Masculinities*. Berkeley: University of California Press.

Connell, R. W., and James W. Messerschmidt. 2005. "Hegemonic Masculinity: Rethinking the Concept." *Gender and Society* 19 (6): 829–59.

Connell, Raewyn. 2007. *Southern Theory: The Global Dynamics of Knowledge in Social Science*. Cambridge, England: Polity.

Connor, J. T. H. 2007. "Exhibit Essay Review: 'Faux Reality' Show? The *Body Worlds* Phenomenon and Its Reinvention of Anatomical Spectacle." *Bulletin of the History of Medicine* 81:848–65.

Connor, Linda H., and Geoffrey Samuel, eds. 2001. *Healing Powers and Modernity: Traditional Medicine, Shamanism, and Science in Asian Societies*. Westport, Conn.: Bergin and Garvey.

Conrad, Peter. 1975. "The Discovery of Hyperkinesis: Notes on the Medicalization of Deviant Behavior." *Social Problems* 23 (1): 12–21.

———. 1992. "Medicalization and Social Control." *Annual Review of Sociology* 18:209–32.

———. 2000. "Medicalization, Genetics, and Human Problems." In *Handbook of Medical Sociology*, ed. C. E. Bird, P. Conrad, and A. Fremont, 322–33. Thousand Oaks, Calif.: Sage.

———. 2005. "The Shifting Engines of Medicalization." *Journal of Health and Social Behavior* 46 (1): 3–14.

———. 2006. "Up, Down, and Sideways." *Society* 43 (6): 19–20.

———. 2007. *The Medicalization of Society: On the Transformation of Human Conditions into Treatable Disorders*. Baltimore: Johns Hopkins University Press.

Conrad, Peter, and Valerie Leiter. 2004. "Medicalization, Markets, and Consumers." *Journal of Health and Social Behavior* 45 (2): 158–76.

Conrad, Peter, and Deborah Potter. 2000. "From Hyperactive Children to ADHD Adults: Observations on the Expansion of Medical Categories." *Social Problems* 47 (4): 559–82.

———. 2004. "Human Growth Hormone and the Temptations of Biomedical Enhancement." *Sociology of Health and Illness* 26 (2): 184–215.

Conrad, Peter, and Joseph Schneider. 1980a/1992. *Deviance and Medicalization: From Badness to Sickness.* St. Louis: Mosby. Expanded ed. Philadelphia: Temple University Press.

———. 1980b. "Looking at Levels of Medicalization: A Comment on Strong's Critique of the Thesis of Medical Imperialism." *Social Science and Medicine* 14A: 75–79.

Cooper, Frederick, and Randall Packard. 2005. "The History and Politics of Development Knowledge." In *The Anthropology of Development and Globalization: From Classical Political Economy to Contemporary Neoliberalism,* ed. Marc Edelman and Angelique Haugerud, 126–39. Malden, Mass.: Blackwell.

Cooper, Melinda. 2006. "Resuscitations: Stem Cells and the Crisis of Old Age." *Body and Society* 12 (1): 1–12.

———. 2008. *Life as Surplus: Biotechnology and Capitalism in the Neoliberal Era.* Seattle: University of Washington Press.

Cooper, Richard, Jeffrey Cutler, Patrice Desvigne-Nickens, Steven P. Fortmann, L. Friedman, R. Havlik, G. Hogelin, J. Marler, P. McGovern, G. Morosco, Lori Mosca, T. Pearson, Jeremiah Stamler, D. Stryer, and T. Thom. 2000. "Trends and Disparities in Coronary Heart Disease, Stroke, and Other Cardiovascular Diseases in the United States: Findings of the National Conference on Cardiovascular Disease Prevention." *Circulation* 102 (25): 3137–47.

Cooter, Roger. 2004. "'Framing' the End of the Social History of Medicine." In *Locating Medical History: The Stories and Their Meanings,* ed. Frank Huisman and John Harley Warner, 309–37. Baltimore: Johns Hopkins University Press.

Coronil, Fernando. 2001. "Toward a Critique of Globalcentrism: Speculations on Capitalism's Nature." In *Millennial Capitalism and the Culture of Neoliberalism,* ed. J. Comaroff and J. Comaroff, 63–87. Durham, N.C.: Duke University Press.

Costa, Alberto. 1993. "Tamoxifen Trial in Healthy Women at Risk of Breast Cancer." *Lancet* 342: 444.

Courtenay, Will H. 2000. "Constructions of Masculinity and Their Influence on Men's Well-Being: A Theory of Gender and Health." *Social Science and Medicine* 50: 1385–401.

Crawford, Robert. 1985. "A Cultural Account of 'Health': Control, Release, and the Social Body." In *Issues in the Political Economy of Health,* ed. J. B. McKinlay, 60–106. New York: Methuen-Tavistock.

———. 1994. "The Boundaries of the Self and the Unhealthy Other: Reflections on Health, Culture, and AIDS." *Social Science and Medicine* 38 (10): 1347–65.

————. 1999. "Transgression for What? A Response to Simon Williams." *Health* 3 (4): 355–66.

————. 2004. "Risk Ritual and the Management of Control and Anxiety in Medical Culture." *Health* 8 (4): 505–28.

Crenner, Christopher. 2005. *Private Practice: In the Early Twentieth-Century Medical Office of Dr. Richard Cabot.* Baltimore: Johns Hopkins University Press.

Cueto, Marcos. 2007. *The Value of Health: A History of the Pan American Health Organization.* Washington: Pan American Health Organization.

Cunningham, Andrew, and Bridie Andrews, eds. 1997. *Western Medicine as Contested Knowledge.* Manchester: Manchester University Press.

Cusack, Tricia. 2007. "Introduction: Riverscapes and the Formation of National Identity." *National Identities* 9 (2): 101–4.

Cussins, Charis. 1996. "Ontological Choreography: Agency through Objectification in Infertility Clinics." *Social Studies of Science* 26 (3): 575–610.

Cutting Edge Information. 2004a. "Cardiovascular Marketing: Budgets, Staffing, and Strategy." http://www.researchandmarkets.com/reportinfo.asp?report_id=53547&t=0&cat_id=16 (accessed February 15, 2005).

————. 2004b. "Clinical Trials Target Specific Ethnic Groups." April 13. http://www.biobn.com/index.cfm?Page=viewnews&NewsID=0002150187 (accessed March 23, 2005).

Daemmrich, Arthur. 2004. *Pharmacopolitics: Drug Regulation in the United States and Germany.* Chapel Hill: University of North Carolina Press.

Daly, Jeanne. 2005. *Evidence-Based Medicine and the Search for a Science of Clinical Care.* Berkeley: University of California Press.

Daniels, Cynthia R. 2006. *Exposing Men: The Science and Politics of Male Reproduction.* Oxford: Oxford University Press.

Daniels, Cynthia R., and J. Golden. 2004. "Procreative Compounds: Popular Eugenics, Artificial Insemination, and the Rise of the American Sperm Banking Industry." *Journal of Social History* 38 (1): 5–27.

Daniels, K., V. Feyles, J. Nisker, M. Perez-y-Perez, C. Newton, J. A. Parker, F. Tekpetey, and J. Haase. 2006. "Sperm Donation: Implications of Canada's Assisted Human Reproduction Act 2004 for Recipients, Donors, Health Professionals, and Institutions." *Journal of Obstetrics and Gynaecology Canada* 28 (7): 608–15.

Danziger, Kurt. 1997. *Naming the Mind: How Psychology Found Its Language.* London: Sage.

Darby, Robert, and J. Steven Svoboda. 2007. "A Rose by Any Other Name? Rethinking the Similarities and Differences between Male and Female Genital Cutting." *Medical Anthropology Quarterly* 21 (3): 301–23.

Daston, Lorraine. 2004. "Speechless." In *Things That Talk: Object Lessons from Art and Science,* ed. Lorraine Daston, 9–27. New York: Zone.

Davidoff, F., C. D. DeAngelis, J. M. Drazen, J. Hoey, L. Højgaard, R. Horton, S. Kotzin, M. G. Nicholls, M. Nylenna, A. J. Overbeke, H. C. Sox, M. B. van der Weyden, and M. S. Wilkes. 2001. "Sponsorship, Authorship, and Accountability." *Journal of the American Medical Association* 286:1232–34.

Davidson, Michael. 2008. *Concerto for the Left Hand: Disability and the Defamiliar Body.* Ann Arbor: University of Michigan Press.

Davies, Kevin. 2005. "Coming Attractions." *Bio-IT World*, March 8. http://www.bio -itworld.com/archive/030805/firstbase.html (Accessed Mar. 28, 2005).

Davis, Audrey D. 1981. "Life Insurance and the Physical Examination: A Chapter in the Rise of American Medical Technology." *Bulletin of the History of Medicine* 55:392–406.

Davis, Joseph E. 2006. "How Medicalization Lost Its Way." *Society* 43 (6): 51–56.

Davis, Kathy. 2007. *The Making of Our Bodies, Ourselves: How Feminism Travels across Borders.* Durham, N.C.: Duke University Press.

Davis-Berman, Jennifer, and Frances G. Pestello. 2005. "The Medicated Self." In *Studies in Symbolic Interaction* 28, ed. Norman K. Denzin, 283–308. Greenwich, Conn.: JAI.

Davis-Floyd, Robbie. 1992. *Birth as an American Rite of Passage.* Berkeley: University of California Press.

Davison, Charlie, George Davey Smith, and Stephen Frankel. 1991. "Lay Epidemiology and the Prevention Paradox: The Implications of Coronary Candidacy for Health Education." *Sociology of Health and Illness* 13 (1): 1–19.

Debate. 2008. "Beyond the Genome: The Challenge of Synthetic Biology." *BioSocieties* 3 (1): 3–20.

de Chadarevian, Soraya, and Harmke Kamminga, eds. 1998. *Molecularizing Biology and Medicine: New Practices and Alliances, 1910s–1970s.* Amsterdam: Harwood Academic.

DeJohn, Paula. 2005. "In Contrast with 1990s, Media Prices Set to Rise." *Hospital Material Management* 30 (5): 1–2.

Deleuze, Gilles. 1990. "Postscript on the Societies of Control." *October* 59:3–7.

Deleuze, Gilles, and Félix Guattari. 1987. *A Thousand Plateaus: Capitalism and Schizophrenia.* Minneapolis: University of Minnesota Press.

Delmonico, Francis L., Robert Arnold, Nancy Scheper-Hughes, Laura A. Siminoff, Jeffrey Kahn, and Stuart J. Younger. 2002. "Sounding Board: Ethical Incentives—Not Payment—for Organ Donation." *New England Journal of Medicine* 346 (25): 2002–5.

De Vault, Marjorie. 1991. *Feeding the Family: The Social Organization of Caring as Gendered Work.* Chicago: University of Chicago Press.

Dicks, Bella, Bambo Soyinka, and Amanda Coffey. 2006. "Multimodal Ethnography." *Qualitative Research* 6 (1): 77–96.

Diez-Roux, Ana V. 1998. "Bringing Context Back into Epidemiology: Variables and Fallacies in Multilevel Analysis." *American Journal of Public Health* 88 (2): 216–22.

———. 2007. "Integrating Social and Biologic Factors in Health Research: A Systems View." *Annals of Epidemiology* 17:569–74.

Dill, Bonnie T., Amy E. McLaughlin, and Angel David Nieves. 2007. "Future Directions of Feminist Research: Intersectionality." In *Handbook of Feminist Research, Theory, and Praxis,* ed. S. N. Hesse-Biber, 629–38. Thousand Oaks, Calif.: Sage.

Dingwall, Robert. 2006a. "The Enduring Relevance of Professional Dominance." *Knowledge, Work, and Society* 4 (2): 77–98.

———. 2006b. "Imperialism or Encirclement." *Society* 43 (6): 30–36.

Dominus, Susan. 2004. "Growing Up with Mom and Mom." *New York Times Magazine,* October 24, 69–74.

Donald, Alasdair. 2001. "The Wal-Marting of American Psychiatry: An Ethnography of Psychiatric Practice in the late 20th Century." *Culture, Medicine, and Psychiatry* 25 (4): 427–39.

Dowsett, Gary W., and Murray Couch. 2007. "Male Circumcision and HIV Prevention: Is There Really Enough of the Right Kind of Evidence?" *Reproductive Health Matters* 15 (29): 33–44.

Doyal, Lesley. 2000. "Gender Equity in Health: Debates and Dilemmas." *Social Science and Medicine* 51 (6): 931–39.

Doyle, Jennifer. 1999. "Sex, Scandal, and Thomas Eakins' *The Gross Clinic.*" *Representations* 68:1–33.

Draper, Jan. 2002. "'It Was a Real Good Show': The Ultrasound Scan, Fathers, and the Power of Visual Knowledge." *Sociology of Health and Illness* 24 (6): 771–95.

Dries, Daniel, Derek Exner, Bernard J. Gersh, et al. 1999. "Racial Differences in the Outcome of Left Ventricular Dysfunction." *New England Journal of Medicine* 340:609–16.

Dubos, Rene. 1959. *Mirage of Health: Utopias, Progress, and Biological Chance.* New York: Harper and Brothers.

Duffin, Jacalyn, and Alison Li. 1995. "Great Moments: Parke, Davis, and Company and the Creation of Medical Art." *Isis* 86:1–29.

Duggan, Lisa. 2002. "The New Homonormativity: The Sexual Politics of Neoliberalism." In *Materializing Democracy,* ed. R. Castronovo and D. Nelson, 173–94. Durham, N.C.: Duke University Press.

———. 2003. *Twilight of Equality? Neoliberalism: Cultural Politics and the Attack on Democracy.* Boston: Beacon.

Duman, Ronald S. 2004. "Depression: A Case of Neuronal Life and Death?" *Biological Psychiatry* 56 (3): 140–45.

Dumit, Joseph. 1997. "A Digital Image of the Category of the Person: PET Scanning and Objective Self-Fashioning." In *Cyborgs and Citadels: Anthropological Interventions in Emerging Sciences, Technologies, and Medicines,* ed. G. L. Downey and J. Dumit, 83–102. Santa Fe: School of American Research Press.

———. 2004. *Picturing Personhood: Brain Scans and Biomedical Identity.* Princeton, N.J.: Princeton University Press.

———. 2006. "Illnesses You Have to Fight to Get: Facts as Forces in Uncertain, Emergent Illnesses." *Social Science and Medicine* 62 (3): 577–90.

Dumit, Joseph, and N. Greenslit. 2006. "Informed Health and Ethical Identity Management." *Culture, Medicine, and Psychiatry* 30 (2): 127–34.

Dunn, Barbara K., S. Barnett Kramer, and Leslie Ford. 1998. "Phase III, Large-Scale Chemoprevention Trials." *Hematology/Oncology Clinics of North America* 12:1019–36.

Duster, Troy. 2003. *Backdoor to Eugenics*. 2nd ed. New York: Routledge.

———. 2005. "Medicine, Race, and Reification in Science." *Science* 307 (5712): 1050–51.

———. 2006. "2005 Presidential Address: Comparative Perspectives and Competing Explanations; Taking on the Newly Reconfigured Reductionist Challenge to Sociology." *American Sociological Review* 71 (1):1–15.

Dworkin, Shari, and A. A. Ehrhardt. 2007. "Going beyond ABC to Include GEM (Gender Relations, Economic Contexts, and Migration Movements): Critical Reflections on Progress in the HIV/AIDS Epidemic." *American Journal of Public Health* 97:13–16.

Edelman, Marc, and Angelique Haugerud, eds. 2005. *The Anthropology of Development and Globalization: From Classical Political Economy to Contemporary Neoliberalism*. Malden, Mass.: Blackwell.

Edgley, Charles, and Dennis Brissett. 1990. "Health Nazis and the Cult of the Perfect Body: Some Polemical Observations." *Symbolic Interaction* 13 (2): 257–80.

Editorial. 2004. "Toward the First Racial Medicine." *New York Times*, November 13.

Edwards, Jeanette, Sarah Franklin, Eric Hirsch, Frances Price, and Marilyn Strathern, eds. 1993/1999. *Technologies of Procreation: Kinship in the Age of Assisted Conception*. 2nd ed. New York: Routledge.

Ehrenreich, Barbara, and Deirdre English. 1978/2005. *For Her Own Good: Two Centuries of the Experts' Advice to Women*. 1st and 2nd eds. New York: Anchor.

Ehrenreich, Barbara, and John Ehrenreich. 1971. *The American Health Empire: Power, Profits, and Politics*. A HealthPAC Book. New York: Vintage.

———. 1978. "Medicine as Social Control." In *The Cultural Crisis of Modern Medicine*, ed. John Ehrenreich, 39–79. New York: Monthly Review Press.

Ehrenreich, John, ed. 1978. *The Cultural Crisis of Modern Medicine*. New York: Monthly Review Press.

Eiseman, Elisa, and Donna Fossum. 2005. *The Challenges of Creating a Global Health Resource Tracking System*. Santa Monica: RAND. www.rand.org/pubs/monographs/2005/RAND_MG317.sum.pdf.

Eisenberg, D. M., R. B. Davis, S. L. Ettner, S. Appel, S. Wilkey, M. van Rompay, and R. C. Kessler. 1998. "Trends in Alternative Medicine Use in the U.S., 1990–1997: Results of a Follow-Up National Survey." *Journal of the American Medical Association* 280:1569–75.

Eisenberg, D. M., R. C. Kessler, C. Foster, F. E. Norlock, D. R. Calkins, and T. L. Delbanco. 1993. "Unconventional Medicine in the U.S.: Prevalence, Costs, and Patterns of Use." *New England Journal of Medicine* 328:346–52.

Elkins, J. 1996. *The Object Stares Back: On the Nature of Seeing*. San Diego: Harcourt Brace.

Elliott, Carl. 2003. *Better than Well: American Medicine Meets the American Dream*. New York: W. W. Norton.

———. 2008a. "Guinea Pig Zero: A Journal for Human Research Subjects." http:// www.guineapigzero.com/ (accessed May 14, 2010).

———. 2008b. "Guinea-Pigging." *New Yorker*, January 7, 36–41.

Elliott, Victoria S. 2004. "Color-Blind? The Value of Racial Data in Medical Research." amednews.com. January 5. http://www.ama-assn.org/amednews/2004/01/05/hlsa0105.htm (accessed February 7, 2007).

Ellrodt, G., D. J. Cook, J. Lee, M. Cho, D. Hunt, and S. Weingarten. 1997. "Evidence-Based Disease Management." *Journal of the American Medical Association* 278:1687–92.

Emery, Alan E. H., and Marcia L. H. Emery. 2003. *Medicine and Art*. London: Royal Society of Medicine Press.

Emslie, Carol, Kate Hunt, and Graham Watt. 2001. "Invisible Women? The Importance of Gender in Lay Beliefs about Heart Problems." *Sociology of Health and Illness* 23 (2): 203–33.

Eng, David, Jose Munoz, and Judith Halberstam, eds. 2005. Special issue: "What's Queer about Queer Studies Now?" *Social Text* 23 (3/4).

Entine, Jon. 2001. *Taboo: Why Black Athletes Dominate Sports and Why We Are Afraid to Talk about It*. New York: Public Affairs.

———. 2004. Welcome and Opening Presentation at the American Enterprise Institute for Public Policy Research Conference: Race, Medicine, and Public Policy, November 12. http://www.aei.org/event/937 (accessed February 9, 2005).

Eoyang, Eugene Chen. 2007. *Two-Way Mirrors: Cross-Cultural Studies in Glocalization*. Lanham, Md.: Lexington.

Epstein, Richard. 2005. "Disparities and Discrimination in Health Care Coverage: A Critique of the Institute of Medicine Study." *Perspectives in Biology and Medicine* 48 (1): S26–S41.

Epstein, Steven. 1995. "The Construction of Lay Expertise: AIDS Activism and the Forging of Credibility in the Reform of Clinical Trials." *Science, Technology, and Human Values* 20 (4): 408–37.

———. 1996. *Impure Science: AIDS, Activism, and the Politics of Knowledge*. Berkeley: University of California Press.

———. 2004. "Bodily Differences and Collective Identities: The Politics of Gender and Race in Biomedical Research in the United States." *Body and Society* 10 (2–3): 183–203.

———. 2007. *Inclusion: The Politics of Difference in Medical Research*. Chicago: University of Chicago Press.

———. 2008. "Patient Groups and Health Movements." In *The Handbook of Science and Technology Studies*, ed. Edward J. Hackett, Olga Amsterdamska, Michael Lynch, and Judy Wajcman, 499–540. Cambridge: MIT Press.

Ernst, Waltraud, ed. 2002. *Plural Medicine: Tradition and Modernity, 1800–2000*. New York: Routledge.

Escobar, Arturo. 2003. "Place, Nature, and Culture in Discourses of Globalization." In *Localizing Knowledge in a Globalizing World: Recasting the Area Studies Debate*, ed. A. Mirsepassi, A. Basu, and F. Weaver. Syracuse: Syracuse University Press.

Essau, Cecilia Ahmoi, and Hans-Ulrich Wittchen. 1993. "An Overview of the Composite International Diagnostic Interview (CIDI)." *International Journal of Methods in Psychiatric Research* 3:79–85.

Estes, Carroll L. 1991. "The Reagan Legacy: Privatization, the Welfare State, and Aging." In *States, Labor Markets, and the Future of Old Age Policy*, ed. J. Myles and J. Quadagno, 59–83. Philadelphia: Temple University Press.

Estes, Carroll L., and Associates. 2001. *Social Policy and Aging: A Critical Perspective.* Thousand Oaks, Calif.: Sage.

Estes, Carroll L., Simon Biggs, and Chris Phillipson. 2003. "Biomedicalization, Ethics, and Ageing." In *Social Theory, Social Policy, and Ageing*, 79–101. Maidenhead, England: Open University Press.

Estes, Carroll L., and Elizabeth A. Binney. 1989. "The Biomedicalization of Aging: Dangers and Dilemmas." *Gerontologist* 29:587–96.

Estes, Carroll L., Charlene Harrington, and David N. Pellow. 2000. "The Medical Industrial Complex." In *The Encyclopedia of Sociology*, ed. E. F. Borgatta and R. V. Montgomery, 1818–32. Farmington Hills, Mich.: Gale.

Estes, Carroll L., and Karen W. Linkins. 1997. "Devolution and Aging Policy: Racing to the Bottom of Long Term Care?" *International Journal of Health Services* 27:427–42.

Eubanks, Mary. 1994. "Biomarkers: The Clues to Genetic Susceptibility." *Environmental Health Perspectives* 102:2–8.

European Magnetic Resonance Forum [EMRF]. 2005. "Frequently Asked Questions." http://www.emrf.org (accessed January 2005).

———. 2007. "How Many MR Systems Are There?" http://www.emrf.org (accessed September 24, 2007).

Evans, Gary W., and Elyse Kantrowitz. 2002. "Socioeconomic Status and Health: The Potential Role of Environmental Risk Exposure." *Annual Review of Public Health* 23:303–31.

Ewald, François. 1990. "Norms, Discipline, and the Law." *Representations* 30:138–61.

Farley, John. 2004. *To Cast Out Disease: A History of the International Health Division of the Rockefeller Foundation (1913–51).* New York: Oxford University Press.

Faust, David, and Richard A. Miner. 1986. "The Empiricist and His New Clothes: DSM-III in Perspective." *American Journal of Psychiatry* 143 (8): 962–67.

Fausto-Sterling, Anne. 2004. "Refashioning Race: DNA and the Politics of Health Care." *Differences: A Journal of Feminist Cultural Studies* 15 (3): 1–37.

Featherstone, Mike. 1991. "The Body in Consumer Culture." In *The Body: Social Process and Cultural Theory*, ed. M. Featherstone, M. Hepworth, and B. S. Turner, 170–96. London: Sage.

———. 2006. "Genealogies of the Global." *Theory, Culture, and Society* 23 (2–3): 387–92.

Feldman, Stephen L., and Katherine Douplitzky. 2002. *Hospital: The Unseen Demands of Delivering Medical Care.* Chicago: Ivan R. Dee.

Felix, Robert. 1949. "Mental Disorders as a Public Health Problem." *American Journal of Psychiatry* 106 (6): 401–6.

Ferguson, James. 2008. "Towards a Left Art of Government: From 'Foucauldian Critique' to Foucauldian Politics." Paper presented at the Conference on Foucault across the Disciplines, University of California, Santa Cruz, March 1.

Ferguson, James, and Akhil Gupta. 2002. "Spatializing States: Toward an Ethnography of Neoliberal Governmentality." *American Ethnologist* 29:981–1002.

Ferguson, Susan J., and Anne S. Kasper, eds. 2002. *Breast Cancer: The Social Construction of an Illness.* New York: Palgrave Macmillan.

Figert, Anne E. 1996. *Women and the Ownership of PMS: The Structuring of a Psychiatric Disorder.* New York: Aldine de Gruyter.

Finkler, Kaja. 2001. "The King of the Gene: The Medicalization of Family and Kinship in American Society." *Current Anthropology* 42 (2): 235–63.

Fischer, Michael. 1999. "Emergent Forms of Life: Anthropologies of Late or Postmodernities." *Annual Review of Anthropology* 28:455–78.

Fisher, Bernard, Joseph Costantino, Carol K. Redmond, R. Poisson, D. Bowman, J. Couture, et al. 1989. "A Randomized Clinical Trial Evaluating Tamoxifen in the Treatment of Patients with Node-Negative Breast Cancer Who Have Estrogen-Receptor-Positive Tumors." *New England Journal of Medicine* 320:479–84.

Fisher, Bernard, Joseph Costantino, D. Lawrence Wickerham, Carol K. Redmond, Maureen Kavanah, Walter M. Kronin, Victor Vogel, Andrè Robidoux, Nikolay Dimitrov, James Atkins, Mary Daly, Samuel Wieand, Elizabeth Tan-Chiu, Leslie Ford, Norman Wolmark, and other NSABP investigators. 1998. "Tamoxifen for Prevention of Breast Cancer: Report of the National Surgical Adjuvant Breast and Bowel Project P-1 Study." *Journal of the National Cancer Institute* 90:1371–88.

Fisher, Jill A. 2009. *Medical Research for Hire: The Political Economy of Pharmaceutical Clinical Trials.* New Brunswick, N.J.: Rutgers University Press.

Fisher, Melissa S., and Greg Downey, eds. 2006. *Frontiers of Capital: Ethnographic Reflections on the New Economy.* Durham, N.C.: Duke University Press.

Fishman, Jennifer R. 2000. "Breast Cancer: Risk, Science, and Environmental Activism in an 'At Risk' Community." In *Ideologies of Breast Cancer: Feminist Perspectives,* ed. L. Potts, 181–204. New York: St. Martin's.

———.2003. "Desire for Profit: Viagra and the Remaking of Sexual Dysfunction." Ph.D. diss. in sociology, Department of Social and Behavioral Sciences, University of California, San Francisco.

———. 2004. "Manufacturing Desire: The Commodification of Female Sexual Dysfunction." *Social Studies of Science* 34 (2): 187–218.

Fishman, Jennifer R., and Laura Mamo. 2002. "What's in a Disorder? A Cultural Analysis of the Medical and Pharmaceutical Constructions of Male and Female Sexual Dysfunction." *Women and Therapy* 24 (2): 179–93.

Fishman, Linda E., and James D. Bentley. 1997. "The Evolution of Support for Safety-Net Hospitals." *Health Affairs* 16:30–47.

Flaherty, Bernard E. 1961. Conclusion to *Psychophysiological Aspects of Space Flight*, ed. Bernard E. Flaherty. New York: Columbia University Press.

Fleck, Ludwik. 1935/1979. *Genesis and Development of a Scientific Fact*. Chicago: University of Chicago Press.

Fleiss, Joseph L., Robert L. Spitzer, Jacob Cohen, and Jean Endicott. 1972. "Three Computer Diagnosis Methods Compared." *Archives of General Psychiatry* 27:643–49.

Food and Drug Administration [FDA]. 1997. Center for Drug Evaluation and Research, Cardiovascular and Renal Drugs Advisory Committee, transcript of meeting, February 27. http://www.fda.gov/ohrms/DOCKETS/ac/97/transcpt/3264 T1.TXT (accessed August 5, 2002).

———. 2003. "Full-Body CT Scans: What You Need to Know." DHHS Publication (FDA) 03-0001.

———. 2004. Division of Drug Marketing, Advertising, and Communication. Correspondence, "Christine Hemler Smith to Mark R. Szewczak, AstraZeneca, LP," December 21. http://www.fda.gov/cder/warn/2004/12779.pdf (accessed March 23, 2005).

———. 2005. "FDA Approves BiDil Heart Failure Drug for Black Patients." *FDA News*, June 23. http://www.fda.gov/bbs/topics/NEWS/2005/NEW01190.html (accessed July 5, 2005).

Forsythe, Diana E. 1992. "Blaming the User in Medical Informatics." *Knowledge and Society: The Anthropology of Science and Technology* 9:95–111.

———. 1996. "New Bodies, Old Wine: Hidden Cultural Assumptions in a Computerized Explanation System for Migraine Sufferers." *Medical Anthropology Quarterly* 10:551–74.

———. 2001. *Studying Those Who Study Us: An Anthropologist in the World of Artificial Intelligence*, ed. David J. Hess. Stanford: Stanford University Press.

Fosket, Jennifer Ruth. 1999a. Field Notes: American Society of Clinical Oncology. Annual Meetings. Atlanta.

———. 1999b. Field Notes: Breast Cancer Risk Communication Workshop, October 3–5, sponsored by the National Cancer Institute and American Cancer Society. Hilton Head, S.C.

———. 1999c. Field Notes: Northern California Cancer Center (NCCC) Twenty-Five-Year Anniversary Cancer Symposium. San Francisco, Northern California Cancer Center.

———. 2002. "Breast Cancer Risk and the Politics of Prevention: Analysis of a Clinical Trial." Ph.D. diss., Department of Social and Behavioral Sciences, University of California, San Francisco.

———. 2004. "Constructing High-Risk Women: The Development and Standardization of a Breast Cancer Risk Assessment Tool." *Science, Technology, and Human Values* 29 (3): 291–313.

Fosket, Jennifer Ruth, Angela Karran, and Christine LaFia. 2000. "Breast Cancer in Popular Women's Magazines from 1913–1997." In *Breast Cancer: Society Constructs an Epidemic,* ed. S. Ferguson and A. Kasper, 303–23. New York: St. Martin's.

Foucault, Michel. 1970. *The Order of Things: An Archaeology of the Human Sciences.* New York: Vintage.

———. 1973. *The Birth of the Clinic: An Archaeology of Medical Perception.* New York: Pantheon.

———. 1978/2000. "Governmentality." In *Michel Foucault: Power,* vol. 3, ed. James D. Faubion, 201–22. New York: The New Press.

———. 1978/1990. *The History of Sexuality.* Vol. 1. *An Introduction.* New York: Vintage.

———. 1979. *Discipline and Punish: The Birth of the Prison.* New York: Vintage.

———. 1980. *Power/Knowledge: Selected Interviews and Other Writings, 1972–1977.* Brighton, England: Harvester.

———. 1984. "Biopower." In *The Foucault Reader,* ed. Paul Rabinow, 258–89. New York: Pantheon.

———. 1988. "Technologies of the Self." In *Technologies of the Self: A Seminar with Michel Foucault,* ed. L. H. Martin, H. Gutman, and P. H. Hutton, 16–49. Amherst: University of Massachusetts Press.

———. 1991a. "Governmentality." In *The Foucault Effect: Studies in Governmentality,* ed. Graham Burchell, Colin Gordon, and Peter Miller, 87–194. Chicago: University of Chicago Press.

———. 1991b. "Questions of Method." In *The Foucault Effect: Studies in Governmentality,* ed. Graham Burchell, Colin Gordon, and Peter Miller, 73–86. Chicago: University of Chicago Press.

———. 2003. *Society Must Be Defended: Lectures at the College de France, 1975–1976.* Trans. David Macey. New York: Picador.

———. 2007. *Security, Territory, Population: Lectures at the College de France, 1977–1978.* Ed. Michel Senellart. Trans. Graham Burchell. Basingstoke, U.K.: Palgrave Macmillan.

———. 2008. *Birth of the Biopolitical: Lectures at the College de France.* Trans. Graham Burchell. New York: Palgrave Macmillan.

Fox, Bonnie, and Diana Worts. 1999. "Revisiting the Critique of Medicalized Childbirth: A Contribution to the Sociology of Birth." *Gender and Society* 13 (3): 326–46.

Fox, Renee C. 1977. "The Medicalization and Demedicalization of American Society." *Daedalus* 106:9–22.

———. 2001. "Medical Uncertainty Revisited." In *Gender, Health, and Healing: The Public/Private Divide,* ed. Gillian Bendelow, Mick Carpenter, Caroline Vautier, and Simon Williams, 236–53. London: Routledge.

Franklin, Sarah. 2000. "Life Itself: Global Nature and the Genetic Imaginary." In *Global Nature, Global Culture,* by S. Franklin, C. Lury, and J. Stacey, 188–227. London: Sage.

———. 2001. "Culturing Biology: Cell Lines for the Second Millennium." *Health* 5 (3): 335–54.

————. 2003. "Ethical Biocapital: New Strategies of Cell Culture." In *Remaking Life and Death: Toward an Anthropology of the Biosciences,* ed. Sarah Franklin and Margaret Lock, 97–128. Santa Fe: School of American Research Advanced Seminar Series.

————. 2005. "Stem Cells R Us: Emergent Life Forms and the Global Biological." In *Global Assemblages: Technology, Politics, and Ethics as Anthropological Problems,* ed. Aihwa Ong and Stephen J. Collier, 59–78. Malden, Mass.: Blackwell.

————. 2006. "Embryonic Economies: The Double Reproductive Value of Stem Cells." *BioSocieties* 1 (1): 71–90.

Franklin, Sarah, and M. Lock. 2003. "Animation and Cessation: The Remaking of Life and Death." In *Remaking Life and Death: Toward an Anthropology of the Biosciences,* ed. Sarah Franklin and Margaret Lock, 3–22. Santa Fe: School of American Research Advanced Seminar Series.

Franklin, Sarah, and Celia Roberts. 2006. *Born and Made: An Ethnography of Preimplantation Genetic Diagnosis.* Princeton, N.J.: Princeton University Press.

Fraser, Laura. 1998. *Losing It: False Hopes and Fat Profits in the Diet Industry.* New York: Plume.

Frayling, Christopher. 2005. *Mad, Bad, and Dangerous: The Scientist and the Cinema.* London: Reaktion.

Frazzetto, Giovanni, Sinead Keenan, and Ilina Singh. 2007. "'I Bambini e le Droghe': The Right to Ritalin vs. the Right to Childhood in Italy." *BioSocieties* 2:393–412.

Freedman, Daniel X. 1984. "Editorial: Psychiatric Epidemiology Counts." *Archives of General Psychiatry* 41:931–33.

Freese, J., and S. Shostak. 2009. "Genetics and Social Inquiry." *Annual Review of Sociology* 35:107–28.

Freidson, Eliot. 1970. *Profession of Medicine: A Study in the Sociology of Applied Knowledge.* Chicago: University of Chicago Press.

————. 2001. *Professionalism: The Third Logic.* Chicago: University of Chicago Press.

Freitas, Anthony J. 1998. "Belongings: Citizenship, Sexuality, and the Market." In *Everyday Inequalities: Critical Inquiries,* ed. J. O'Brien and J. A. Howard, 361–84. Malden, Mass.: Blackwell.

Freudenheim, Milt. 2006. "Attention Shoppers: Low Prices on Shots in the Clinic off Aisle 7." *New York Times,* May 14, 1, 18.

Freudenheim, Milt, and Melody Petersen. 2001. "The Drug-Price Express Runs into a Wall." *New York Times,* December 23, BU 1, 13.

Freund, Peter E. S. 1982. *The Civilized Body: Social Domination, Control, and Health.* Philadelphia: Temple University Press.

Friedman, Lester D., ed. 2004. *Cultural Sutures: Medicine and Media.* Durham, N.C.: Duke University Press.

Friedman, Meyer, and Ray H. Rosenman. 1959. "Association of Specific Overt Behavior Pattern with Blood and Cardiovascular Findings. *Journal of the American Medical Association* 169:1286–96.

———. 1960. "Overt Behavior Pattern in Coronary Disease." *Journal of the American Medical Association* 173:1320–25.

———. 1974. *Type A Behavior and Your Heart.* New York: Alfred A. Knopf.

Friese, Carrie. 2006. "Rethinking the Biological Clock: Eleventh-Hour Moms, Miracle Moms, and Meanings of Age-Related Infertility." *Social Science and Medicine* 63 (6): 1550–60.

———. 2007. "Enacting Conservation and Biomedicine: Cloning Animals of Endangered Species in the Cultures of Late Modernity." Ph.D. diss., University of California, San Francisco.

Fu, Daiwie. 2007. "How Far Can East Asian STS Go?" *EASTS: East Asian Science and Technology Studies Journal* 1 (1): 1–14.

Fujimura, Joan H. 1988/1997. "The Molecular Biology Bandwagon in Cancer Research: Where Social Worlds Meet." *Social Studies of Science* 35:261–83.

———. 1999. "The Practices and Politics of Producing Meaning in the Human Genome Project." *Sociology of Science Yearbook.* 21 (1): 49–87.

Fujimura, Joan H., Troy Duster, and Ramya Rajagopalan. 2008. "Introduction: Race, Genetics, and Disease: Questions of Evidence, Questions of Consequence." *Social Studies of Science* 38 (5): 643–56.

Fullwiley, Duana. 2008. "The Biologistical Construction of Race: 'Admixture' Technology and the New Genetic Medicine." *Social Studies of Science* 38 (5): 695–735.

Furedi, Frank. 2006. "The End of Professional Dominance." *Society* 43 (6): 14–18.

Gaesser, Glenn. 2002. *Big Fat Lies: The Truth about Your Weight and Your Health.* Carlsbad, Calif.: Gurze.

Gaffikin, Lynne, Harshad Sanghvi, Ricky Lu, Paul D. Blumenthal, and Anne Szarewski. 2008. "Trying to Resolve a Dispute over the Best Way to Diagnose Cervical Neoplasia in a Developing Country: Letters from the Global Literature; Unresolved Issues and Responses." *Medscape Journal of Medicine,* January 10. http://www.medscape.com/viewarticle/567881.

Gail, Mitchell H., Joseph Costantino, John Bryant, Robert Croyle, Laurence Freedman, Kathy Helzlsouer, and Victor Vogel. 1999. "Weighing the Risks and Benefits of Tamoxifen Treatment for Preventing Breast Cancer." *Journal of the National Cancer Institute* 91:1829–46.

Gail, Mitchell H., Louise A. Brinton, David P. Byar, Donald K. Corle, Sylvan B. Green, Catharine Schairer, and John J. Mulvihill. 1989. "Projecting Individualized Probabilities of Developing Breast Cancer for White Females Who Are Being Examined Annually." *Journal of the National Cancer Institute* 81:1879–86.

Ganchoff, Chris. 2004. "Regenerating Movements: Embryonic Stem Cells and the Politics of Potentiality." *Sociology of Health and Illness* 26:757–74.

———. 2008. "Speaking for Stem Cells: Biomedical Activism and Emerging Forms of Patienthood." *Advances in Medical Sociology.* Vol. 10. Bingley, England: JAI.

Gannett News Service. 2003. "AIDS Vaccine Protects Asians, Blacks." February 24. http://www.tucsoncitizen.com/national/2_24_03aids.html (accessed March 6, 2003).

Garcia-Closas, Montserrat, Karl T. Kelsey, John K. Wienke, Xiping Xu, John C. Wain, and David C. Christiani. 1997. "A Case-Control Study of Cytochrome P450 1A1, Glutathione S-Transferase M1, Cigarette Smoking, and Lung Cancer Susceptibility." *Cancer Causes and Control* 8:544–53.

Gardner, Alan. 2004. "Gender, Ethnic Gaps Found in Health Care." September 10. http://www.medicinenet.com (accessed February 23, 2005).

Garrett, Laurie. 2007. "The Challenge of Global Health." *Foreign Affairs* 86 (1).

Garza, Cynthia. 2005. "Using Prenatal Ultrasounds as Keepsakes Grows More Popular and Controversial: FDA and Others Question Ethics of the Procedure in Nonmedical Uses." *Houston Chronicle*, August 2, A1.

Gaudillière, Jean Paul, and Ilana Löwy, eds. 1998. *The Invisible Industrialist: Manufactures and the Production of Scientific Knowledge.* New York: St Martin's.

Geertz, Clifford. 1979. "From the Native's Point of View: On the Nature of Anthropological Understanding." In *Interpretive Social Science*, ed. Paul Rabinow and William M. Sullivan, 225–41. Berkeley: University of California Press.

Genteric Inc. 2001. "Genteric Announces Breakthrough Gene Therapy in a Pill: First Patent Ever Awarded for Oral Delivery of Non-viral Gene Therapy." Press release.

Gentleman, Amelia. 2008. "Kidney Theft Ring Preys on India's Poorest Laborers." *New York Times*, January 30, A3.

George, L. K. 1998. "Self and Identity in Later Life: Protecting and Enhancing the Self." *Journal of Aging and Identity* 3 (3): 133–52.

Gevitz, Norman, ed. 1988. *Other Healers: Unorthodox Medicine in America.* Baltimore: Johns Hopkins University Press.

Ghanem, H., T. Sherif, T. Abdel-Gawad, and T. Asaad. 1998. "Short Term Use of Intracavernous Vasoactive Drugs in the Treatment of Persistent Psychogenic Erectile Dysfunction." *International Journal of Impotence Research* 10:211–14.

Gibbon, Sahra. 2007. *Breast Cancer Genes and the Gendering of Knowledge: Science and Citizenship in the Cultural Context of the "New" Genetics.* New York: Palgrave Macmillan.

Gibbon, Sahra, and Carlos Novas. 2008. *Biosocialities, Genetics, and the Social Sciences.* London: Routledge.

Giddens, Anthony. 1991. *Modernity and Self Identity.* Oxford: Polity.

Gifford, Sandra. 1986. "The Meaning of Lumps: A Case Study of the Ambiguities of Risk." In *Anthropology and Epidemiology: Interdisciplinary Approaches to the Study of Health and Disease*, ed. C. R. Janes, R. Stall, and S. M. Gifford, 213–46. Boston: Reidel.

Gillum, Robert F. 1987. "Heart Failure in the United States, 1970–1985." *American Heart Journal* 113:1043–45.

Gilman, Sander L. 1995. *Picturing Health and Illness: Images of Identity and Difference.* Baltimore: Johns Hopkins University Press.

———. 2001. *Making the Body Beautiful: A Cultural History of Aesthetic Surgery.* Princeton, N.J.: Princeton University Press.

———. 2008. *Race in Contemporary Medicine.* London: Routledge.

Ginsburg, Faye. 1989. *Contested Lives: The Abortion Debate in an American Community.* Berkeley: University of California Press.

Ginsburg, Faye, and Rayna Rapp, eds. 1995. *Conceiving the New World Order: The Global Politics of Reproduction.* Berkeley: University of California Press.

Gitelman, Donald, and Bruce Fligg. 1992. "Diversified Technique." In *Principles and Practice of Chiropractic,* 2nd ed., ed. Scott Haldeman, 483–502. Norwalk, Conn.: Appleton and Lange.

Glassner, Barry. 2000. *The Culture of Fear: Why Americans Are Afraid of the Wrong Things.* New York: Basic.

Glenn, Evelyn Nakano. 1992/1996. "From Servitude to Service Work: Historical Continuities in the Racial Division of Paid Reproductive Labor." In *The Second Signs Reader: Feminist Scholarship, 1983–1996,* ed. Ruth-Ellen B. Joeres and Barbara Laslett, 27–69. Chicago: University of Chicago Press.

Goffman, Erving. 1963. *Stigma: Notes on the Management of Spoiled Identity.* Englewood Cliffs, N.J.: Spectrum.

———. 1976. *Gender Advertisements.* New York: Harper and Row.

Golan, Tal. 1998. "The Authority of Shadows: The Legal Embrace of the X-Ray." *Historical Reflections* 24 (3): 437–58.

Golden, Janet, and Charles Rosenberg. 1991. *Pictures of Health: A Photographic History of Health Care in Philadelphia, 1860–1945.* Philadelphia: University of Pennsylvania Press.

Goldstein, Diane E. 2000. "'When Ovaries Retire': Contrasting Women's Experiences with Feminist and Medical Models of Menopause." *Health* 4 (3): 309–23.

Goode, Erich, and Nachman Ben-Yehuda. 1994. *Moral Panics: The Social Construction of Deviance.* Oxford: Blackwell.

Goodman, Jordan. 2003. "Pharmaceutical Industry." In *Companion to Medicine in the Twentieth Century,* ed. Roger Cooter and John Pickstone, 141–54. London: Routledge.

Goodman, Jordan, Anthony McElligott, and Lara Marks. 2003. *Useful Bodies: Humans in the Service of Medical Science in the Twentieth Century.* Baltimore: Johns Hopkins University Press.

Gordon, Colin. 2004. *Dead on Arrival: The Politics of Health Care in Twentieth Century America.* Princeton, N.J.: Princeton University Press.

Gordon, Deborah R. 1988. "Tenacious Assumptions of Western Medicine." In *Biomedicine Examined,* ed. Margaret Lock and Deborah R. Gordon, 19–56. Dordrecht, Netherlands: Kluwer.

Gossel, Patricia Peck. 1999. "Packaging the Pill." In *Manifesting Medicine: Bodies and Machines,* ed. Robert Bud, Bernard Finn, and Helmuth Trischler, 105–22. Sydney: Harwood Academic.

Gottweis, Herbert. 2005. "Governing Genomics in the 21st Century: Between Risk and Uncertainty." *New Genetics and Society* 24 (2): 175–93.

Gottweis, Herbert, and Alan R. Petersen, eds. 2008. *Biobanks: Governance in Comparative Perspective.* New York: Routledge.

Gravlee, Clarence C., and Elizabeth Sweet. 2008. "Race, Ethnicity, and Racism in Medical Anthropology, 1977–2002." *Medical Anthropology Quarterly* 22 (1): 27–51.

Gray, Chris Hables, Heidi J. Figueroa-Sarriera, and Steven Mentor, eds. 1995. *The Cyborg Handbook*. New York: Routledge.

Green, R. M. 2007. *Babies by Design: The Ethics of Genetic Choice*. New Haven: Yale University Press.

Greene, Jeremy A. 2004. "Attention to 'Details': Etiquette and Pharmaceutical Salesmen in Postwar America." *Social Studies of Science* 34:271–92.

————. 2006. *Prescribing by Numbers: Drugs and the Definition of Disease*. Baltimore: Johns Hopkins University Press.

Greenhalgh, Susan. 2001. *Under the Medical Gaze: Facts and Fictions of Chronic Pain*. Berkeley: University of California Press.

————. 2008. *Just One Child: Science and Policy in Deng's China*. Berkeley: University of California Press.

Grewal, Inderpal, and Caren Kaplan, eds. 1994. *Scattered Hegemonies: Postmodernity and Transnational Feminist Practices*. Minneapolis: University of Minnesota Press.

Grey, Michael R. 2002. *New Deal Medicine: The Rural Health Programs of the Farm Security Administration*. Baltimore: Johns Hopkins University Press.

Grissom, Grant, and Kenneth Howard. 2000. "Directions and COMPASS-PC." In *Handbook of Psychological Assessment in Primary Care Settings*, ed. Mark E. Marvish, 255–75. Mahwah, N.J.: Lawrence Erlbaum.

Groopman, Jerome. 2008. "Faith and Healing: Review of Anne Harrington, *The Cure Within: A History of Mind-Body Medicine*." *New York Times Book Review*, January 27, 14.

Grosz, Elizabeth. 1994. *Volatile Bodies: Toward a Corporeal Feminism*. Bloomington: Indiana University Press.

Guillen, Mauro. 2001. "Is Globalization Civilizing, Destructive, or Feeble? A Critique of Five Key Debates in the Social Science Literature." *Annual Review of Sociology* 27:236–60.

Gussow, Mel. 1979. "The Time of the Wounded Hero." *New York Times*, April 15, section 2, p. 1.

Hackett, Edward J., Olga Amsterdamska, Michael Lynch, and Judy Wajcman, eds. 2008. *The Handbook of Science and Technology Studies*. 3rd ed. Cambridge: MIT Press.

Hacking, Ian. 1983. *Representing and Intervening: Introductory Topics in the Philosophy of Natural Science*. Cambridge: Cambridge University Press.

————. 1999. *The Social Construction of What?* Cambridge, Mass.: Harvard University Press.

————. 2006. "The Cartesian Body." *BioSocieties* 1 (1): 13–15.

Hafferty, Frederick W. 2006. "Medicalization Reconsidered." *Society* 43 (6): 41–46.

Hajdu, David. 2008. *The Ten-Cent Plague: The Great Comic-Book Scare and How It Changed America*. New York: Farrar, Straus and Giroux.

Halfon, Saul. 2007. *The Cairo Consensus: Demographic Surveys, Women's Empowerment, and Regime Change in Population Policy*. Lanham, Md.: Rowman and Littlefield.

Hall, Stephen S. 2005. "The Short of It." *New York Times Magazine,* October 16, 54–59.

Hall, Stuart. 1996a. "The West and the Rest." In *Modernity: An Introduction to Modern Societies,* ed. Stuart Hall, David Held, Don Hubert, and Ken Thompson, 184–201. Malden, Mass.: Blackwell.

———. 1996b. "When was 'The Post-colonial'? Thinking at the Limit." In *The Post-colonial Question: Common Skies, Divided Horizons,* ed. Iain Chambers and Lidia Curti, 262–90. London: Routledge.

———. 1997. *Representation: Cultural Representations and Signifying Practices.* London: Blackwell.

Halpern, Sydney A. 1990. "Medicalization as a Professional Process: Post War Trends in Pediatrics." *Journal of Health and Social Behavior* 31:28–42.

———. 2004. *Lesser Harms: The Morality of Risk in Medical Research.* Chicago: Chicago University Press.

Hamilton, David. 1986. *The Monkey Gland Affair.* London: Chatto and Windus.

Han, Gil Soo. 2000. "Traditional Herbal Medicine in the Korean Community in Australia: A Strategy to Cope with Health Demands of Immigrant Life." *Health* 4 (4): 426–54.

Hannerz, Ulf. 1996. *Transnational Connections: Culture, People, Places.* London: Routledge.

Hansen, Bert. 1997. "The Image and Advocacy of Public Health in American Caricature and Cartoons from 1860–1900." *American Journal of Public Health* 87:1798–807.

———. 1999. "New Images of a New Medicine: Visual Evidence for the Widespread Popularity of Therapeutic Discoveries in America after 1885." *Bulletin of the History of Medicine* 73:629–78.

———. 2004. "Medical History for the Masses: How American Comic Books Celebrated Heroes of Medicine in the 1940s." *Bulletin of the History of Medicine* 78:148–91.

———. 2009. *Picturing Medical Progress from Pasteur to Polio: A History of Mass Media Images and Popular Attitudes in America.* New Brunswick, N.J.: Rutgers University Press.

Haraway, Donna. 1981. "The High Cost of Information in Post–World War II Evolutionary Biology: Ergonomics, Semiotics, and the Sociobiology of Communication Systems." *Philosophical Forum* 13 (2–3): 244–78.

———. 1983. "Signs of Dominance: From a Physiology to a Cybernetics of Primate Society, C. R. Carpenter, 1930–1970." In *Studies in History of Biology,* vol. 6, ed. William Coleman and Camille Limuges, 129–219. Baltimore: Johns Hopkins University Press.

———. 1985. "Manifesto for Cyborgs: Science, Technology, and Socialist Feminism in the 1980s." *Socialist Review* 80:65–108.

———. 1991. *Simians, Cyborgs, and Women: The Reinvention of Nature.* New York: Routledge.

———. 1997. *Modest_Witness@Second_Millennium.FemaleMan©_Meets_Onco-Mouse™: Feminism and Technoscience.* New York: Routledge.

————. 2004. *The Haraway Reader*. New York: Routledge.

————. 2006. "When We Have Never Been Human, What Is to Be Done: Interview with Donna Haraway." *Theory, Culture, and Society* 23 (7–8): 135–58.

————. 2007. *When Species Meet*. Minneapolis: University of Minnesota Press.

Hardey, Michael. 1999. "Doctor in the House: The Internet as a Source of Lay Health Knowledge and the Challenge to Expertise." *Sociology of Health and Illness* 21 (6): 820–35.

Hardt, Michael, and Antonio Negri. 2000. *Empire*. Cambridge: Harvard University Press.

Harrington, Anne. 2008. *The Cure Within: A History of Mind-Body Medicine*. New York: W. W. Norton.

Harrington, Anne, Nikolas Rose, and Ilina Singh. 2006. "Editors' Introduction." *BioSocieties* 1 (1): 1–5.

Harris, J. 2007. *Enhancing Evolution: The Ethical Case for Making Better People*. Princeton, N.J.: Princeton University Press.

Harrison, Barbara. 2002. "Seeing Health and Illness Worlds — Using Visual Methodologies in Sociology of Health and Illness: A Methodological Review." *Sociology of Health and Illness* 24 (6): 856–72.

Harrison, Barbara, and K. Aranda. 1999. "Photography, Power, and Resistance: The Case of Health and Medicine." In *Consuming Cultures: Power and Resistance*, ed. J. Hearn and S. Roseneil. London: Macmillan.

Hartley, Heather, and Cynthia-Lou Coleman. 2007. "News Media Coverage of Direct-to Consumer Pharmaceutical Advertising: Implications for Countervailing Powers Theory." *Health* 12 (1): 107–32.

Hartouni, Valerie. 1997. *Cultural Conceptions: On Reproductive Technologies and the Remaking of Life*. Minneapolis: University of Minnesota Press.

Harvey, A. McGehee, and Susan L. Abrams. 1986. *"For the Welfare of Mankind": The Commonwealth Fund and American Medicine*. Baltimore: Johns Hopkins University Press.

Harvey, David. 1989. *The Condition of Postmodernity*. Oxford: Blackwell.

Hatch, Anthony. 2009. "The Politics of Metabolism: The Metabolic Syndrome and the Reproduction of Race." Ph.D. diss., Department of Sociology, University of Maryland.

Hattis, Dale. 1996. "Variability in Susceptibility: How Big, How Often, for What Responses to What Agents?" *Environmental Toxicology and Pharmacology* 2:135–45.

Hatty, Suzanne E., and James Hatty. 1999. *The Disordered Body, Epidemic Disease, and Cultural Transformation*. Albany: SUNY Press.

Hatzichristou, D. G. 2002. "Sildenafil Citrate: Lessons Learned from Three Years of Clinical Experience." *International Journal of Impotence Research* 14:S42–S52.

Hayden, Cori. 2003. *When Nature Goes Public: The Making and Unmaking of Bioprospecting in Mexico*. Princeton, N.J.: Princeton University Press.

————. 2005. "Bioprospecting's Representational Dilemma." *Science as Culture* 14 (2): 185–200.

———. 2007. "Taking as Giving: Bioscience, Exchange, and the Politics of Benefit-Sharing." *Social Studies of Science* 37 (5): 729–58.

Hayles, N. Katherine. 1999. *How We Became Posthuman: Virtual Bodies in Cybernetics, Literature, and Informatics.* Chicago: University of Chicago Press.

Healy, David. 1996. "Psychopharmacology in the New Medical State." In *Psychotropic Drug Development: Social, Economic, and Pharmacological Aspects,* ed. David Healy and Declan P. Doogan, 13–40. London: Chapman and Hall Medical.

———. 1997. *The Anti-depressant Era.* Cambridge: Harvard University Press.

———. 1998. *The Psychopharmacologists II: Interviews by Dr. David Healy.* London: Chapman and Hall.

———. 2002. *The Creation of Psychopharmacology.* Cambridge: Harvard University Press.

Hedgecoe, Adam. 2001. "Schizophrenia and the Narrative of Enlightened Geneticization." *Social Studies of Science* 31:375–411.

———. 2004. *The Politics of Personalised Medicine: Pharmacogenetics in the Clinic.* Cambridge: Cambridge University Press.

———. 2008. "From Resistance to Usefulness: Sociology and the Clinical Use of Genetics Tests." *BioSocieties* 3 (2): 183–94.

Heimer, Matthew. 2002. "Club Med." *Smart Money,* July, 82.

Helfand, William H. 2002. *Quack, Quack, Quack: The Sellers of Nostrums in Prints, Posters, Ephemera, and Books.* New York: Grolier Club.

Hemminki, Kari, E. Grzybowska, P. Widlak, and M. Chorazy. 1996. "DNA Adducts in Environmental, Occupational, and Life-Style Studies in Human Biomonitoring." *Acta Biochimica Polonica* 43:305–12.

Henderson, Saras, and Alan Petersen, eds. 2002. *Consuming Health: The Commodification of Health Care.* London: Routledge.

Henry, Michael, Jennifer R. Fishman, and Stuart J. Youngner. 2007. "Propranolol and the Prevention of Post-traumatic Stress Disorder: Is It Wrong to Erase the 'Sting' of Bad Memories?" *American Journal of Bioethics* 7 (9): 12–20.

Hess, David J. 2003. "CAM Cancer Therapies in Twentieth-Century North America: Examining Continuities and Change." In *The Politics of Healing: Histories of Alternative Medicine in Twentieth Century North America,* ed. Robert Johnston, 231–44. New York: Routledge.

———. 2005. "Complementary and Alternative Medicine." In *Oxford Encyclopedia of Science, Technology, and Society,* ed. Sal Restivo, 67–72. New York: Oxford University Press.

Hewa, Soma, and Philo Hove, eds. 1997. *Philanthropy and Cultural Context: Western Philanthropy in South, East, and Southeast Asia in the 20th Century.* Lanham, Md.: University Press of America.

Hilgartner, Stephen. 2007. "Making the Bioeconomy Measurable: Politics of an Emerging Machinery." *BioSocieties* 2 (3): 382–86.

Hisashige, A. 1994. "MR Imaging in Japan and the United States: Analysis of Utilization and Economics." *American Journal of Roentgenology* 162:507–10.

Hite, Sheri. 1976. *The Hite Report*. New York: Macmillan.

Hoberman, John. 2006. *Testosterone Dreams: Rejuvenation, Aphrodesia, Doping*. Berkeley: University of California Press.

Hodgetts, Darrin, and Kerry Chamberlain. 1999. "Medicalization and the Depiction of Lay People in Television Health Documentary." *Health* 3:317–33.

Hoffman, Jan. 2005. "Awash in Information, Patients Face a Lonely, Uncertain Road." *New York Times*, August 14, national edition, 1, 16.

Hog, Erling, and Elisabeth Hsu. 2002. "Introduction to Special Issue: Countervailing Creativity; Patient Agency in the Globalization of Asian Medicines." *Anthropology and Medicine* 9:205–21.

Hogle, Linda F. 1999. *Recovering the Nation's Body: Cultural Memory, Medicine, and the Politics of Redemption*. New Brunswick, N.J.: Rutgers University Press.

———. 2000. "Regulating Human Tissue Innovations: Hybrid Forms of Nature and Governmentality." *Sciences Sociales et Santé* 18:53–74.

———. 2003. "Life/Time Warranty: Rechargeable Cells and Extendable Lives." In *Remaking Life and Death: Toward an Anthropology of the Biosciences*, ed. Sarah Franklin and Margaret Lock, 61–96. Santa Fe: School of American Research Advanced Seminar Series.

———. 2006. "Enhancement Technologies and the Body." In *Annual Review of Anthropology*, vol. 34, ed. W. Durham, J. Comaroff, and J. Hill, 695–716. Palo Alto: Annual Reviews.

———. 2008. "Emerging Medical Technologies." In *The Handbook of Science and Technology Studies*, ed. Edward J. Hackett, Olga Amsterdamska, Michael Lynch, and Judy Wajcman, 841–74. 3rd ed. Cambridge: MIT Press.

Holmes, T. 2004. "Vigilante Awarded BiDil Ad Campaign." Blackenterprise.com, December 27. http://www.blackenterprise.com (accessed January 26, 2005).

hooks, bell. 1994. *Sisters of the Yam: Black Women and Self-Recovery*. Boston: South End.

Horton, Richard. 2000. "How Sick Is Modern Medicine?" *New York Review*, November 2, 46–50.

Horwitz, Allan V. 2002. *Creating Mental Illness*. Chicago: University of Chicago Press.

Horwitz, Allan V., and Jerome C. Wakefield. 2007. *The Loss of Sadness: How Psychiatry Transformed Normal Sorrow into Depressive Disorder*. New York: Oxford University Press.

Hotz, Robert. 2001. "Byte by Byte: A Map of the Brain." *Los Angeles Times*, July 2, A12.

Houck, Judith. 2006. *Hot and Bothered: Women, Medicine, and Menopause in Modern America*. Cambridge: Harvard University Press.

Howson, Alexandra. 1998a. "Embodied Obligation: The Female Body and Health Surveillance." In *The Body in Everyday Life*, ed. S. Nettleton and J. Watson, 218–41. New York: Routledge.

———. 1998b. "Surveillance, Knowledge, and Risk: The Embodied Experience of Cervical Screening." *Health* 2:195–215.

———. 2004. *The Body in Society: An Introduction*. Cambridge: Polity.

Huget, Jennifer. 2003. "ADHD Made Visible." *Washington Post*, December 9, HE1.

Hughes, Everett C. 1958. *Men and Their Work*. Glencoe, Ill.: Free Press.

———. 1971. *The Sociological Eye*. Chicago: Aldine Atherton.

Hunt, Kelly J. 2002. "Mexican American More Likely to Die of Heart Disease Than Caucasians." American Heart Association press release, from the Asia Pacific Scientific Forum, April 25.

Hunt, Linda M., and Nedal H. Arar. 2001. "An Analytical Framework for Contrasting Patient and Provider Views of the Process of Chronic Disease Management." *Medical Anthropology Quarterly* 15:347–67.

Hunt, Nancy Rose. 1999. *A Colonial Lexicon of Birth Ritual, Medicalization, and Mobility in the Congo*. Durham, N.C.: Duke University Press.

Hutchins, J. 1998. "Bringing International Patients to American Hospitals: The Johns Hopkins Perspective." *Managed Care Quarterly* 6:22–27.

Huyard, Caroline. 2009. "Who Rules Rare Disease Associations? A Framework to Understand their Actions." *Sociology of Health and Illness* 31 (7): 979–993.

Illich, Ivan. 1976/1982. *Medical Nemesis: The Expropriation of Health*. New York: Pantheon.

———. 1992. *The Mirror of the Past: Lectures and Addresses, 1978–1990*. New York: Marion Boyars.

Incrocci, L., W. C. J. Hop, and A. K. Slob. 1996. "Visual Erotic and Vibrotactile Stimulation and Intracavernous Injection in Screening Men with Erectile Dysfunction: A 3 Year Experience with 406 Cases." *International Journal of Impotence Research* 8:227–32.

Information Means Value [IMV]. 2003. "Latest IMV Study Shows Continued Strength in MRI Market as Product Portfolio Broadens." http://www.imvlimited.com (accessed October 15, 2005).

———. 2005. "Non-hospital Sites Perform Over 40% of All Procedures." http://www.imvlimited.com (accessed October 15, 2005).

———. 2007. "Latest IMV Market Report Shows Continued High Demand for High Field MRI Systems." http://www.imvlimited.com (accessed September 21, 2007).

Ingstad, Benedicte, and Susan Reynolds Whyte, eds. 2007. *Disability in Local and Global Worlds*. Berkeley: University of California Press.

Inhorn, Marcia C. 2006. "Making Muslim Babies: IVF and Gamete Donation in Sunni versus Shi'a Islam." *Culture, Medicine, and Psychiatry* 30 (4): 427–50.

———. 2007. "Medical Anthropology at the Intersections." *Medical Anthropology Quarterly* 21 (3): 249–55.

Institute of Medicine [IOM]. 1999. *To Err Is Human: Building a Safer Health System*. Washington: National Academy of the Sciences.

———. 2002. *Unequal Treatment: Confronting Racial and Ethnic Disparities in Health Care*. Washington: National Academies Press.

International Society on Hypertension in Blacks [ISHIB]. 2005. "Examining the Evidence: Optimizing Heart Failure Management in African American Patients." Lunch Symposium, Annual Meeting of the International Society on Hypertension in Blacks, July 17. http://www.ishib.org/ishib2005/program.htm (accessed July 13, 2005).

Iriye, Akira. 2002. *Global Community: The Role of International Organizations in the Making of the Contemporary World*. Berkeley: University of California Press.

Irwin, Alan, and Brian Wynne, eds. 1996. *Misunderstanding Science: The Public Reconstruction of Science and Technology*. New York: Cambridge University Press.

Ivry, T., and E. Teman. 2008. "Expectant Israeli Fathers and the Medicalized Pregnancy: Ambivalent Compliance and Critical Pragmatism." *Culture, Medicine, and Psychiatry* 32 (3): 358–85.

Jackson, Beth E. 2003. "Situating Epidemiology." In *Gender Perspectives on Health and Medicine: Key Themes*, ed. Marcia Texler Segal, Vasilikie Demos, and Jennie Jacobs Kronenfeld, 11–57. Boston: JAI.

Jackson, Stevi, and Sue Scott. 1997. "Gut Reactions to Matters of the Heart: Reflections on Rationality, Irrationality, and Sexuality." *Sociological Review* 45 (4): 551–75.

Jackson, Sue, and Graham Scambler. 2007. "Perceptions of Evidence-Based Medicine: Traditional Acupuncturists in the U.K. and Resistance to Biomedical Modes of Evaluation." *Sociology of Health and Illness* 29 (3): 412–29.

Jain, Sarah Lochlan. 1998. "Mysterious Delicacies and Ambiguous Agents: Lennart Nilsson in *National Geographic*." *Configurations* 6:373–95.

JAMA. 1891. "Andrology as a Specialty." *JAMA* 17:691.

Jasanoff, Sheila. 2000. "Reconstructing the Past, Constructing the Present: Can Science Studies and the History of Science Live Happily Ever After?" *Social Studies of Science* 30:621–31.

———, ed. 2004. *States of Knowledge: The Co-production of Science and the Social Order*. London: Taylor and Francis.

———. 2007. *Designs on Nature: Science and Democracy in Europe and the United States*. Princeton, N.J.: Princeton University Press.

Jasso-Aquilar, Rebeca, Howard Waitzkin, and Angela Landwehr. 2004. "Multinational Corporations and Health Care in the United States and Latin America." *Journal of Health and Social Behavior* 45 (extra issue): 136–57.

Jenkins, H. W. 1998. "Is Advertising the New Wonder Drug?" *Wall Street Journal*, March 25, A23.

Jensen, Casper Bruun. 2005. "An Experiment in Performative History: Electronic Patient Records as a Future Generating Device." *Social Studies of Science* 35 (2): 241–67.

Jobson, Kenneth. 2004. "Algorithms Assisted by Informatics." *Psychiatric Times*, April 15, 26–27.

Johnson, Carolyn. 2004. "High-Tech Images Open Window into the Skull." *Boston Globe*, September 28, C4.

Johnson, Ericka. 2005. "The Ghost of Anatomies Past: Simulating the One-Sex Body in Modern Medical Training." *Feminist Theory* 6 (2): 141–59.

Johnson, Paul Christopher. 2007. *Diaspora Conversions: Black Carib Religion and the Recovery of Africa*. Berkeley: University of California Press.

Johnson, Robert D. 2004a. "Contemporary Anti-vaccination Movements in Historical Perspective." In *The Politics of Healing: Histories of Alternative Medicine in Twentieth Century North America*, ed. Robert D. Johnson, 259–86. New York: Routledge.

————. 2004b. *The Politics of Healing: Histories of Alternative Medicine in Twentieth
Century North America*. New York: Routledge.

Jones, Anne Hudson. 1981. "Medicine and the Physician in Popular Culture." In *Handbook of American Popular Culture*, vol. 3, ed. M. Thomas Inge, 183–203. Westport,
Conn.: Greenwood.

Jones, David S. 2002. "The Health Care Experiments at Many Farms: The Navajo,
Tuberculosis, and the Limits of Modern Medicine, 1952–1962." *Bulletin of the History of Medicine* 76:749–90.

Jones, James H. 1981/1993. *Bad Blood: The Tuskegee Syphilis Experiment*. New York:
Free Press.

Jordan, V. Craig. 1994. "The Development of Tamoxifen for Breast Cancer Therapy."
In *Long-Term Tamoxifen Treatment for Breast Cancer*, ed. V. C. Jordan, 3–26. Madison: University of Wisconsin Press.

————. 1995. "Tamoxifen for Breast Cancer Prevention." *Proceedings of the Society for
Experimental Biology and Medicine* 208:144–49.

Jordanova, Ludmilla. 1993. "Gender and the Historiography of Science." *British Journal
for the History of Science* 26:469–83.

————. 1998. "Medicine and the Genres of Display." In *Visual Display: Culture beyond
Appearances*, ed. Lynne Cooke and Peter Wollen, 202–17. New York: New Press.

Joyce, Kelly. 2005a. "Appealing Images: Magnetic Resonance Imaging and the Production of Authoritative Knowledge." *Social Studies of Science* 35 (3): 437–62.

————. 2005b. "Cultures of Medicine: Perceptions of Magnetic Resonance Imaging Technology in Japan and the United States." Paper presented at the American
Sociological Association 100th Annual Meeting. Philadelphia, August 13–16.

————. 2008. *Magnetic Appeal: MRI and the Myth of Transparency*. Ithaca, N.Y.: Cornell University Press.

Joyce, Kelly, and Laura Mamo. 2006. "Graying the Cyborg: New Directions in Feminist Analyses of Aging, Science, and Technology." In *Age Matters: Realigning Feminist Thinking*, ed. Toni Calasanti and Katherine Allen, 99–122. New York: Routledge.

Justice, Judith. 1986. *Policies, Plans, and People: Foreign Aid and Health Development*.
Berkeley: University of California Press.

Jutel, Annemarie. 2009. "Sociology of Diagnosis: A Preliminary Review." *Sociology of
Health and Illness* 31 (2): 278–99.

Kahn, Jennifer. 2001. "Let's Make Your Head Interactive." *Wired*, April 9. http://www
.wired.com/wired/archive/9.08/brain.html (accessed October 15, 2005).

Kahn, Jonathan. 2003. "Getting the Numbers Right: Statistical Mischief and Racial
Profiling in Heart Failure Research." *Perspectives in Biology and Medicine* 46:473–83.

————. 2004. "How a Drug Becomes 'Ethnic': Law, Commerce, and the Production of Racial Categories in Medicine." *Yale Journal of Health Policy, Law, and Ethics*
4:1–46.

———. 2005. "'Ethnic' Drugs." *Hastings Center Report* 35 (1): 49.

———. 2006. "Race, Pharmacogenomics, and Marketing: Putting BiDil in Context." *American Journal of Bioethics* 6 (5): W1–W5.

———. 2008. "Exploiting Race in Drug Development: BiDil's Interim Model of Pharmacogenomics." *Social Studies of Science* 38:737–58.

Kahn, Susan Martha. 2000. *Reproducing Jews: A Cultural Account of Assisted Conception in Israel.* Durham, N.C.: Duke University Press.

Kaiser Family Foundation. 2001. *Understanding the Effects of Direct-to-Consumer Prescription Drug Advertising.* Menlo Park, Calif.: Henry J. Kaiser Family Foundation.

Kakigi, R., and H. Shibasaki. 1991. "Effects of Age, Gender, and Stimulus Side on Scalp Topography of Evoked Potentials Following Median Nerve Stimulation." *Journal of Clinical Neurophysiology* 8 (3): 320–30.

Kamps, Louise. 2006. "The Medical Vacation." *Travel and Leisure,* July, 108–10.

Kaplan, Helen Singer. 1974. *The New Sex Therapy: Active Treatments of Dysfunctions.* New York: Brunner-Mazel.

Kaplan, Marcie. 1983. "A Woman's View of DSM-III." *American Psychologist,* July, 786–92.

Karlberg, Kristen. 2000. "The Work of Genetic Care Providers: Managing Uncertainty and Ambiguity." In *Research in the Sociology of Health Care,* vol. 17, ed. J. J. Kronenfeld, 81–97. Stamford, Conn.: JAI.

Karpf, Anne. 1988. *Doctoring the Media: The Reporting of Health and Medicine.* London: Routledge.

Katz, Michael B. 1989. *The Undeserving Poor: From the War on Poverty to the War on Welfare.* New York: Pantheon.

Katz, N. M., B. J. Gersh, and J. L. Cox. 1998. "Changing Practice of Coronary Bypass Surgery and Its Impact on Early Risk and Long-Term Survival." *Current Opinion in Cardiology* 13 (6): 465–75.

Katz, Stephen, and Barbara Marshall. 2003. "New Sex for Old: Lifestyle, Consumerism, and the Ethics of Aging Well." *Journal of Aging Studies* 17:3–16.

Kaufman, Jay S., and Richard S. Cooper. 1999. "Seeking Causal Explanations in Social Epidemiology." *American Journal of Epidemiology* 150 (2): 113–20.

Kaufman, Martin. 1988. "Homeopathy in America: The Rise, Fall, and Persistence of a Medical Heresy." In *Other Healers: Unorthodox Medicine in America,* ed. Norman Gevetz, 99–123. Baltimore: Johns Hopkins University Press.

Kaufman, Sharon R. 2005. *. . . And a Time to Die: How American Hospitals Shape the End of Life.* New York: Scribner.

Kaufman, Sharon R., and Lynne Morgan. 2005. "The Anthropology of the Beginnings and Ends of Life." *Annual Review of Anthropology* 34:317–41.

Kaufman, Sharon R., Ann J. Russ, and Janet K. Shim. 2006. "Aged Bodies and Kinship Matters: The Ethical Field of Kidney Transplant." *American Ethnologist* 33 (1): 81–99.

Kaufman, Sharon R., Janet K. Shim, and Ann J. Russ. 2004. "Revisiting the Biomedicalization of Aging: Clinical Trends and Ethical Challenges." *Gerontologist* 44 (6): 731–38.

———. 2006. "Old Age, Life Extension, and the Character of Medical Choice." *Journal of Gerontology, Social Sciences* 61B (4): S175–S84.

Kawachi, Ichiro, and Bruce P. Kennedy. 2001. "How Income Inequality Affects Health: Evidence from Research in the United States." In *Income, Socioeconomic Status, and Health: Exploring the Relationships*, ed. James A. Auerbach and Barbara Kivimae Krimgold. Washington: National Policy Association.

Kay, Lily E. 1993a. "Life as Technology: Representing, Intervening and Molecularizing." *Rivista di Storia Scienza*, n.s., 1 (2): 85–103.

———. 1993b. *The Molecular Vision of Life: Caltech, the Rockefeller Foundation, and the New Biology*. New York: Oxford University Press.

———. 2000. *Who Wrote the Book of Life? A History of the Genetic Code*. Stanford: Stanford University Press.

Keane, Helen. 2005. "Diagnosing the Male Steroid User: Drug Use, Body Image, and Disordered Masculinity." *Health* 9 (2): 189–209.

Kempner, Joanna. 2006. "Uncovering the Man in Medicine: Lessons Learned from a Case Study of Cluster Headache." *Gender and Society* 20 (5): 632–56.

Kerber, Ross. 2005. "Drug Makers Adding X-Rays, MRIs to Arsenal: Imaging May Help Predict Pharmaceuticals' Effects." *Boston Globe*, January 3, C1, C3.

Kerr, Anne. 2005. "Understanding Genetic Disease in a Socio-historical Context: A Case Study of Cystic Fibrosis." *Sociology of Health and Illness* 27 (6): 873–96.

Kerr, Anne, and Sarah Franklin. 2006. "Genetic Ambivalence: Expertise, Uncertainty, and Communication in the Context of New Genetic Technologies." In *New Technologies in Healthcare: Challenge, Change, and Innovation*, ed. Andrew Webster, 40–56. London: Palgrave.

Kessler, Ronald, Katherine McGonagle, Shanyang Zhao, Christopher Nelson, Michael Hughes, Suzann Eshleman, Hans-Ulrich Wittchen, and Kenneth Kendler. 1994. "Lifetime and 12-Month Prevalence of *DSM-III-R* Psychiatric Disorders in the United States." *Archives of General Psychiatry* 51 (January): 8–19.

Kevles, Bettyann H. 1997. *Naked to the Bone: Medical Imaging in the 20th Century*. New Brunswick, N.J.: Rutgers University Press.

Kilbourne, Edwin D. 1973. "The Molecular Epidemiology of Influenza." *Journal of Infectious Diseases* 127:478–87.

Kim, Jongyoung. 2007. "Alternative Medicine's Encounter with Laboratory Science: The Scientific Construction of Korean Medicine in a Global Age." *Social Studies of Science* 37 (6): 855–80.

Kimmel, Michael S. 2000. *Gendered Society*. New York: Oxford University Press.

Kim-Puri, H. J. 2005. "Conceptualizing Gender-Sexuality-State-Nation: An Introduction." *Gender and Society* 19 (2): 137–59.

King, Nicholas B. 2001. "Infectious Disease in a World of Goods." Ph.D. diss., History of Science, Harvard University.

————. 2002. "Security, Disease, Commerce: Ideologies of Post-colonial Global Health." *Social Studies of Science* 32 (56): 763–89.

————. 2003. "The Influence of Anxiety: September 11th, Bioterrorism, and American Public Health." *Journal of the History of Medicine and the Allied Sciences* 58 (4): 433–41.

————. 2005. "The Ethics of Biodefense." *Bioethics* 19 (4): 432–46.

Kinsey, Alfred C., Wardell B. Pomeroy, Clyde E. Martin, and Paul H. Gebhard. 1953. *Sexual Behavior in the Human Male.* New York: Pocket Books.

Kiple, Kenneth F., ed. 1993. *The Cambridge World History of Human Disease.* Cambridge: Cambridge University Press.

Kirk, Stuart, and Herb Kutchins. 1992. *The Selling of DSM: The Rhetoric of Science in Psychiatry.* New York: Aldine de Gruyter.

Kirschmann, Anne Taylor. 2004. "Making Friends for 'Pure' Homeopathy: Hahnemannians and the Twentieth-Century Preservation and Transformation of Homeopathy." In *The Politics of Healing: Histories of Alternative Medicine in Twentieth Century North America*, ed. Robert D. Johnson, 29–42. New York: Routledge.

Klawiter, Maren. 2008. *The Biopolitics of Breast Cancer: Changing Cultures of Disease and Activism.* Minneapolis: University of Minnesota Press.

Kleinman, A., V. Das, and M. Lock, eds. 1997. *Social Suffering.* Berkeley: University of California Press.

Klerman, Gerald L. 1984. "The Advantages of DSM-III." *American Journal of Psychiatry* 141 (4): 539–42.

Klerman, Gerald L., James H. Coleman, and Robert Purpura. 1986. "The Design and Conduct of the Upjohn Cross-National Collaborative Panic Study." *Psychopharmacology Bulletin* 22 (1): 59–64.

Kline, Nathan S., ed. 1956. *Psychopharmacology.* Washington: American Association for the Advancement of Science.

Kline, Nathan S., and Manfred Clynes. 1961. "Drugs, Space, and Cybernetics: Evolution to Cyborgs." In *Psychopharmacological Aspects of Space Flight*, ed. Bernard E. Flaherty, 345–71. New York: Columbia University Press.

Kline, Nathan S., and Eugene Laska. 1968. Preface to *Computers and Electronic Devices in Psychiatry*, ed. Nathan S. Kline and Eugene Laska. New York: Grune and Stratton.

Kline, Nathan S., and George Logemann. 1967. "The Use of Computers in Office Practice of Psychiatry." *Comprehensive Psychiatry* 8 (6): 544–51.

Knorr Cetina, Karen. 2005. "The Rise of a Culture of Life: The Biological Sciences Are Encouraging the Move Away from the Ideals of the Enlightenment towards an Idea of Individual Perfectibility and Enhancement." *EMBO Reports* 6:S76–S80.

Koenig, Barbara. 1988. "The Technological Imperative in Medical Practice: The Social Creation of a Routine Treatment." In *Biomedicine Examined*, ed. M. Lock and D. Gordon, 456–96. Dordrecht, Netherlands: Kluwer Academic.

Koenig, Barbara, Sandra Soo-Jin Lee, and Sarah Richardson, eds. 2008. *Revisiting Race in a Genomic Age.* New Brunswick, N.J.: Rutgers University Press.

Kohler, Robert E. 1982. *From Medical Chemistry to Biochemistry: The Making of a Biomedical Discipline.* Cambridge: Cambridge University Press.

———. 1991. *Partners in Science: Foundations and Natural Scientists, 1900–1945.* Chicago: University of Chicago Press.

Kolata, Gina. 1995. "Man's World, Woman's World? Brain Studies Point to Differences." *New York Times,* February 28, C1, C7.

———. 2002. "Fertility Inc.: Clinics Race to Lure Clients." *New York Times,* January 1, D1, D7.

———. 2006. "If You've Got a Pulse, You're Sick." *New York Times,* May 21, WK1, 5.

———. 2009. "Mammogram Debate Took Group by Surprise." *New York Times,* November 20, A22.

Kolko, Beth E., Lisa Nakamura, and Gilbert B. Rodman, eds. 2000. *Race in Cyberspace.* New York: Routledge.

Komesaroff, Linda, ed. 2007. *Surgical Consent: Bioethics and Cochlear Implantation.* Washington: Gallaudet University Press.

Korogi, Y., and M. Takahashi. 1997. "Cost Containment and Diffusion of MRI: Oil and Water? Japanese Experience." *European Radiology* 7 (5): S256–S258.

Kowalczyk, Liz. 2002. "Rush for Medical Scans Raises Concerns on Costs." Excite Media Group. http://www.excitepr.com (accessed March 2005).

Kraidy, Marwan M. 2002. "The Global, the Local, and the Hybrid: A Native Ethnography of Glocalization." In *Ethnographic Research: A Reader,* ed. Stephanie Taylor, 187–209. London: Sage.

Kramer, Morton. 1975. "Some Perspectives on the Role of Biostatistics and Epidemiology in the Prevention and Control of Mental Disorders." *Milbank Memorial Fund Quarterly* (summer): 279–336.

Krane, R. J., I. Goldstein, and I. S. DeTejada. 1989. "Impotence." *New England Journal of Medicine* 321:1649–59.

Krieger, Nancy. 1994. "Epidemiology and the Web of Causation: Has Anyone Seen the Spider?" *Social Science and Medicine* 39 (7): 887–903.

———. 1997. "The Need for Epidemiologic Theory." *Epidemiology* 8 (2): 212–14.

———. 1999. "Embodying Inequality: A Review of Concepts, Measures, and Methods for Studying Health Consequences Of Discrimination." *International Journal of Health Services* 29 (2): 295–352.

———. 2000. "Epidemiology and Social Sciences: Towards a Critical Reengagement in the 21st Century." *Epidemiological Reviews* 22 (1): 155–63.

———. 2003. "Genders, Sexes, and Health: What Are the Connections and Why Does It Matter?" *International Journal of Epidemiology* 32:652–57.

———. 2008. "Hormone Therapy and the Rise and Perhaps Fall of U.S. Breast Cancer Incidence Rates: Critical Reflections." *International Journal of Epidemiology* 37 (3): 638–40.

Krieger, Nancy, Ilana Löwy, Robert Aronowitz, Judyann Bigby, Kay Dickersin, Elizabeth Garner, Jean-Paul Gaudillière, Carolina Hinestrosa, Ruth Hubbard, Paula A. Johnson, et al. 2005. "Hormone Replacement Therapy, Cancer, and

Women's Health: Historical, Epidemiological, Biological, Clinical, and Advocacy Perspectives. An Interdisciplinary Analysis of the Hormone Therapy Replacement Saga." *Journal of Epidemiology and Community Health* 59:740–48.

Krieger, Nancy, Diane L. Rowley, Allen A. Herman, Byllye Avery, and Mona T. Phillips. 1993. "Racism, Sexism, and Social Class: Implications for Studies of Health, Disease, and Well-Being." *American Journal of Preventive Medicine* 9 (6): 82–122.

Krimsky, Sheldon. 1999. "The Profit of Scientific Discovery and Its Normative Implications." *Chicago-Kent Law Review* 75:15–40.

Kroll-Smith, Steve, and H. Hugh Floyd. 1997. *Bodies in Protest: Environmental Illness and the Struggle over Medical Knowledge.* New York: New York University Press.

Kuczynski, Alex. 2006. *Beauty Junkies: Inside Our $15 Billion Obsession with Cosmetic Surgery.* New York: Doubleday.

Kugelmann, Robert. 1992. *Stress: The Nature and History of Engineered Grief.* Westport, Conn.: Praeger.

Kuhlmann, Ellen, and Birgit Babitsch. 2002. "Bodies, Health, Gender: Bridging Feminist Theories and Women's Health." *Women's Studies International Forum* 25 (4): 433–42.

Kuhn, Maggie, with Christina Long and Laura Quinn. 1991. *No Stone Unturned: The Life and Times of Maggie Kuhn.* New York: Ballantine.

Kuo, Wen-Hua. 2007. "Race at the Frontier of Pharmaceutical Regulation: Deciphering the Racial Difference Debate in the ICH." Paper presented at the Second Conference of *EASTS* Journal, NTU, Taipei, August.

Kutchins, Herb, and Stuart A. Kirk. 1988. "The Business of Diagnosis: DSM-III and Clinical Social Work." *Social Work* 33 (4): 215–20.

Laird, Pamela Walker. 1998. *Advertising Progress: American Business and the Rise of Consumer Marketing.* Baltimore: Johns Hopkins University Press.

Lakoff, Andrew. 2004. "The Anxieties of Globalization: Antidepressant Sales and Economic Crisis in Argentina." *Social Studies of Science* 34 (2): 247–69.

———. 2006. *Pharmaceutical Reason: Knowledge and Value in Global Psychiatry.* Cambridge: Cambridge University Press.

———. 2008. "The Right Patients for the Drug: Pharmaceutical Circuits and the Codification of Illness." In *The Handbook of Science and Technology Studies*, ed. Edward J. Hackett, Olga Amsterdamska, Michael Lynch, and Judy Wajcman, 741–60. 3rd ed. Cambridge: MIT Press.

Landecker, Hannah. 1999. "Between Beneficence and Chattel: The Human Biological in Law and Science." *Science in Context* 12 (1): 203–25.

Landzelius, K. 2006. "Introduction: Patient Organization Movements and New Metamorphoses in Patienthood." *Social Science and Medicine* 62 (3): 529–37.

Langford, Jean M. 2002. *Fluent Bodies: Ayurvedic Remedies for Postcolonial Imbalance.* Durham, N.C.: Duke University Press.

Laqueur, Thomas W. 1989. "Bodies, Details and the Humanitarian Narrative." In *The New Cultural History*, ed. L. Hunt. Berkeley: University of California Press.

Lash, Scott. 2001. "Technological Forms of Life." *Theory, Culture, and Society* 18 (1): 105–20.

———. 2007. "Power after Hegemony: Cultural Studies in Mutation?" *Theory, Culture, and Society* 24 (3): 55–78.

Laska, Eugene, and Nathan S. Kline. 1968. "Computers at the Rockland Research Center." In *Computers and Electronic Devices in Psychiatry*, ed. Nathan S. Kline and Eugene Laska, 4–11. New York: Grune and Stratton.

Laska, Eugene, D. Morrill, Nathan S. Kline, E. Hackett, and George Simpson. 1967. "SCRIBE—A Method for Producing Automated Narrative Psychiatric Case Histories." *American Journal of Psychiatry* 124 (1): 82–84.

Latimer, J. 2007. "Diagnosis, Dysmorphology, and the Family: Knowledge, Motility, Choice." *Medical Anthropology* 26:53–94.

Latour, Bruno. 1987. *Science in Action: How to Follow Scientists and Engineers through Society*. Cambridge: Harvard University Press.

———. 1993. *We Have Never Been Modern*. Trans. Catherine Porter. Cambridge: Harvard University Press.

Latour, Bruno, and Steve Woolgar. 1979/1987. *Laboratory Life: The Social Construction of Scientific Facts*. Princeton, N.J.: Princeton University Press.

Lavayssière, Robert, and Anne-Elizabeth Cabée. 2000. "MRI in France: The French Paradox." *Journal of Magnetic Resonance Imaging* 13 (4): 528–33.

Leach, Melissa, and James Fairhead. 2007. *Vaccine Anxieties: Global Science, Child Health, and Society*. London: Earthscan.

Leake, Chauncey D., ed. 1975. *Percival's Medical Ethics*. 2nd ed. New York: Krieger.

Leavitt, Judith Walzer, and Ronald L. Numbers. 1978. Introduction to *Sickness and Health in America: Readings in the History of Medicine and Public Health*, ed. Judith Walzer Leavitt and Ronald L. Numbers. Madison: University of Wisconsin Press.

Lederer, Susan E., and Naomi Rogers. 2003. "Media." In *Companion to Medicine in the Twentieth Century*, ed. Roger Cooter and John Pickstone, 487–502. London: Routledge.

Lee, Ellie. 2006. "Medicalizing Motherhood." *Society* 43 (6): 47–50.

Lee, S., and A. Mysyk. 2004. "The Medicalization of Compulsive Buying." *Social Studies of Medicine* 58:1709–18.

Lee, Sandra Soo-Jin. 2005. "Racializing Drug Design: Implications of Pharmacogenomics for Health Disparities." *American Journal of Public Health* 95 (12): 2133–38.

Lenoir, Timothy. 1997. *Instituting Science: The Cultural Production of Scientific Disciplines*. Stanford: Stanford University Press.

Lentzos, Filippa. 2006. "Rationality, Risk, and Response: A Research Agenda for Biosecurity." *BioSocieties* 1 (4): 453–64.

Leonhardt, David. 2001. "Health Care as Main Engine: Is That So Bad?" *New York Times*, Money and Business, November 11, 1, 12.

Lerner, Baron H. 1998. "Fighting the War on Breast Cancer: Debates over Early Detection, 1945 to the Present." *Annals of Internal Medicine* 129:74–78.

———. 1999. "Great Expectations: Historical Perspectives on Genetic Breast Cancer Testing." *American Journal of Public Health* 89:938–44.

———. 2000. "Inventing a Curable Disease: Historical Perspectives on Breast Cancer." In *Breast Cancer: Society Constructs an Epidemic*, ed. S. Ferguson and A. Kasper, 25–49. New York: St. Martin's.

———. 2002. "What's Behind It All." *Washington Post*, March 3, B01.

———. 2006. *When Illness Goes Public: Celebrity Patients and How We Look at Medicine*. Baltimore: Johns Hopkins University Press.

Leslie, Charles, ed. 1978. *Theoretical Foundations for the Comparative Study of Medical Systems*. New York: Pergamon.

Leslie, Stuart W., and Robert Kargon. 2006. "Exporting MIT: Science, Technology, and Nation-Building in India and Iran." *Osiris* 21:110–32.

Letiche, Hugo. 2002. "Viagra(ization) or Technoromanticism." *Consumption, Markets, and Culture* 5 (3): 247–60.

Leung, Angela Ki Che, and Charlotte Furth, eds. Forthcoming. *Health and Hygiene in Modern Chinese East Asia*. Durham, N.C.: Duke University Press.

Leungwattanakji, Somboon, Vincent Flynn, and Wayne J. G. Hellstrom. 2001. "Intracavernosal Injection and Intraurethral Therapy for Erectile Dysfunction." *Urologic Clinics of North America* 28:343–54.

Lewin, Ellen. 1993. *Lesbian Mothers: Accounts of Gender in American Culture*. Ithaca, N.Y.: Cornell University Press.

Lewis, Bradley. 2006. *Moving beyond Prozac, DSM, and the New Psychiatry: The Birth of Postpsychiatry*. Ann Arbor: University of Michigan Press.

Lewis, Carol. 2001. "Full-Body CT Scans: What You Need to Know." *FDA Consumer Magazine* 35 (6): 33.

Lewis, Michael. 2000. *The New New Thing: A Silicon Valley Story*. New York: W. W. Norton.

Lewis, Oscar. 1959. *Five Families: Mexican Case Studies in the Culture of Poverty*. New York: Basic.

———. 1966. *La Vida: A Puerto Rican Family in the Culture of Poverty—San Juan and New York*. New York: Random House.

Lew-Ting, Chih-Yin. 2005. "Antibiomedicine Belief and Integrative Health Seeking in Taiwan." *Social Science and Medicine* 60 (9): 2111–16.

Lexchin, Joel. 2006. "Bigger and Better: How Pfizer Redefined Erectile Dysfunction." *PLoS Medicine* 3:429–32.

Lezaun, Javier. 2007. "Towards a Bioeconomy." *BioSocieties* 2 (3): 381.

Ley, Barbara L. 2009. *From Pink to Green: Disease Prevention and the Environmental Breast Cancer Movement*. New Brunswick, N.J.: Rutgers University Press.

Liao, D., L. Cooper, J. Cai, J. Toole, N. Bryan, G. Burke, E. Shahar, and J. Neito. 1997. "The Prevalence and Severity of White Matter Lesions, Their Relationship to Ethnicity, Gender, and Cardiovascular Disease Risk Factors." *Neuroepidemiology* 16 (3): 149–62.

Light, Donald W. 2000a. "The Medical Profession and Organizational Change: From Professional Dominance to Countervailing Power." In *Handbook of Medical Sociology*, ed. C. E. Bird, P. Conrad, and A. M. Fremont, 201–16. Upper Saddle River, N.J.: Prentice Hall.

———. 2000b. "The Sociological Character of Health Care Markets." In *Handbook of Social Studies in Health and Medicine*, ed. G. L. Albrecht, R. Fitzpatrick, and S. C. Scrimshaw, 394–408. Thousand Oaks, Calif.: Sage.

———. 2004. "Introduction: Ironies of Success—A New History of the American Health Care 'System.'" *Journal of Health and Social Behavior* 45 (Extra Issue): 1–24.

Lillienfield, David E., and Paul D. Stolley. 1994. *Foundations of Epidemiology*. New York: Oxford University Press.

Lim, E. C., and R. C. Seet. 2008. "In-house Medical Education: Redefining Tele-education." *Teach Learn Med* 20 (2): 193–95.

Lim, Michelle. 2002. "Fetal Photos and Body Scans in Parking Lots: The Implications of Bypassing Physicians in the Medical Marketplace." American Medical Association, Virtual Mentor 4 (11): http://virtualmentor.ama-assn.org/2002/11/ebyt1-0211 .html (accessed October 2005).

Lindee, Susan. 2005. *Moments of Truth in Genetic Medicine*. Baltimore: Johns Hopkins University Press.

Lippman, Abby. 1992. "Led (Astray) by Genetic Maps: The Cartography of the Human Genome and Health Care." *Social Science and Medicine* 35 (12): 1469–76.

Lippman, Scott M., and Powel H. Brown. 1999. "Tamoxifen Prevention of Breast Cancer: An Instance of the Fingerpost." *Journal of the National Cancer Institute* 91: 1809–19.

Litt, Jacqueline S. 2000. *Medicalized Motherhood: Perspectives from the Lives of African American Women and Jewish Women*. New Brunswick, N.J.: Rutgers University Press.

Litwin, Mark S., Robert J. Nied, and Nasreen Dhanani. 1998. "Health-Related Quality of Life in Men with Erectile Dysfunction." *Journal of General Internal Medicine* 13: 159–69.

Liu, Lydia, ed. 1999. *Tokens of Exchange: The Problem of Translation in Global Circulations*. Durham, N.C.: Duke University Press.

Lloyd, Stephanie. 2006. "The Clinical Clash over Social Phobia: The Americanization of French Experience?" *BioSocieties* 1: 229–49.

Loberg, Michael. 2005a. "NitroMed Web Cast at 23d Annual JP Morgan Healthcare Conference." January 10. http://www.mapdigital.com/jpmorgan/healthcare05/ ondemand.html#n (accessed January 25, 2005).

———. 2005b. "NitroMed Webcast Presentation at UBS Global Life Sciences Conference." September 26. http://event.streamx.us/event/default.asp?event= ubs20050926 (accessed September 27, 2005).

Lock, Margaret M. 1993. *Encounters with Aging: Mythologies of Menopause in Japan and North America*. Berkeley: University of California Press.

———. 1998. "Situating Women in the Politics of Health." In *The Politics of Women's*

Health: Exploring Agency and Autonomy, ed. S. Sherwin, 178–204. Philadelphia: Temple University Press.

———. 2002. "Biomedical Technologies: Anthropological Approaches." In *Encyclopedia of Medical Anthropology: Health and Illness in the World's Cultures*, 86–95. Norwell, Mass.: Kluwer Academic Plenum.

———. 2005. "Eclipse of the Gene and the Return of Divination." *Current Anthropology* 46:S47–S70. Suppl. S.

———. 2006. "The Molecularized Mind and the Search for Incipient Dementia." *Sciences Sociales et Santé* 24 (1): 21–56.

———. 2008. "Biomedical Technologies, Cultural Horizons, and Contested Boundaries." In *The Handbook of Science and Technology Studies*, ed. Edward J. Hackett, Olga Amsterdamska, Michael Lynch, and Judy Wajcman, 875–900. 3rd ed. Cambridge: MIT Press.

Lock, Margaret, Julia Freeman, Rosemarie Sharpies, and Stephanie Lloyd. 2006. "When It Runs in the Family: Putting Susceptibility Genes in Perspective." *Public Understanding of Science* 15 (3): 277–300.

Lock, Margaret, and Deborah Gordon, eds. 1988. *Biomedicine Examined*. Boston: Kluwer Academic.

Lock, Margaret, and Patricia A. Kaufert, eds. 1998. *Pragmatic Women and Body Politics*. New York: Cambridge University Press.

Lock, Margaret, and Mark Nichter. 2002. "Introduction: From Documenting Medical Pluralism to Critical Interpretations of Globalized Health Knowledge, Policies, and Practices." In *New Horizons in Medical Anthropology: Essays in Honour of Charles Leslie*, ed. Mark Nichter and Margaret Lock, 1–34. London: Routledge.

Loe, Meika. 2001. "Fixing Broken Masculinity: Viagra as a Technology for the Production of Gender and Sexuality." *Sexuality and Culture* 5 (3): 97–125.

———. 2004a. *The Rise of Viagra: How the Little Blue Pill Changed Sex in America*. New York: New York University Press.

———. 2004b. "Sex and the Senior Woman: Pleasure and Danger in the Viagra Era." *Sexualities* 7 (3): 303–26.

———. 2006. "'The Viagra Blues': Embracing or Resisting the Viagra Body." In *Medicalized Masculinities*, ed. D. Rosenfeld and C. A. Faircloth, 21–44. Philadelphia: Temple University Press.

Lohr, Steve. 2000. "Welcome to the Internet, the First Global Colony." *New York Times*, January 9, section 4, p. 1.

Loomis, Dana, and Steven Wing. 1990. "Is Molecular Epidemiology a Germ Theory for the End of the Twentieth Century?" *International Journal of Epidemiology* 19:1–3.

LoPiccolo, J. 1978. "Direct Treatment of Sexual Dysfunction." In *Handbook of Sex Therapy*, ed. J. LoPiccolo and P. LoPiccolo. New York: Plenum.

Lorber, Judith, and Lisa Jean Moore. 2002. *Gender and the Social Construction of Illness*. 2nd ed. Walnut Creek, Calif.: Rowman and Littlefield.

Loudon, Irvine, ed. 1997. *Western Medicine: An Illustrated History*. Oxford: Oxford University Press.

Love, R. R. 1995. "Tamoxifen Chemoprevention: Public Health Goals, Toxicities for All, and Benefits to a Few." *Annals of Oncology* 6:127–28.

Low, Morris. 2005. *Building a Modern Japan: Science, Technology, and Medicine in the Meiji Era and Beyond.* New York: Palgrave Macmillan.

Lowe, Donald M. 1995. *The Body in Late-Capitalist U.S.A.* Durham, N.C.: Duke University Press.

Löwy, Ilana, and Jean-Paul Gaudillière. 2008. "Localizing the Global: Testing for Hereditary Risks of Breast Cancer." *Science, Technology, and Human Values* 33 (3): 299–325.

Luker, Kristin. 1996. *Dubious Conceptions: The Politics of Teenage Pregnancy.* Cambridge: Harvard University Press.

Lupton, Deborah. 1993. "Risk as Moral Danger: The Social and Political Functions of Risk Discourse in Public Health." *International Journal of Health Services* 23 (3): 425–35.

———. 1994/2003. *Medicine as Culture: Illness, Disease, and the Body in Western Society.* 2nd ed. London: Sage.

———. 1995. *The Imperative of Health: Public Health and the Regulated Body.* Thousand Oaks, Calif.: Sage.

———. 1997. "Foucault and the Medicalization Critique." In *Foucault: Health and Medicine,* ed. A. Petersen and R. Bunton, 94–110. London: Routledge.

———. 1999. *Risk.* London: Routledge.

———. 2000. "The Social Construction of Medicine and the Body." In *Handbook of Social Studies in Health and Medicine,* ed. G. L. Albrecht, R. Fitzpatrick, and S. C. Scrimshaw, 50–63. London: Sage.

Lury, Celia. 1996. *Consumer Culture.* New Brunswick, N.J.: Rutgers University Press.

Lyman, Karen A. 1989. "Bringing the Social Back In: A Critique of the Biomedicalization of Dementia." *Gerontologist* 29:597–605.

Ma, EunJung. 2007. "Reconfiguring Korean Medicine in Postcolonial Korea." Paper presented at the Second Conference of EASTS Journal, NTU, Taipei, August.

MacKenzie, Donald. 2001. *Mechanizing Proof: Computing, Risk, and Trust.* Cambridge: MIT Press.

Madigan, M. P., R. G. Ziegler, J. Benichou, C. Byrne, and R. N. Hoover. 1995. "Proportion of Breast Cancer Cases in the United States Explained by Well Established Risk Factors." *Journal of the National Cancer Institute* 87:1681–85.

Makoul, Gregory, and Limor Peer. 2004. "Dissecting the Doctor Shows: A Content Analysis of *ER* and *Chicago Hope.*" In *Cultural Sutures: Medicine and Media,* ed. Lester D. Friedman, 244–60. Durham, N.C.: Duke University Press.

Malacrida, Claudia. 2004. "Medicalization, Ambivalence, and Social Control: Mothers' Descriptions of Educators and ADD/ADHD." *Health* 8 (1): 61–80.

Mamo, Laura. 2002. "Sexuality, Reproduction, and Biomedical Negotiations: An Analysis of Achieving Pregnancy in the Absence of Heterosexuality." Ph.D. diss., Department of Social and Behavioral Sciences, University of California, San Francisco.

———. 2007a. "Negotiating Conception: Lesbians' Hybrid-Technological Practices." *Science, Technology, and Human Values* 32 (3): 369–93.

———. 2007b. *Queering Reproduction: Achieving Pregnancy in the Age of Technoscience.* Durham, N.C.: Duke University Press.

Mamo, Laura, and Jennifer Fishman. 2001. "Potency in All the Right Places: Viagra as a Technology of the Gendered Body." *Body and Society* 7:13–35.

Mamo, Laura, and Jennifer R. Fosket. 2009. "Scripting the Body: Pharmaceuticals and the (Re)making of Menstruation." *Signs: Journal of Women in Culture and Society* 34 (4): 925–49.

Mamo, Laura, and Vrushali Patil. n.d. "The New Citizen Queer: Theorizing U.S. Reproduction." Manuscript.

Manderson, Lenore. 1996. *Sickness and the State: Health and Illness in Colonial Malaya, 1870–1940.* Cambridge: Cambridge University Press.

Marchessault, Janine. 2000. "David Suzuki's *The Secret of Life*: Informatics and the Popular Discourse of the Life Code." In *Wild Science: Reading Feminism, Medicine, and the Media*, ed. Janine Marchessault and Kim Sawchuk, 55–65. London: Routledge.

Marchessault, Janine, and Kim Sawchuk. 2000. Introduction to *Wild Science: Reading Feminism, Medicine, and the Media*, ed. Janine Marchessault and Kim Sawchuk, 1–8. London: Routledge.

Marcus, George E., and Erkan Saka. 2006. "Assemblage." *Theory, Culture, and Society* 23 (2–3): 101–9.

Marks, Harry M. 1993. "Medical Technologies: Social Contexts and Consequences." In *Companion Encyclopedia of the History of Medicine*, vol. 1, ed. W. F. Bynum and R. Porter, 1592–1618. New York: Routledge.

———. 1997. *The Progress of Experiment: Science and Therapeutic Reform in the United States, 1900–1990.* New York: Cambridge University Press.

Marshall, Eliot. 1995. "Tamoxifen's Trials and Tribulations." *Science* 270:910.

Marshall, Barbara L. 2002. "'Hard Science': Gendered Constructions of Sexual Dysfunction in the 'Viagra Age.'" *Sexualities* 5 (2): 131–58.

———. 2007. "Climacteric Redux? (Re)medicalizing the Male Menopause." *Men and Masculinities* 9 (4): 509–29.

Marshall, Barbara L., and Stephen Katz. 2002. "Forever Functional: Sexual Fitness and the Ageing Male Body." *Body and Society* 8:43–70.

Martin, Emily. 1987. *The Woman in the Body: A Cultural Analysis of Reproduction.* Boston: Beacon.

———. 1994. *Flexible Bodies: The Role of Immunity in American Culture from the Days of Polio to the Age of AIDS.* Boston: Beacon.

———. 2006. "The Pharmaceutical Person." *BioSocieties* 1 (3): 273–88.

———. 2007. *Bipolar Expeditions: Mania and Depression in American Culture.* Princeton, N.J.: Princeton University Press.

Martinez, Airin. In prep. "Comiendo Bien: A Situational Analysis of Healthy Eating among Latino Immigrant Families in San Francisco." Ph.D. diss., University of California, San Francisco.

Marzano, Marco. 2009. "Comment on Biomedicalization." *Salute e Società 8 (2)*: 259–263 [in Italian]. *Salute e Società 8 (2)*: 243–247 [in English].

Massumi, Brian. 2006. "National Enterprise Emergency: Refiguring Political Decision." Presented at the conference "Beyond Biopower: State Racism and the Politics of Life and Death," March 2006, City University of New York Graduate Center, New York. http://web.gc.cuny.edu/womenstudies/biopolitics.html.

Masters, William H., and Virginia E. Johnson. 1966. *Human Sexual Response*. Boston: Little, Brown.

Maternowska, M. Catherine. 2006. *Reproducing Inequities: Poverty and the Politics of Population in Haiti*. New Brunswick, N.J.: Rutgers University Press.

Maugh, Thomas. 2004. "Drug for Blacks Only Stirs Hope, Concern." *Los Angeles Times*, November 9, A1.

May, C., T. Rapley, T. Moreira, T. Finch, and B. Heaven. 2006. "Technogovernance: Evidence, Subjectivity, and the Clinical Encounter in Primary Care Medicine." *Social Science and Medicine 62 (4)*: 1022–30.

Maynard, Ronald J. 2006. "Controlling Death—Compromising Life: Chronic Disease, Prognostication, and the New Biotechnologies." *Medical Anthropology Quarterly 20 (2)*: 212–34.

Mbembe, Achille. 2003. "Necropolitics." *Public Culture 15 (1)*: 11–40.

McCall, Leslie. 2001. *Complex Inequality: Gender, Class, and Race in the New Economy*. New York: Routledge.

McCarthy, Doyle. 1996. *Knowledge as Culture: The New Sociology of Knowledge*. New York: Routledge.

McEwen, Bruce S. 2000. "The Neurobiology of Stress: From Serendipity to Clinical Relevance." *Brain Research 886 (1–2)*: 172–89.

———. 2003. "Mood Disorders and Allostatic Load." *Biological Psychiatry 54*:200–207.

McInaney, Maureen. 2000. "Group Appointments Can Benefit Busy Doctors and Chronically Ill Patients." UCSF *News Service Release*, October 18.

McKeown, Thomas. 1979. *The Role of Medicine: Dream, Mirage or Nemesis?* Princeton, N.J.: Princeton University Press.

McKinlay, John B. 1974/1994. "A Case for Refocussing Upstream: The Political Economy of Illness." In *The Sociology of Health and Illness*, ed. P. Conrad and R. Kern, 509–23. 4th ed. New York: St. Martin's.

McKinlay, John B., and L. D. Marceau. 2002. "The End of the Golden Age of Doctoring." *International Journal of Health Services 32 (2)*: 379–416.

McKinlay, John B., and John D. Stoeckle. 1988. "Corporatization and the Social Transformation of Doctoring." *International Journal of Health Services 18*:191–205.

McMichael, Anthony J. 1995. "The Health of Persons, Populations, and Planets: Epidemiology Comes Full Circle." *Epidemiology 6 (6)*: 633–36.

Mechanic, David. 2002. "Socio-cultural Implications of Changing Organizational Technologies in the Provision of Care." *Social Science and Medicine* 54:459–67.

Meckler, Laura. 2003. "Uninsured Population Continues to Expand." *San Francisco Chronicle*, March 5, A4.

Mensah, Maria Nengeh. 2000. "Screening Bodies, Assigning Meaning: *ER* and the Technology of HIV Testing." In *Wild Science: Reading Feminism, Medicine, and the Media*, ed. Janine Marchessault and Kim Sawchuk, 139–50. London: Routledge.

Merkin, Donald H. 1976. *Pregnancy as a Disease: The Pill in Society*. Port Washington, N.Y.: Kennikat.

Messer-Davidow, Ellen, David R. Shumway, and David J. Sylvan. 1993. *Knowledges: Historical and Critical Studies in Disciplinarity*. Charlottesville: University Press of Virginia.

Metzl, Jonathan. 2003. *Prozac on the Couch: Prescribing Gender in the Era of Wonder Drugs*. Durham, N.C.: Duke University Press.

———. 2004. "The Pharmaceutical Gaze: Psychiatry, Scopophilia, and Psychotropic Medication Advertising, 1964–1985." In *Cultural Sutures: Medicine and Media*, ed. Lester D. Friedman, 15–54. Durham, N.C.: Duke University Press.

Midanik, Lorraine T. 2006. *Biomedicalization of Alcohol Studies*. Piscataway, N.J.: Transaction.

Miller, Daniel, ed. 1998. *Material Cultures — Why Some Things Matter*. London: University College of London Press.

Miller, Peter, and Nikolas Rose. 1997. "Mobilizing the Consumer: Assembling the Subject of Consumption." *Theory, Culture, and Society* 14 (1): 1–36.

Millman, Marcia. 1981. *Such a Pretty Face: Being Fat in America*. New York: Berkley.

Mills, C. Wright. 1959. *The Sociological Imagination*. New York: Oxford University Press.

Mills, Robert J. 2002. "Health Insurance Coverage: 2001." U.S. Census Bureau. http://www.census.gov/prod/2002pubs/p60-220.pdf (accessed September 15, 2008).

Mirken, Bruce. 1996. "Ask Your Doctor." *San Francisco Bay Guardian*, October 23, 47–61.

Mirowski, Philip. 2007. "Johnny's in the Basement Mixin' Up the Medicine: Review of Angell, Avorn, and Daemmrich on the Modern Pharmaceutical Predicament." *Social Studies of Science* 37 (2): 311–27.

Mirzoeff, Nicholas, ed. 1998. *Visual Culture Reader*. New York: Routledge.

Mitchell, W. J. T. 1995. *Picture Theory: Essays on Verbal and Visual Representation*. Chicago: University of Chicago Press.

Mitman, Gregg. 1999. *Reel Nature: America's Romance with Wildlife on Film*. Cambridge: Harvard University Press.

Mol, Annemarie. 2002. *The Body Multiple: Ontology in Medical Practice*. Durham, N.C.: Duke University Press.

Montini, Theresa. 1996. "Gender and Emotions in the Advocacy for Breast Cancer Informed Consent Legislation." *Gender and Society* 10:9–23.

Montini, Theresa, and Sheryl Ruzek. 1989. "Overturning Orthodoxy: The Emergence of Breast Cancer Treatment Policy." In *Research in the Sociology of Health Care*, vol. 8, ed. Dorothy Wertz, 3–32. Greenwich, Conn.: JAI.

Montoya, Michael. 2007. "Bioethnic Conscription: Genes, Race, and Mexicana/o Ethnicity in Diabetes Research." *Cultural Anthropology* 22 (1): 94–128.

Moore, J. Stuart. 1993. *Chiropractic in America: The History of a Medical Alternative.* Baltimore: Johns Hopkins University Press.

Moore, Jeremy. 2003. "Hypertension Treatment among Blacks: Should It Be Different?" *Today in Cardiology*, October. http://www.cardiologytoday.com (accessed February 8, 2008).

Moore, Lisa Jean. 1997. "'It's Like You Use Pots and Pans to Cook, It's the Tool': The Technologies of Safer Sex." *Science, Technology, and Human Values* 22 (4): 434–71.

Moore, Lisa Jean, and Adele E. Clarke. 2001. "The Traffic in Cyberanatomies: Sex/Gender/Sexuality in Local and Global Formations." *Body and Society* 7 (1): 57–96.

Morgan, Kathryn Pauly. 1998. "Contested Bodies, Contested Knowledges: Women, Health, and the Politics of Medicalization." In *The Politics of Women's Health: Exploring Agency and Autonomy*, ed. Feminist Health Care Ethics Research Network, Susan Sherwin, Coordinator, 83–121. Philadelphia: Temple University Press.

Morgen, Sandra. 2002. *Into Our Own Hands: The Women's Health Movement in the United States, 1969–1990.* New Brunswick, N.J.: Rutgers University Press.

———. 2006. "Movement-Grounded Theory: Intersectional Analyses of Health Inequalities in the U.S." In *Gender, Race, Class, and Health: Intersectional Approaches*, ed. Amy J. Schulz and Leith Mullings, 394–423. San Francisco: Jossey-Bass.

Morrison, Linda J. 2005. *Talking Back to Psychiatry: The Psychiatric Consumer/Survivor/Ex-Patient Movement.* New York: Routledge.

Mosca, L., W. K. Jones, K. B. King, P. Ouyang, R. F. Redberg, and M. N. Hill. 2000. "Awareness, Perception, and Knowledge of Heart Disease Risk and Prevention among Women in the United States." *Archives of Family Medicine* 9 (6): 506–15.

Moss, Nancy E. 2002. "Gender Equity and Socioeconomic Inequality: A Framework for the Patterning of Women's Health." *Social Science and Medicine* 54:649–61.

Moss, Pamela, and Isabel Dyck. 2002. *Women, Body, Illness: Space and Identity in the Everyday Lives of Women with Chronic Illness.* Lanham, Md.: Rowman and Littlefield.

Moynihan, Ray, and Alan Cassels. 2005. *Selling Sickness: How the World's Biggest Pharmaceutical Companies Are Turning Us All into Patients.* New York: Nation Books.

Moynihan, Ray, and David Henry. 2006. "The Fight against Disease Mongering: Generating Knowledge for Action." *PLoS Medicine* 3:425–28.

Mudur, Ganapati. 2004. "Hospitals in India Woo Foreign Patients." *British Medical Journal* 328:1338. http://www.bmj.com/cgi/content/extract/328/7452/1338.

Mueller, Mary-Rose. 1997. "Science versus Care: Physicians, Nurses, and the Dilemma of Clinical Research." In *The Sociology of Medical Science and Technology*, ed. M. A. Elston, 57–78. Malden, Mass.: Blackwell.

Mueller, Mary Rose, and Laura Mamo. 2000. "Changes in Medicine, Changes in Nursing: Career Contingencies and the Movement of Nurses into Clinical Trial Coordi-nation." *Sociological Perspectives* 43:S43–S57.

Mundy, Alicia. 2001. *Dispensing with the Truth: The Victims, the Drug Companies, and the Dramatic Story behind the Battle over Fen-Phen.* New York: St. Martin's.

Murcott, Anne, ed. 2006. *Sociology and Medicine: Selected Papers by P. M. Strong.* Burlington, Vt.: Ashgate.

Murphy, Michelle. 2003. "Liberation through Control in the Body Politics of U.S. Radical Feminism." In *The Moral Authority of Nature,* ed. Lorraine Daston and Fernando Vidal, 331–56. Chicago: University of Chicago Press.

——. 2004. "Immodest Witnessing: Vaginal Self-Examination and the Evidence of Experience in the U.S. Feminist Self Help Movement." *Feminist Studies* 30 (1): 115–47.

——. 2006. *Sick Building Syndrome and the Problem of Uncertainty: Environmental Politics, Technoscience, and Women Workers.* Durham, N.C.: Duke University Press.

——. Forthcoming. *Seizing the Means of Reproduction.* Durham, N.C.: Duke University Press.

Murray, Stuart J., and Dave Holmes, eds. 2009. *Critical Interventions in the Ethics of Healthcare: Challenging the Principle of Autonomy in Bioethics.* Surrey, England: Ashgate.

Myers, David. 2003. *Psychology.* 7th ed. New York: Worth.

Mykhalovskiy, Eric, Liza McCoy, and Michael Bresalier. 2004. "Compliance/Adherence, HIV, and the Critique of Medical Power." *Social Theory and Health* 2 (4): 315–40.

Mykhalovskiy, Eric, and Lorna Weir. 2004. "The Problem of Evidence-Based Medicine: Directions for Social Science." *Social Science and Medicine* 59 (5): 1059–69.

Mykytyn, Courtney Everts. 2006. "Anti-aging Medicine: A Patient/Practitioner Movement to Redefine Aging." *Social Science and Medicine* 62 (3): 643–53.

Naik, Gautam, and Antonio Regalado. 2006. "Trail of a Killer: A Fitness Mogul, Stricken by Illness, Hunts for Genes: Study of Lou Gehrig's Disease Pinpoints DNA Variations Common to Its Sufferers." *Wall Street Journal,* November 30, A1.

Napoli, Maryann. 2001. "Another View of Tamoxifen." *American Journal of Nursing* 101:33.

National Institutes of Health. 1976. *NIH Factbook: Guide to National Institutes of Health Programs and Activities.* Bethesda: Marquis Academic Media.

——. 1992. "Consensus Development Conference Statement on Impotence." *International Journal of Impotence Research* 5:181–99.

——. 2000a. "NIH Obligations and Amounts Obligated for Grants and Direct Operations." http://www.nih.gov/about/almanac/index.html (accessed July 10, 2005).

——. 2000b. "NIH Overview." http://www.nih.gov/about/NIHoverview.html #goal (accessed July 10, 2005).

National Research Council. 2000. *Networking Health: Prescriptions for the Internet.* Washington: National Academy Press.

Navarro, Vicente. 1976. *Medicine under Capitalism*. New York: Prodist.

———. 1986. *Crisis, Health, and Medicine: A Social Critique*. New York: Tavistock.

———. 1999. "Health and Equity in the World in the Era of Globalization." *International Journal of Health Services* 29 (2): 215–26.

———. 2007. *Neoliberalism, Globalization, and Inequalities: Consequences for Health and Quality of Life*. Amityville, N.Y.: Baywood.

NCHS. 1998. *Health, United States, 1998, with Socioeconomic Status and Health Chartbook*. Hyattsville, Md.: National Center for Health Statistics.

———. 2001. *Health, United States, 2001, with Rural and Urban Health Chartbook*. Hyattsville, Md.: National Center for Health Statistics.

Nelkin, Dorothy. 1995. "Scientific Controversies." In *Handbook of Science and Technology Studies*, ed. Sheila Jasanoff et al., 444–56. Thousand Oaks, Calif.: Sage.

Nelson, Alondra. 2008a. "Bio Science: Genetic Genealogy Testing and the Pursuit of African Ancestry." *Social Studies of Science* 38 (5): 759–83.

———. 2008b. "Of Race and Kin: Diasporic Subjectivity, Genetic Genealogy, and the Pursuit of African Ancestry." In *Revisiting Race in a Genomic Age*, ed. Barbara Koenig, Sandra Soo-Jin Lee, and Sarah Richardson, 253–68. New Brunswick, N.J.: Rutgers University Press.

———. Forthcoming. *Body and Soul: The Black Panther Party and the Politics of Race and Health*. Berkeley: University of California Press.

Nelson, Fiona. 2000. "Lesbian Families: Achieving Motherhood." *Journal of Gay and Lesbian Social Services* 10 (1): 27–46.

Nelson, Jennifer. 2003. *Women of Color and the Reproductive Rights Movement*. New York: New York University Press.

Newman, Karen. 1996. *Fetal Positions: Individualism, Science, Visuality*. Stanford, Calif.: Stanford University Press.

Newman, Lisa A., Henry M. Kuerer, Kelly K. Hunt, Georges Vlastos, Frederick Ames, Merrick I. Ross, and Eva Singletary. 2000. "Prophylactic Mastectomy." *Journal of the American College of Surgeons* 191:322–30.

Nguyen, Vinh-kim. 2005. "Antiretroviral Globalism, Biopolitics, and Therapeutic Citizenship." In *Global Assemblages: Technology, Politics, and Ethics as Anthropological Problems*, ed. Aihwa Ong and Stephen J. Collier, 124–44. Malden, Mass.: Blackwell.

Nichter, Mark. 2008. *Global Health: Why Cultural Perceptions, Social Representations, and Biopolitics Matter*. Tucson: University of Arizona Press.

Nichter, Mark, and Jennifer Jo Thompson. 2006. "'For My Wellness, Not Just My Illness': North Americans' Use of Dietary Supplements." *Culture, Medicine, and Psychiatry* 30 (2): 175–222.

Nichter, Mark, and Nancy Vuckovic. 1994. "Agenda for an Anthropology of Pharmaceutical Practice." *Social Science and Medicine* 39 (11): 1509–25.

NIH Consensus Development Panel on Impotence. 1993. "Impotence: NIH Consensus Conference." *Journal of the American Medical Association* 270:83–90.

NitroMed. 1999. "NitroMed Acquires BiDil New Drug Application for Treatment of Congestive Heart Failure." September 10. [now http://www.bidil.com] http://www.nitromed.com/newsindex.html (accessed December 7, 2003).

—. 2003. "SEC Filing, Form S-1/A." October 2, 11.

—. 2004. "NitroMed Stops Heart Failure Study (A-HEFT) in African Americans Due to Significant Survival Benefit of BiDil." July 19. [now www.bidil.com] http://www.nitromed.com/07_19_04a.asp (accessed March 21, 2005).

—. 2005. "BiDil Named to American Heart Association's 2004 'Top 10 Advances' List." January 11. [now http://www.bidil.com] http://www.nitromed.com/01_11_05.asp (accessed March 21, 2005).

Nobles, Melissa. 2000. *Shades of Citizenship*. Stanford: Stanford University Press.

Novas, Carlos. 2006. "The Political Economy of Hope: Patients' Organizations, Science, and Biovalue." *BioSocieties* 1 (3): 289–306.

Novas, Carlos, and Nikolas Rose. 2000. "Genetic Risk and the Birth of the Somatic Individual." *Economy and Society* 29 (4): 485–513.

Novello, Antonia. 2002. "DOH Medicaid Update: Magnetic Resonance Imaging—Physician and Clinic Billing." New York State, Office of Medicaid Management. http://www.health.state.ny.us/health_care/medicaid/program/update/2002/may2002.htm#mri (accessed June 27, 2006).

Novotny, Thomas E., and Vincanne Adams. 2007. "Global Health Diplomacy: A Call for a New Field of Teaching and Research." *San Francisco Medicine* 80 (2): 22–23. http://globalhealthsciences.ucsf.edu/pdf/SFMedicineHD_Article.pdf.

Nye, Robert A. 2003. "The Evolution of the Concept of Medicalization in the Late Twentieth Century." *Journal of History of the Behavioral Sciences* 39 (2): 115–29.

Ogburn, William Fielding. 1922/1964. *Social Change with Respect to Culture and Original Nature*. New York: B. W. Huebsch, 1922. Reprint, Gloucester, Mass.: Peter Smith, 1964.

Oh, Eun-Hwan, Yulchi Imanaka, and Edward Evans. 2005. "Determinants of the Diffusion of Computer Tomography and Magnetic Resonance Imaging." *International Journal of Technology Assessment in Health Care* 21 (1): 73–80.

Olesen, Virginia L. 2000. "Emotions and Gender in U.S. Health Care Contexts: Implications for Change and Stasis in the Division of Labour." In *Theorizing Medical Sociology*, ed. S. Williams, J. Gabe, and M. Calnan, 315–32. London: Routledge.

—. 2002. "Resisting 'Fatal Unclutteredness': Conceptualizing the Sociology of Health and Illness into the Millennium." In *Gender, Health, and Healing: The Public/Private Divide*, ed. Gillian Bendelow, Mick Carpenter, Caroline Vautier, and Simon Williams, 254–66. London: Routledge.

Olesen, Virginia, and Debora Bone. 1998. "Emotions in Rationalizing Organizations: Conceptual Notes from Professional Nursing in the U.S.A." In *Emotions in Social Life: Critical Themes and Contemporary Issues*, ed. G. Bendelow and S. Williams, 313–29. London: Routledge.

Olesen, Virginia, and Ellen Lewin. 1985. "Women, Health, and Healing." In *Women, Health, and Healing: Toward a New Perspective*, ed. E. Lewin and V. Olesen, 1–24. New York: Tavistock.

Olney, Buster. 2002. "Pro Football: Giants Notebook; For Hilliard, Serenity Vanishes in a Second." *New York Times*, October 30, D2.

Olsen, Ole, and Peter C. Gøtzsche. 2001. "Cochrane Review in Screening for Breast Cancer with Mammography." *Lancet* 358:1340–42.

Omenn, Gilbert S. 1991. "Future Research Directions in Cancer Ecogenetics." *Mutation Research* 247 (2): 283–91.

——. 2000. "Public Health Genetics: An Emerging Interdisciplinary Field for the Post-genomic Era." *Annual Review of Public Health* 21:1–13.

Omi, Michael, and Howard Winant. 1994. *Racial Formation in the United States: From the 1960s to the 1990s*. 2nd ed. New York: Routledge.

O'Neil, Edward. 2006. *A Practical Guide to Global Health Service*. Chicago: American Medical Association.

Ong, Aihwa. 2008. "Scales of Exception: Experiments with Knowledge and Sheer Life in Tropical Southeast Asia." *Singapore Journal of Tropical Geography* 29:1–13.

Ong, Aihwa, and Stephen J. Collier, eds. 2005. *Global Assemblages: Technology, Politics, and Ethics as Anthropological Problems*. Malden, Mass.: Blackwell.

Orbach, Susie. 1978. *Fat Is a Feminist Issue*. New York: Berkley.

Organs Watch. 2001. Organs Watch Website. http://sunsite.berkeley.edu/biotech/organswatch (accessed November 15, 2001).

Orr, Jackie. 2006. *Panic Diaries: A Genealogy of Panic Disorder*. Durham, N.C.: Duke University Press.

Ottman, Ruth. 1996. "Gene-Environment Interaction: Definitions and Study Designs." *Preventive Medicine* 125 (6): 764–70.

Oudshoorn, Nelly. 1990. "On the Making of Sex Hormones: Research Materials and the Production of Knowledge." *Social Studies of Science* 20:5–33.

——. 1994. *Beyond the Natural Body: An Archeology of Sex Hormones*. New York: Routledge.

——. 2002. "Drugs for Healthy People: The Culture of Testing Hormonal Contraceptives for Women and Men." In *Biographies of Remedies: Drugs, Medicines, and Contraceptives in Dutch and Anglo-American Healing Cultures*, ed. G. M. van Heteren, M. Gijswijt-Horstra, and E. M. Tansey, 79–92. Atlanta: Rodopi.

——. 2003. *The Male Pill: A Biography of a Technology in the Making*. Durham, N.C.: Duke University Press.

Oudshoorn, Nelly, M. Brouns, and E. van Oost. 2005. "Diversity and Distributed Agency in the Design and Use of Medical Video-Communication Technologies." In *Inside the Politics of Technology*, ed. Hans Harbers, 85–109. Amsterdam: Amsterdam University Press.

Oudshoorn, Nelly, and Trevor Pinch, eds. 2003. *How Users Matter: The Co-construction of Users and Technology*. Cambridge: MIT Press.

Oudshoorn, Nelly, and Andre Somers. 2007. "Constructing the Digital Patient: Patient Organizations and the Development of Health Websites." In *Biomedicine as Culture: Instrumental Practices, Technoscientific Knowledge, and New Modes of Life*, ed. Valerie Burri and Joseph Dumit, 223–28. New York: Routledge.

Overmoyer, Beth A. 1999. "The Role of Tamoxifen in Preventing Breast Cancer." *Cleveland Clinic Journal of Medicine* 66:33–40.

Packard, Randall. 1998. *White Plague, Black Labor: Tuberculosis and the Political Economy of Health and Disease in South Africa.* Berkeley: University of California Press.

———. 2003. "Post-colonial Medicine." In *Companion to Medicine in the Twentieth Century*, ed. Roger Cooter and John Pickstone, 97–112. London: Routledge.

Packard, Randall M. P., J. Brown, R. L. Berkelman, and H. Frumkin, eds. 2004. *Emerging Illnesses and Society: Negotiating the Public Health.* Baltimore: Johns Hopkins University Press.

Page, Benjamin B., and David A. Valone, eds. 2007. *Philanthropic Foundations and the Globalization of Scientific Medicine and Public Health.* Lanham, Md.: University Press of America.

Pandey, Sanjay K., John J. Hart, and Sheela Tiwary. 2003. "Women's Health and the Internet: Understanding Emerging Trends and Implications." *Social Science and Medicine* 56 (1): 179–91.

Pantilat, S. Z., A. Alpers, and R. M. Wachter. 1999. "A New Doctor in the House: Ethical Issues in Hospitalist Systems." *Journal of the American Medical Association* 282:171–74.

Parens, Erik. 2006. *Surgically Shaping Children: Technology, Ethics, and the Pursuit of Normality.* Baltimore: Johns Hopkins University Press.

Parry, Bronwyn. 2004. *Trading the Genome: Investigating the Commodification of Bioinformation.* New York: Columbia University Press.

———. 2007. "Cornering the Futures Market in Bio-epistemology." *BioSocieties* 2 (3): 386–89.

Parsons, Talcott. 1951. *The Social System.* New York: Free Press.

Parthasarathy, Shobita. 2007. *Building Genetic Medicine: Breast Cancer, Technology, and the Comparative Politics of Health Care.* Cambridge: MIT Press.

Pasveer, Bernike. 1989. "Knowledge of Shadows: The Introduction of X-ray Images in Medicine." *Sociology of Health and Illness* 11 (4): 360–81.

Patterson, James T. 1987. *The Dread Disease: Cancer and Modern American Culture.* Cambridge: Harvard University Press.

Patton, Cindy. 2002. *Globalizing AIDS.* Minneapolis: University of Minnesota Press.

Pauly, Philip J. 1987. *Controlling Life: Jacques Loeb and the Engineering Ideal in Biology.* New York: Oxford University Press.

Payer, Lynn. 1988/1996. *Medicine and Culture: Varieties of Treatment in the United States, England, West Germany, and France.* New York: Henry Holt.

Payne, S. 2006. *The Health of Men and Women.* Cambridge, England: Polity.

Pearce, Neil. 1996. "Traditional Epidemiology, Modern Epidemiology, and Public Health." *American Journal of Public Health* 86 (5): 678–83.

Pearson, G. A. 1996. "Of Sex and Gender." *Science* 274:328–29.

Pelling, Margaret, and Scott Mandelbrote, eds. 2006. *The Practice of Reform in Health, Medicine, and Science, 1500–2000.* Farnham, England: Ashgate.

Perera, Frederica P. 1987. "Molecular Cancer Epidemiology: A New Tool in Cancer Prevention." *Journal of the National Cancer Institute* 78:887–98.

———. 1997. "Environment and Cancer: Who Are Susceptible?" *Science* 278:1068–73.

———. 2000. "Molecular Epidemiology: On the Path to Prevention?" *Journal of the National Cancer Institute* 92:602–12.

Perera, Frederica P., and J. Bernard Weinstein. 1982. "Molecular Epidemiology and Carcinogen-DNA Adduct Detection: New Approaches to Studies of Human Cancer Causation." *Journal of Chronic Disease* 35:581–600.

———. 2000. "Molecular Epidemiology: Recent Advances and Future Directions." *Carcinogenesis* 213:517–24.

Pescosolido, Bernice A. 2006. "Professional Dominance and the Limits of Erosion." *Society* 43 (6): 21–29.

Pescosolido, Bernice A., and Jack K. Martin. 2004. "Cultural Authority and the Sovereignty of American Medicine: The Role of Networks, Class, and Community." *Journal of Health, Health Policy, and Law* 29 (4–5): 735–55.

Petchesky, Rosalind Pollack. 2003. *Global Prescriptions: Gendering Health and Human Rights*. London: Zed.

Petersen, Alan. 1997. "Risk, Governance, and the New Public Health." In *Foucault, Health, and Medicine*, ed. A. Petersen and R. Bunton, 189–223. New York: Routledge.

———. 2007. *The Body in Question: A Socio-cultural Approach*. London: Routledge.

Petersen, Alan, and Robin Bunton, eds. 1997. *Foucault, Health, and Medicine*. London: Routledge.

Petersen, Alan, and Deborah Lupton. 1996. *The New Public Health: Health and Self in the Age of Risk*. London: Sage.

Petryna, Adriana. 2002. *Life Exposed: Biological Citizens after Chernobyl*. Princeton, N.J.: Princeton University Press.

———. 2006. "Globalizing Human Subjects Research." In *Global Pharmaceuticals: Ethics, Markets, Practices*, ed. Adriana Petryna, Andrew Lakoff, and Arthur Kleinman, 61–84. Durham, N.C.: Duke University Press.

———. 2007. "Clinical Trials Offshored: On Private Sector Science and Public Health." *BioSocieties* 2 (1): 21–40.

Petryna, Adriana, and Arthur Kleinman. 2006. "The Pharmaceutical Nexus." In *Global Pharmaceuticals: Ethics, Markets, Practices*, ed. Adriana Petryna, Andrew Lakoff, and Arthur Kleinman, 1–32. Durham, N.C.: Duke University Press.

Petryna, Adriana, Andrew Lakoff, and Arthur Kleinman, eds. 2006. *Global Pharmaceuticals: Ethics, Markets, Practices*. Durham, N.C.: Duke University Press.

Pfeffer, Naomi. 1985. "The Hidden Pathology of the Male Reproductive System." In *The Sexual Politics of Reproduction*, ed. H. Homans, 30–44. Aldershot, UK: Gower.

Pfeffer, Naomi, and Julie Kent. 2007. "Framing Women, Framing Fetuses: How Britain Regulates Arrangements for the Collection and Use of Aborted Fetuses in Stem Cell Research and Therapies." *BioSocieties* 2 (4): 429–47.

Pfizer Incorporated. 2003. "Welcome to Viagra." Pfizer Incorporated.

Pfohl, Stephen J. 1985. *Images of Deviance and Social Control: A Sociological History.* New York: McGraw Hill.

Pickstone, John V. 1993. "The Biographical and the Analytical: Towards a Historical Model of Science and Practice in Modern Medicine." In *Medicine and Change: Historical and Sociological Studies of Medical Innovation*, ed. I. Löwy, 23–47. Montrouge, France: John Libbey Eurotext.

Piercy, Marge. 1976. *Woman on the Edge of Time.* New York: Fawcett Crest.

Pigg, Stacy Leigh. 2005. "Globalizing the Facts of Life." In *Sex in Development: Science, Sexuality, and Morality in Global Perspective*, ed. Vincanne Adams and Stacy Leigh Pigg, 39–67. Durham, N.C.: Duke University Press.

Pitts, Victoria. 2004. "Illness and Internet Empowerment: Writing and Reading Breast Cancer in Cyberspace." *Health* 8 (1): 33–59.

Piven, Frances Fox, and Richard A. Cloward. 1993. *Regulating the Poor: The Functions of Public Welfare.* Updated ed. New York: Vintage.

Plotnick, Rod. 2001. *Introduction to Psychology (with Infotrac).* Belmont, Calif.: Wadsworth.

Poling, Travis. 2004. "The Big Buzz Is Magnets; Once Rare, MRI Machines Have Become Common and Profitable." *San Antonio Express-News*, June 19, D1.

Pollack, Andrew. 2003. "AIDS Vaccine Numbers Off, Statistician Says: Effectiveness for Minorities May Be Overstated." *San Francisco Chronicle*, February 27, A1.

Pollack, Andrew, and Lawrence Altman. 2003. "Large Trial Finds AIDS Vaccine Fails to Stop Infection." *New York Times*, February 24.

Pollan, Michael. 2007. *The Omnivore's Dilemma: A Natural History of Four Meals.* New York: Penguin.

Poole, Charles, and Kenneth J. Rothman. 1998. "Our Conscientious Objection to the Epidemiology Wars." *Journal of Epidemiology and Community Health* 52 (10): 613–14.

Popay, Jennie, and Gareth Williams. 1996. "Public Health Research and Lay Knowledge." *Social Science and Medicine* 42 (5): 759–68.

Popay, Jennie, Gareth Williams, Carol Thomas, and Tony Gatrell. 1998. "Theorising Inequalities in Health: The Place of Lay Knowledge." *Sociology of Health and Illness* 20 (5): 619–44.

Pope, Catherine. 2003. "Resisting Evidence: The Study of Evidence-Based Medicine as a Contemporary Social Movement." *Journal for the Social Study of Health, Illness, and Medicine* 7 (3): 267–82.

Porter, Roy. 1985. "The Patient's View: Doing Medical History from Below." *Theory and Society* 14:175–98.

———. 1995. *Trust in Numbers: The Pursuit of Objectivity in Science and Public Life.* Princeton, N.J.: Princeton University Press.

———. 1999. *The Greatest Benefit to Mankind: A Medical History of Humanity.* New York: W. W. Norton.

Potts, Annie. 2000. "'The Essence of the Hard On': Hegemonic Masculinity and the Cultural Construction of 'Erectile Dysfunction.'" *Men and Masculinities* 3 (1): 85–103.

—————. 2004. "Deluxe on Viagra (or What Can a 'Viagra-Body' Do?)" *Body and Society* 10 (1): 17–36.

Potts, Annie, Victoria M. Grace, Tiina Vares, and Nicola Gavey. 2006. "'Sex for Life'? Men's Counterstories on 'Erectile Dysfunction,' Male Sexuality, and Ageing." *Sociology of Health and Illness* 28 (3): 306–29.

Potts, Annie, Nicola Gavey, Victoria M. Grace, and Tiina Vares. 2003. "The Downside of Viagra: Women's Experiences and Concerns." *Sociology of Health and Illness* 25 (7): 697–719.

Potts, Annie, Victoria Grace, Nicola Gavey, and Tiina Vares. 2004. "'Viagra Stories': Challenging 'Erectile Dysfunction.'" *Social Science and Medicine* 59 (3): 489–99.

Poulton, Terry. 1997. *No Fat Chicks: How Big Business Profits from Making Women Hate Their Bodies and How to Fight Back*. New York: Birch Lane.

Powles, Trevor, Ros Eeles, Sue Ashley, Doug Easton, Jenny Chang, Mitch Dowsett, Alwynne Tidy, Jenny Viggers, and Jane Davey. 1998. "Interim Analysis of the Incidence of Breast Cancer in the Royal Marsden Hospital Tamoxifen Randomised Chemoprevention Study." *Lancet* 352:98–101.

Prainsack, Barbara. 2006. "'Negotiating Life': The Regulation of Human Cloning and Embryonic Stem Cell Research in Israel." *Social Studies of Science* 36 (2): 173–205.

Prasad, Amit. 2006. "Social Adoption of a Technology: Magnetic Resonance Imaging (MRI) in India." *International Journal of Contemporary Sociology* 43 (2): 327–55.

Prasad, Amit, and Srirupa Prasad. 2007. "Citizenship, Identity, and Power within Neoliberal Globalization: Drug Testing and Medical Transcription in India." Annual Transnational Workshop, Department of Sociology, University of Illinois, Urbana-Champaign, April.

Preda, Alex. 2005. *AIDS, Rhetoric, and Medical Knowledge*. New York: Cambridge University Press.

President's Commission on Mental Health. 1978. Report to the President from the President's Commission on Mental Health. 4 vols. Washington: U.S. Government Printing Office.

Press, Nancy, Jennifer Fishman, and Barbara Koenig. 2000. "Collective Fear, Individualized Risk: The Social and Cultural Context of Genetic Testing for Breast Cancer." *Nursing Ethics* 7 (3): 237–49.

Pritchard, Kathleen I. 1998. "Is Tamoxifen Effective in Prevention of Breast Cancer?" *Lancet* 352:80–81.

Psillidis, Lori, Jennifer Flach, and Rose Mary Padberg. 1997. "Participants Strengthen Clinical Trial Research: The Vital Role of Participant Advisors in the Breast Cancer Prevention Trial." *Journal of Women's Health* 6:227–32.

Quadagno, Jill. 2004. "Why the United States Has No National Health Insurance: Stakeholder Mobilization Against the Welfare State, 1945–1996." *Journal of Health and Social Behavior* 45 (extra issue): 25–44.

Rabeharisoa, Vololona. 2008. "A Journey in America's Biopolitical Landscape." *BioSocieties* 3 (2): 227–33.

Rabeharisoa, Vololona, and Michel Callon. 1998. "The Participation of Patients in the

Process of Production of Knowledge: The Case of the French Muscular Dystrophy Association" (in French). *Sciences Sociale et Santé* 16:41–65.

Rabinow, Paul. 1992. "Artificiality and Enlightenment: From Sociobiology to Biosociality." In *Incorporations*, ed. J. Crary and S. Kwinter, 234–52. New York: Zone.

———. 1996. *Making PCR: A Story of Biotechnology.* Chicago: University of Chicago Press.

———. 1999. *French DNA: Trouble in Purgatory.* Chicago: University of Chicago Press.

———. 2003. *Anthropos Today: Reflections on Modern Equipment.* Princeton, N.J.: Princeton University Press.

Rabinow, Paul, and Nikolas Rose. 2003/2006. "Biopower Today." 2006. *BioSocieties* 1:195–217. Earlier version at http://www.lse.ac.uk/collections/sociology/pdf/RabinowandRose-BiopowerToday03.pdf.

Radiological Society of North America [RSNA]. 2003. "Salaries Rise for Radiologists in 2002." http://www.rsna.org/publications/rsnanews/oct03/salaries-1.htm (accessed March 2005).

Radley, Alan. 2002. "Portrayals of Suffering: On Looking Away, Looking At, and the Comprehension of Illness Experience." *Body and Society* 8 (3): 1–23.

Radley, Alan, and Susan E. Bell. 2007. "Artworks, Collective Experience, and Claims for Social Justice: The Case of Women Living with Breast Cancer." *Sociology of Health and Illness* 29 (3): 366–90.

Radley, Alan, Deborah Lupton, and Christian Ritter. 1997. "Health: An Invitation and an Introduction." *Health* 1 (1): 5–21.

Rafalovich, Adam. 2005. "Relational Troubles and Semiofficial Suspicion: Educators and the Medicalization of 'Unruly' Children." *Symbolic Interaction* 28 (1): 25–46.

Rajan, Kaushik Sunder. 2005. "Subjects of Speculation: Emergent Life Sciences and Market Logics in the United States and India." *American Anthropologist* 107 (1): 19–30.

———. 2006. *Biocapital: The Constitution of Postgenomic Life.* Durham, N.C.: Duke University Press.

———. 2007. "Biocapital Downstream: The Experimental Machinery of Global Clinical Trials." Paper presented at University of California, San Francisco, November 15.

Rajfer, J., W. J. Aronson, P. A. Bush, F. J. Dorey, and L. J. Ignarro. 1992. "Nitric Oxide as a Mediator of Relaxation of the Corpus Cavernosum in Response to Nonadrenergic, Noncholinergic Neurotransmission." *New England Journal of Medicine* 326:90–94.

Ramirez de Arellano, A. 2007. "Patients without Borders: The Emergence of Medical Tourism." *International Journal of Health Services* 37:193–98.

Randeria, S. 2007. "The State of Globalization: Legal Plurality, Overlapping Sovereignties, and Ambiguous Alliances between Civil Society and the Cunning State in India." *Theory, Culture, and Society* 24 (1): 1–33.

Rankin, Tracy L. 2005. "Andrology as the Medical Specialty to Focus Medical Training on Men's Health?" *Journal of Men's Health and Gender* 2 (1): 45–48.

Rapp, Rayna. 1999. *Testing Women, Testing the Fetus: The Social Impact of Amniocentesis in America.* New York: Routledge.

———. 2001. "Gender, Body, Biomedicine: How Some Feminist Concerns Dragged Reproduction into the Center of Social Theory." *Medical Anthropology Quarterly* 15 (4): 466–477.

———. 2003. "Cell Life and Death, Child Life and Death: Genomic Horizons, Genetic Diseases, Family Stories." In *Remaking Life and Death: Towards an Anthropology of the Biosciences,* ed. S. Franklin and M. Lock. Santa Fe: School of American Research Advanced Seminar Series.

Ratcliff, Kathryn Strother, ed. 2002. *Women and Health: Power, Technology, Inequality, and Conflict in a Gendered World.* Boston: Allyn and Bacon.

Ravdin, Peter M., Kathleen A. Cronin, Nadia Howlader, Christine D. Berg, Rowan T. Chlebowski, Eric J. Feuer, Brenda K. Edwards, and Donald A. Berry. 2007. "The Decrease in Breast-Cancer Incidence in 2003 in the United States." *New England Journal of Medicine* 356:1670–74.

Raz, Aviad E. 1999. "Glocalization and Symbolic Interactionism." *Studies in Symbolic Interaction* 22:3–16.

Reagan, Leslie J., Nancy Tomes, and Paula A. Treichler, eds. 2007. *Medicine's Moving Pictures: Medicine, Health, and Bodies in American Film and Television.* Rochester, N.Y.: University of Rochester Press.

Reardon, Jenny. 2004. *Race to the Finish: Identity and Governance in an Age of Genomics.* Princeton, N.J.: Princeton University Press.

Redfield, Peter. 2008. "Vital Mobility and the Humanitarian Kit." In *Biosecurity Interventions: Global Health and Security in Question,* ed. Andrew Lakoff and Stephen Collier, 147–71. New York: Columbia University Press.

Regier, Darrel A., Jerome K. Myers, Morton Kramer, Lee N. Robins, Dan Blazer, Richard Hough, William Eaton, and Ben Locke. 1984. "The NIMH Epidemiologic Catchment Area Program." *Archives of General Psychiatry* 41:934–48.

Reich, Michael R., ed. 2002. *Public-Private Health Partnerships for Public Health.* Cambridge: Harvard University Press for Harvard Series on Population and International Health.

Reichertz, Jo. 2007. "Abduction: The Logic of Discovery of Grounded Theory." In *Handbook of Grounded Theory,* ed. A. Bryant and K. Charmaz, 214–28. London: Sage.

Reid, Lynette, Natalie Ram, and R. Blake Brown. 2007. "Compensation for Gamete Donation: The Analogy with Jury Duty." *Cambridge Quarterly of Healthcare Ethics* 16:35–43.

Reid, Roddey. 1998. "Unsafe at Any Distance: Todd Haynes' Visual Culture of Health and Risk." *Film Quarterly* 51 (3): 32–44.

———. 2005. *Globalizing Tobacco Control: Anti-smoking Campaigns in California, France, and Japan.* Bloomington: Indiana University Press.

Relman, Arnold S. 1980. "The Medical-Industrial Complex." *New England Journal of Medicine* 303:963–70.

Renaud, Marc. 1995. "Le concept de médicalisation, a-t-il toujours la même perti-nence?" [The Concept of Medicalization: Is It Still Salient?]. In *Medicalization and Social Control*, ed. L. Bouchard and D. Cohen, 167–73. Paris: Actas.

Report on Presidential Commission on Mental Health. 1978.

Reverby, Susan. 1981. "Stealing the Golden Eggs: Ernest Amory Codman and the Sci-ence and Management of Medicine." *Bulletin of the History of Medicine* 55:156–71.

———. 2000. *Tuskegee's Truths: Rethinking the Tuskegee Syphilis Study*. Chapel Hill: University of North Carolina Press.

Reverby, Susan M., and David Rosner. 1979. "Beyond 'the Great Doctors.'" In *Health Care in America: Essays in Social History*, ed. Susan M. Reverby and David Rosner. Philadelphia: Temple University Press.

———. 2004. "'Beyond the Great Doctors' Revisited: A Generation of the New So-cial History of Medicine." In *Locating Medical History: The Stories and Their Mean-ings*, ed. Frank Huisman and John Harley Warner, 167–93. Baltimore: Johns Hop-kins University Press.

Reynolds, P. Preston. 1997. "The Federal Government's Use of Title VI and Medicare to Racially Integrate Hospitals in the United States, 1963 through 1967." *American Journal of Public Health* 87 (11): 1850–58.

Reznik, David L., John W. Murphy, and Lisa Liska Belgrave. 2007. "Globalization and Medicine in Trinidad." *Sociology of Health and Illness* 29 (4): 536–50.

Rheinberger, Hans-Jorg. 2000. "Beyond Nature and Culture: Modes of Reasoning in the Age of Molecular Biology and Medicine." In *Living and Working with the New Medical Technologies*, ed. Margaret Lock, Allan Young, and Alberto Cambrosio, 19–30. Cambridge: Cambridge University Press.

Riccardi, Nicholas, and Terence Monmaney. 2000. "King/Drew Medical Research Suspended." *Los Angeles Times*, April 27, 1.

Richards, Evelleen. 1991. *Vitamin C and Cancer: Medicine or Politics?* London: Mac-millan.

Richardson, Diane. 2001. "Extending Citizenship: Cultural Citizenship and Sexuality." In *Culture and Citizenship*, ed. N. Stevenson, 153–66. London: Sage.

Riessman, Catherine Kohler. 1983. "Women and Medicalization: A New Perspective." *Social Policy* 14:3–18.

Riska, Elianne. 2000. "The Rise and Fall of Type A Man." *Social Science and Medicine* 51 (11): 1665–74.

———. 2002. "From Type A Man to the Hardy Man: Masculinity and Health." *Soci-ology of Health and Illness* 24 (3): 347–58.

———. 2003. "Gendering the Medicalization Thesis." In *Gender Perspectives on Health and Medicine*, vol. 7, ed. Marcia Texler Segal, Vasilikie Demos, and Jennie Jacobs Kronenfeld, 59–87. Boston: JAI.

———. 2004. *Masculinity and Men's Health: Coronary Heart Disease in Medical and Public Discourse*. Lanham, Md.: Rowman and Littlefield.

Risse, Guenter B. 1999. *Mending Bodies, Saving Souls: A History of Hospitals*. New York: Oxford University Press.

Risse, Guenter B., Ronald L. Numbers, and Judith Walzer Leavitt, eds. 1977. *Medicine without Doctors: Home Health Care in American History.* New York: Science History.

Roberts, Celia. 2007. *Messengers of Sex: Hormones, Biomedicine, and Feminism.* Cambridge: Cambridge University Press.

Robertson, Rose Marie. 2001. "Women and Cardiovascular Disease: The Risks of Misperception and the Need for Action." *Circulation* 103 (19): 2318–20.

Robertson, Steve. 2007. *Understanding Men and Health: Masculinities, Health, and Wellbeing.* Maidenhead, England: Open University Press.

Robins, Lee N., and Darrel A. Regier, eds. 1991. *Psychiatric Disorders in America: The Epidemiologic Catchment Area Study.* New York: Free Press.

Robins, Lee N., John E. Helzer, Jack Croughan, and K. S. Ratcliff. 1981. "National Institute of Mental Health Diagnostic Interview Schedule." *Archives of General Psychiatry* 38:381–89.

Robins, Lee N., John Wing, Hans Ulrich Wittchen, John E. Helzer, Thomas Babor, Jay Burke, Anne Farmer, Assen Jablenski, Roy Pickens, Darrel A. Regier, Norman Sartorius, and Leland Towle. 1988. "The Composite International Diagnostic Interview." *Archives of General Psychiatry* 45:1069–77.

Robinson, James C. 1999. *The Corporate Practice of Medicine: Competition and Innovation in Health Care.* Berkeley: University of California Press.

Robinson, Sally. 2000. *Marked Men: White Masculinity in Crisis.* New York: Columbia University Press.

———. 2002. "Men's Liberation, Men's Wounds: Emotion, Sexuality, and Reconstruction of Masculinity in the 1970s." In *Boys Don't Cry? Rethinking Narratives of Masculinity and Emotion in the U.S.,* ed. M. Shamir and J. Travis, 205–29. New York: Columbia University Press.

Rock, M., E. Mykhalovskiy, and T. Schlich. 2007. "People, Other Animals, and Health Knowledges: Towards a Research Agenda." *Social Science and Medicine* 64 (9): 1970–76.

Rockhill, Beverly, Donna Spiegelman, Celia Byrne, David J. Hunter, and Graham Colditz. 2001. "Validation of the Gail et al. Model of Breast Cancer Risk Prediction and Implications for Chemoprevention." *Journal of the National Cancer Institute* 93:358–66.

Rodriguez-Ocana, Esteban, ed. 2002. *The Politics of the Healthy Life: An International Perspective.* Sheffield, England: European Association for the History of Medicine and Public Health.

Rogers, Naomi. 1998. *An Alternative Path: The Making and Remaking of Hahnemann Medical College of Philadelphia.* New Brunswick, N.J.: Rutgers University Press.

Rogers, Richard, ed. 2000. *Preferred Placement: Knowledge Politics on the Web.* Maastricht, Netherlands: Jan van Eyck.

Rosaldo, Renato. 2003. "Introduction: The Borders of Belonging: Nation and Citizen in the Hinterlands." In *Cultural Citizenship in Island Southeast Asia: Nation and Belonging in the Hinterlands,* ed. Renato Rosaldo, 1–15. Berkeley: University of California Press.

Rose, Nikolas. 1990. *Governing the Soul: The Shaping of the Private Self.* London: Routledge.

———. 1992. "Engineering the Human Soul: Analyzing Psychological Expertise." *Science in Context* 5 (2): 351–69.

———. 1994. "Medicine, History, and the Present." In *Reassessing Foucault: Power, Medicine, and the Body,* ed. C. Jones and R. Porter, 48–72. London: Routledge.

———. 1996. *Inventing Our Selves: Psychology, Power, and Personhood.* Cambridge: Cambridge University Press.

———. 2001. "The Politics of Life Itself." *Theory, Culture, and Society* 18:1–30.

———. 2007. *The Politics of Life Itself: Biomedicine, Power, and Subjectivity in the Twenty-first Century.* Princeton, N.J.: Princeton University Press.

Rose, Nikolas, and Carlos Novas. 2004. "Biological Citizenship." In *Global Assemblages: Technology, Politics, and Ethics as Anthropological Problems,* ed. Aihwa Ong and Stephen J. Collier, 439–63. Malden, Mass.: Blackwell.

Rosenberg, Charles E. 1962. *The Cholera Years: The United States in 1832, 1849, and 1866.* Chicago: University of Chicago Press.

———. 1963/1976. "Martin Arrowsmith: The Scientist as Hero." *American Quarterly* 15:447–58. Reprinted in *No Other Gods: Science and American Social Thought* (Baltimore: Johns Hopkins University Press, 1976), 123–32.

———. 1992. *Explaining Epidemics and Other Studies in the History of Medicine.* Cambridge: Cambridge University Press.

———. 1995. *The Care of Strangers: The Rise of America's Hospital System.* Baltimore: Johns Hopkins University Press.

———. 2006. "Contested Boundaries: Psychiatry, Disease, and Diagnosis." *Perspectives in Biology and Medicine* 49 (3): 407–24.

———. 2007. *Our Present Complaint: American Medicine, Then and Now.* Baltimore: Johns Hopkins University Press.

Rosenfeld, Dana, and Christopher A. Faircloth, eds. 2006. *Medicalized Masculinities.* Philadelphia: Temple University Press.

Rothman, Barbara Katz. 1998. *Genetic Maps and Human Imaginations: The Limits of Science in Understanding Who We Are.* New York: W. W. Norton.

Rothman, David J. 2000. "The Shame of Medical Research." *New York Review,* November 30, 60–64.

Rothman, Sheila M., and David J. Rothman. 2003. *The Pursuit of Perfection: The Promise and Perils of Medical Enhancement.* New York: Pantheon.

Rothschild, Joan. 2005. *The Dream of the Perfect Child.* Bloomington: Indiana University Press.

Rowland, D. L., H. S. M. Boedhoe, G. Dohle, and A. K. Slob. 1999. "Intracavernosal Self-Injection Therapy in Men with Erectile Dysfunction: Satisfaction and Attribution in 119 Patients." *International Journal of Impotence Research* 11:145–51.

Russell, Sabin, and Tom Abate. 2001. "Shutdown Puts Spotlight on Human Research: Experts Say John's Hopkins Case Reflects Problems across the U.S." *San Francisco Chronicle,* July 21, A1.

Ruzek, Sheryl. 1978. *The Women's Health Movement: Feminist Alternatives to Medical Control.* New York: Praeger.

———. 1980. "Medical Response to Women's Health Activities: Conflict, Cooperation, Accommodation, and Cooptation." In *Research in the Sociology of Health Care,* ed. J. A. Roth, 325–54. Stamford, Conn.: JAI.

Ruzek, Sheryl D., and J. Hill. 1986. "Promoting Women's Health: Redefining the Knowledge Base and Strategies for Change." *Health Promotion* 1:301–9.

Ruzek, Sheryl B., Virginia L. Olesen, and Adele E. Clarke, eds. 1997. *Women's Health: Complexities and Differences.* Columbus: Ohio State University Press.

Sabo, Donald, and David Frederick Gordon, eds. 1995. *Men's Health and Illness: Gender, Power, and the Body.* London: Sage.

Saetnan, Anne, Nelly Oudshoorn, and Marta Kirejczyk, eds. 2000. *Bodies of Technology: Women's Involvement with Reproductive Medicine.* Columbus: Ohio State University Press.

Saillant, Francine, and Serge Genest, eds. 2007. *Medical Anthropology: Regional Perspectives and Shared Concerns.* Malden, Mass.: Blackwell.

Salmon, J. Warren. 1990. "Profit and Health Care: Trends in Corporatization and Proprietarization." In *The Corporate Transformation of Health Care: Issues and Directions,* ed. J. W. Salmon, 55–77. Amityville, N.Y.: Baywood.

Santos, Fernando. 2007. "Evidence from Bite Marks, It Turns Out, Is Not So Elementary." *New York Times,* January 28, 7.

Sapolsky, Robert M. 1998. *Why Zebras Don't Get Ulcers: A Guide to Stress, Stress-Related Diseases, and Coping.* New York: W. H. Freeman.

Sassen, Saskia. 2002. *Global Networks: Linked Cities.* New York: Routledge.

Satel, Sally. 2002. "I Am a Racially Profiling Doctor." *New York Times Magazine,* May 5.

———. 2004a. "Race and Medicine Can Mix without Prejudice: How the Story of BiDil Illuminates the Future of Medicine." *Medical Progress Today,* December 10. http://www.medicalprogresstoday.com/spotlight/spotlight_indarchive.php?id=449 (accessed March 23, 2005).

———. 2004b. Transcript of Remarks Presented at American Enterprise Institute for Public Policy Research Conference: Race, Medicine, and Public Policy, November 12. http://www.aei.org/include/event_print.asp?eventID=937 (accessed February 9, 2005).

Satel, Sally, and Jonathan Klick. 2005. "The Institute of Medicine Report: Too Quick to Diagnose Bias." *Perspectives in Biology and Medicine* 48 (1): S15–S25.

Saul, Stephanie. 2005. "Drug Makers to Police Consumer Campaigns." *New York Times,* August 3, C7.

Savitz, David A. 1997. "The Alternative to Epidemiologic Theory: Whatever Works." *Epidemiology* 8 (2): 210–12.

Sawchuk, Kim. 2000. "Biotourism, *Fantastic Voyage,* and Sublime Inner Space." In *Wild Science: Reading Feminism, Medicine, and the Media,* ed. Janine Marchessault and Kim Sawchuk, 9–23. London: Routledge.

Schacht, Thomas. 1985. "DSM-III and the Politics of Truth." *American Psychologist* 40 (5): 513–21.

Scheper-Hughes, Nancy. 2000. "The Global Traffic in Human Organs." *Current Anthropology* 41 (2): 191–224.

———. 2005. "The Last Commodity: Posthuman Ethics and the Global Traffic in 'Fresh' Organs." In *Global Assemblages: Technology, Politics, and Ethics as Anthropological Problems*, ed. Aihwa Ong and Stephen J. Collier, 145–67. Malden, Mass.: Blackwell.

Schiebinger, Londa. 2004. *Plants and Empire: Colonial Bioprospecting and the Improvement of the World*. Cambridge: Harvard University Press.

Schiller, Dan. 1999. *Digital Capitalism: Networking the Global Market System*. Cambridge: MIT Press.

Schiller, Joyce K., Heather Campbell Coyle, Molly S. Hutton, and Susan Fillin-Yeh, eds. 2007. *John Sloan's New York*. Wilmington: Delaware Art Museum.

Schirren, Carl. 1985. "Andrology: Origin and Development of a Special Discipline in Medicine." *Andrologia* 17 (2): 117–25.

———. 1996. "Andrology—Development and Future: Critical Remarks after 45 Years of Medical Practice." *Andrologia* 28 (3): 137–40.

Schmid, Randolph E. 2006. "Doctors Frown on Unneeded Fetal Scans." *San Francisco Chronicle*, August 8, A4.

Schmidt, Matthew, and Lisa J. Moore. 1998. "Constructing a 'Good Catch,' Picking a Winner: The Development of Technosemen and the Deconstruction of the Monolithic Male." In *Cyborg Babies: From Techno-sex to Techno-tots*, ed. R. Davis-Floyd and J. Dumit, 21–39. New York: Routledge.

Schneider, Joseph W., and Peter Conrad. 1980. "The Medical Control of Deviance: Contests and Consequences." In *Research in the Sociology of Health Care*, ed. J. A. Roth, 1–53. Stamford, Conn.: JAI.

Schneider, William H., ed. 2002. *Rockefeller Philanthropy and Modern Biomedicine: International Initiatives from World War I to the Cold War*. Bloomington: Indiana University Press.

Schneirov, Matthew, and Jonathan David Geczik. 2003. *A Diagnosis for Our Times: Alternative Health, from Lifeworld to Politics*. Albany: SUNY Press.

Schroeder, Doris. 2007. "Public Health, Ethics, and Functional Foods." *Journal of Agricultural and Environmental Ethics* 20 (3): 247–50.

Schulte, Paul A. 1993. "A Conceptual and Historical Framework for Molecular Epidemiology." In *Molecular Epidemiology: Principles and Practices*, ed. Paul A. Schulte and Frederica P. Perera. San Diego: Academic.

Schulte, Paul A., and Frederica P. Perera. 1993. *Molecular Epidemiology: Principles and Practices*. San Diego: Academic.

Schulte-Sasse, Linda. 2006. "Advise and Consent: On the Americanization of Body Worlds." *BioSocieties* 1 (4): 369–84.

Schulz, Amy J., and Leith Mullings, eds. 2006. *Gender, Race, Class, and Health: Intersectional Approaches*. San Francisco: Jossey-Bass.

Schwartz, Hillel. 1986. *Never Satisfied: A Cultural History of Diets, Fantasies, and Fat.* London: Collier Macmillan.

Scott, Susia. 2006. "The Medicalization of Shyness: From Social Misfits to Social Fitness." *Sociology of Health and Illness* 28 (2): 133–53.

Scott, W. Richard, Martin Ruef, Peter J. Mendel, and Carol A. Coronna. 2000. *Institutional Change and Healthcare Organizations: From Professional Dominance to Managed Care.* Chicago: University of Chicago Press.

Seale, Clive. 2004. *Health and Media.* Oxford: Blackwell.

———. 2005. "New Directions for Critical Internet Health Studies: Representing Cancer Experience on the Web." *Sociology of Health and Illness* 27 (4): 515–40.

———. 2008. "Mapping the Field of Medical Sociology: A Comparative Analysis of Journals." *Sociology of Health and Illness* 30 (5): 677–95.

Seale, Clive, Debbie Cavers, and Mary Dixon-Woods. 2006. "Commodification of Body Parts: By Medicine or Media?" *Body and Society* 12 (1): 25–42.

Sears, R. Bradley, Gary J. Gates, and William B. Rubenstein. 2005. *Same-Sex Couples and Same-Sex Couples Raising Children in the United States: Data from Census 2000.* Los Angeles: Williams Project on Sexual Orientation Law and Public Policy, UCLA School of Law.

Segal, Marcia Texler, and Vasilikie Demos, with Jennie Jacobs Kronenfeld. 2003. "Gendered Perspectives on Medicine: An Introduction." In *Gendered Perspectives on Health and Medicine: Key Themes*, vol. 7, ed. Marcia Texler Segal and Vasilikie Demos with Jennie Jacobs Kronenfeld, 1–9. Oxford: Elsevier.

Seidman, S. N., S. P. Roose, M. A. Menza, R. Shabsigh, and R. C. Rosen. 2002. "Sildenafil Improved Erectile Dysfunction and Quality of Life in Men with Comorbid Mild-to-Moderate Depression." *ACP Journal Club* 137:21.

Sen, Gita, Asha George, and Piroska Ostlin, eds. 2002. *Engendering International Health: The Challenge of Equity.* Cambridge: MIT Press.

Sengoopta, Chandak. 2006. *The Most Secret Quintessence of Life: Sex, Glands, and Hormones, 1850–1950.* Chicago: University of Chicago Press.

Sensenig, James. 2007. "The Need for Diversity in Models of Medicine." In *Philanthropic Foundations and the Globalization of Scientific Medicine and Public Health*, ed. Benjamin B. Page and David A. Valone, 125–29. Lanham, Md.: University Press of America.

Serlin, David. 2004. *Replaceable You: Engineering the Body in Postwar America.* Chicago: University of Chicago Press.

Sharf, Barbara F. 1997. "Communicating Breast Cancer On-line: Support and Empowerment on the Internet." *Women and Health* 26:65–84.

Sheehan, H. E., and B. P. Brenton, eds. 2002. "Preface to Special Issue: Global Perspectives on Complementary and Alternative Medicine." *Annals of the American Academy of Political and Social Science* 583:6–11.

Shepard, Peggy, Mary E. Northridge, Swati Prakash, and Gabriel Stover. 2002. "Advancing Environmental Justice through Community Based Participatory Research." *Environmental Health Perspectives* 110:139–40. Suppl. 2.

Sherman, Janette D. 1998. "Tamoxifen and Prevention of Breast Cancer." *Toxicology and Industrial Health* 14:485–99.

Shields, Peter G., and Curt C. Harris. 1991. "Molecular Epidemiology and the Genetics of Environmental Cancer." *Journal of the American Medical Association* 246:681–87.

Shilling, Chris. 2005. *The Body in Culture, Technology, and Society.* London: Sage.

Shim, Janet K. 2000. "Bio-power and Racial, Class, and Gender Formation in Biomedical Knowledge Production." In *Research in the Sociology of Health Care*, vol. 17, ed. J. J. Kronenfeld, 173–95. Stamford, Conn.: JAI.

———. 2002a. "Race, Class, and Gender across the Science-Lay Divide: Expertise, Experience, and 'Difference' in Cardiovascular Disease." Ph.D. diss., Department of Social and Behavioral Sciences, University of California, San Francisco.

———. 2002b. "Understanding the Routinised Inclusion of Race, Socioeconomic Status, and Sex in Epidemiology: The Utility of Concepts from Technoscience Studies." *Sociology of Health and Illness* 24:138–50.

———. 2005. "Constructing 'Race' across the Science-Lay Divide: Racial Formation in the Epidemiology and Experience of Cardiovascular Disease." *Social Studies of Science* 35 (3): 405–36.

———. 2010. "Cultural Health Capital: A Theoretical Approach to Understanding Health Care Interactions and the Dynamics of Unequal Treatment." *Journal of Health and Social Behavior* 51 (1): 1–15.

Shim, Janet K., Ann J. Russ, and Sharon R. Kaufman. 2006. "Risk, Life Extension, and the Pursuit of Medical Possibility." *Sociology of Health and Illness* 28 (4): 479–502.

———. 2007. "Clinical Life: Expectation and the Double Edge of Medical Promise." *Health* 11 (2): 245–64.

Shostak, Sara. 2001. "Locating Molecular Biomarkers, Relocating Risk." Paper presented at the annual meeting of the Society for Social Studies of Science, November 1–4, Cambridge, Mass.

———. 2003. "Locating Gene-Environment Interaction: At the Intersections of Genetics and Public Health." *Social Science and Medicine* 56 (11): 2327–42.

———. 2004. "Environmental Justice and Genomics: Acting on the Futures of Environmental Health." *Science as Culture* 13 (4): 539–62.

———. 2005. "The Emergence of Toxicogenomics: A Case Study of Molecularization." *Social Studies of Science* 35 (3): 367–404.

Shostak, Sara, and Erin Rehel. 2007. "Changing the Subject: Science, Subjectivity, and the Structure of 'Ethical Problems.'" In *Advances in Medical Sociology: Sociological Perspectives on Bioethical Issues*, ed. Barbara Katz Rothman, Elizabeth Armstrong, and Rebecca Tiger, 323–46. Oxford: Elsevier.

Shy, Carl M. 1997. "The Failure of Academic Epidemiology: Witness for the Prosecution." *American Journal of Epidemiology* 145 (6): 479–84.

Silver, Lee M. 1997. *Remaking Eden: How Genetic Engineering and Cloning Will Transform the American Family.* New York: Avon.

Silverman, Chloe. 2008. "Brains, Pedigrees, and Promises: Lessons from the Politics of

Autism Genetics." In *Biosocialities, Genetics, and the Social Sciences: Making Biologies and Identities*, ed. S. Gibbon and C. Novas, 38–55. London: Routledge.

Singh, Ilina, and Nikolas Rose. 2006. "Neuro-forum: An Introduction." *BioSocieties* 1 (1): 97–102.

Singh, Jennifer, J. Illes, L. Lazzeroni, and J. Hallmayer. 2009. "Trends in U.S. Autism Research." *Journal of Autism and Developmental Disorders* 39 (5): 788–95. Epub 2009 Jan 16.

Sismondo, Sergio. 2009. "Ghosts in the Machine: Publication Planning in the Medical Sciences." *Social Studies of Science* 39 (2): 171–198.

Skocpol, Theda. 1992. *Protecting Soldiers and Mothers: The Political Origins of Social Policy in the United States*. Cambridge: Harvard University Press.

Slaughter, Sheila, and Gary Rhoades. 2004. *Academic Capitalism: Politics, Policies, and the Entrepreneurial University*. Baltimore: Johns Hopkins University Press.

Sloane, David Charles, and Beverlie Conant Sloane. 2003. *Medicine Moves to the Mall*. Baltimore: Johns Hopkins University Press.

Smart, Andrew, Richard Tutton, Paul Martin, George T. H. Ellison, and Richard Ashcroft. 2008. "The Standardization of Race and Ethnicity in Biomedical Science Editorials and U.K. Biobanks." *Social Studies of Science* 38 (3): 407–23.

Smith, Merritt Roe, and Leo Marx, eds. 1994. *Does Technology Drive History? The Dilemma of Technological Determinism*. Cambridge: MIT Press.

Smith, Vicki. 1997. "New Forms of Work Organizations." *Annual Review of Sociology* 23:315–39.

Sobal, Jeffery. 1995. "The Medicalization and Demedicalization of Obesity." In *Eating Agendas: Food and Nutrition as Social Problems*, ed. Jeffery Sobal and Donna Maurer, 67–90. New York: Aldine de Gruyter.

———. 1999. "The Size Acceptance Movement and the Social Construction of Body Weight." In *Weighty Issues: Fatness and Thinness as Social Problems*, ed. D. Maurer and J. Sobal, 231–48. New York: Aldine de Gruyter.

Solomon, Deborah. 2006. "A Life in Somber Tones: Review of *The Revenge of Thomas Eakins*, by Sidney D. Kirkpatrick." *New York Times Book Review*, April 2, 18.

Spark, Richard F., Robert A. White, and Peter B. Connolly. 1980. "Impotence Is Not Always Psychogenic." *Journal of the American Medical Association* 243:750–55.

Spector, M., and John Kitsuse. 1977. *Constructing Social Problems*. Menlo Park, Calif.: Cummings.

Spitzack, Carole. 1990. *Confessing Excess: Women and the Politics of Body Reduction*. Albany: SUNY Press.

Spitzer, Robert L. 1984. "Psychiatric Diagnosis: Are Clinicians Still Necessary?" In *Psychotherapy Research: Where Are We and Where Should We Go*, ed. Robert L. Spitzer and Janet B. Williams, 273–92. New York: Guilford.

Spitzer, Robert L., and Jean Endicott. 1968. "DIAGNO: A Computer Program for Psychiatric Diagnosis Utilizing the Differential Diagnostic Procedure." *Archives of General Psychiatry* 18:746–56.

———. 1969. "DIAGNO II: Further Developments in a Computer Program for Psychiatric Diagnosis." *American Journal of Psychiatry* 125:12–21. Suppl. 8.

———. 1971. "An Integrated Group of Forms for Automated Psychiatric Case Records." *Archives of General Psychiatry* 24:540–47.

———. 1974. "Can the Computer Assist Clinicians in Psychiatric Diagnosis?" *American Journal of Psychiatry* 131 (5): 523–30.

———. 1975. "Computer Applications in Psychiatry." In *American Handbook of Psychiatry*, ed. Silvano Arieti, 811–39. 2nd ed. New York: Basic.

Spitzer, Robert L., Jean Endicott, Jacob Cohen, and Joseph Fleiss. 1974. "Constraints on the Validity of Computer Diagnosis." *Archives of General Psychiatry* 31:197–203.

Spitzer, Robert L., Jean Endicott, and Eli Robins. 1978. "Research Diagnostic Criteria: Rationale and Reliability." *Archives of General Psychiatry* 35:773–82.

Spivak, Gayatri Chakravorty. 1988. *In Other Worlds: Essays in Cultural Politics*. New York: Routledge.

———. 1993. *Outside in the Teaching Machine*. New York: Routledge.

Squier, Susan Merrill. 2004. *Liminal Lives: Imagining the Human at the Frontiers of Biomedicine*. Durham, N.C.: Duke University Press.

Squiers, Carol. 2005. *The Body at Risk: Photography of Disorder, Illness, and Healing*. New York and Berkeley: International Center of Photography and University of California Press.

Stafford, Barbara Maria. 2007. *Echo Objects: The Cognitive Work of Images*. Chicago: University of Chicago Press.

Stanford Law School Securities Class Action Clearing House. 2005. "Vaxgen." http://securities.stanford.edu/1027/VXGN03-01 (accessed March 25, 2005).

Stanton, Jennifer, ed. 2002. *Innovations in Health and Medicine: Diffusion and Resistance in the Twentieth Century*. London: Routledge.

Star, S. Leigh. 1995. *The Cultures of Computing*. Oxford: Basil Blackwell.

Starr, Paul. 1982. *The Social Transformation of American Medicine: The Rise of a Sovereign Profession and the Making of a Vast Industry*. New York: Basic.

Stearns, Peter. 1997. *Fat History: Bodies and Beauty in the Modern West*. New York: New York University Press.

Steel, T. G. W., H. van Langen, H. Wijktra, and E. J. Meuleman. 2003. "Penile Duplex Pharmaco-Ultrasonography Revisited: Revalidation of the Parameters of the Cavernous Arterial Response." *Journal of Urology* 169:216–20.

Steichen, Edward. 1955. *The Family of Man: The Photographic Exhibition Created by Edward Steichen for the Museum of Modern Art*. New York: Simon and Schuster.

Stein, Rob. 2004. "Bar Code Implant Calls Up Medical Data." *San Francisco Chronicle*, October 14, A1, A16.

Stevens, Rosemary. 1998. *American Medicine and the Public Interest: A History of Specialization*. Berkeley: University of California Press.

Stevenson, Fiona, and Graham Scambler. 2005. "Rejoinder to 'The Myth of Concordance': A Response to Armstrong." *Health* 9 (1): 29–30.

Stimson, Gerry V. 2007. "'Harm Reduction—Coming of Age': A Local Movement with Global Impact." *International Journal of Drug Policy* 18:67–69.

Stoeckle, J. D., and G. A. White. 1985. *Plain Pictures of Plain Doctoring: Vernacular Expression and New Deal Medicine.* Cambridge: MIT Press.

Stoler, Ann Laura, and Frederick Cooper. 1997. "Between Metropole and Colony: Rethinking a Research Agenda." In *Tensions of Empire: Colonial Cultures in a Bourgeois World,* ed. Frederick Cooper and Ann L. Stoler, 1–56. Berkeley: University of California Press.

Stoler, Ann Laura, Carole McGranahan, and Peter C. Perdue, eds. 2008. *Imperial Formations.* Santa Fe: School for Advanced Research Press.

Strategic Research Institute. 2005a. "Sixth Annual Multicultural Pharmaceutical Market Development and Outreach Conference." March 17–18. http://www.srinstitute .com/CustomerFiles/upload/brochure/CM453 brochure.pdf (accessed March 30, 2005).

Strategic Research Institute. 2005b. (No title.) http://www.srinstitute.com/Applica tionFiles/web/WebFrame.cfm?web_id=324 (accessed March 30, 2005).

Strathern, Marilyn. 1992. *Reproducing the Future: Essays on Anthropology, Kinship, and the New Reproductive Technologies.* New York: Routledge.

Strauman, Elena, and Bethany C. Goodier. 2008. "Not Your Grandmother's Doctor Show: A Review of *Grey's Anatomy, House,* and *Nip/Tuck.*" *Journal of Medical Humanities* 29 (2): 127–31.

Strauss, Anselm L. 1978. *Negotiations: Varieties, Processes, Contexts, and Social Order.* San Francisco: Jossey-Bass.

———. 1987. *Qualitative Analysis for Social Scientists.* New York: Cambridge University Press.

———. 1993. *Continual Permutations of Action.* New York: Aldine de Gruyter.

Strauss, Anselm L., and Juliet Corbin. 1988. *Shaping a New Health Care System: The Explosion of Chronic Illness as a Catalyst for Change.* San Francisco: Jossey-Bass.

———. 1998. *Basics of Qualitative Research: Techniques and Procedures for Developing Grounded Theory,* 2nd ed. Newbury Park, Calif.: Sage.

Strauss, Anselm, Juliet Corbin, Fagerhaugh Shizuko, Barney G. Glazer, David Maines, Barbara Suczek, and Carolyn L. Weiner, eds. 1984. *Chronic Illness and the Quality of Life.* 2nd ed. St. Louis: C. V. Mosby.

Strauss, Anselm L., and Barney Glaser. 1964. "The Social Loss of Dying Patients." *American Journal of Nursing* 64:119–21.

———. 1975. *Chronic Illness and the Quality of Life.* St. Louis: C. V. Mosby.

———. 1993. *Continual Permutations of Action.* New York: Aldine de Gruyter.

Strauss, Anselm, Leonard Schatzman, Rue Bucher, Danuta Erlich, and Melvin Sabshin. 1964. *Psychiatric Ideologies and Institutions.* Glencoe, Ill.: Free Press.

Strong, Phil. 1979. "Sociological Imperialism and the Profession of Medicine: A Critical Examination of the Thesis of Medical Imperialism." *Social Science and Medicine* 13A:199–215.

————. 1984. "The Academic Encirclement of Medicine." *Sociology of Health and Illness* 6 (3): 339–58.

Sturken, Marita, and Lisa Cartwright. 2001. *Practices of Looking: An Introduction to Visual Culture.* New York: Oxford University Press.

Stux, Gabriel, and Richard Hammerschlag, eds. 2001. *Clinical Acupuncture: Scientific Basis.* New York: Springer.

Sulik, Gayle A. 2009. "Managing Biomedical Uncertainty: The Technoscientific Illness Identity." *Sociology of Health and Illness* 31 (7): 1059–76.

Sullivan, Jean. 2000. *MassHealth Diagnostic and Surgical Bulletin 5.* Massachusetts Health and Human Services. http://www.mass.gov/Eeohhs2/docs/masshealth/bull_2000/dsf-5.pdf (accessed June 27, 2006).

Summerfield, Derek. 1999. "A Critique of Seven Assumptions behind Psychological Trauma Programmes in War-Affected Areas." *Social Science and Medicine* 48 (10): 1449–62.

Susser, Mervyn. 1989. "Epidemiology Today: 'A Thought-Tormented World.'" *International Journal of Epidemiology* 18 (3): 481–88.

————. 1998. "Does Risk Factor Epidemiology Put Epidemiology at Risk?" *Journal of Epidemiology and Community Health* 52:608–11.

Susser, Mervyn, and Ezra Susser. 1996a. "Choosing a Future for Epidemiology I: Eras and Paradigms." *American Journal of Public Health* 86:668–73.

————. 1996b. "Choosing a Future for Epidemiology II: From Black Box to Chinese Boxes and Eco-epidemiology." *American Journal of Public Health* 86 (5): 674–77.

Swan, John P. 1990. "Universities, Industry, and the Rise of Biomedical Collaboration in America." In *Pill Peddlers: Essays on the History of the Pharmaceutical Industry,* ed. J. Liebman, G. J. Higby, and E. C. Stroud, 73–90. Madison: American Institute of the History of Pharmacy.

Sydie, R. A. 1987. *Natural Women, Cultural Men: A Feminist Perspective on Sociological Theory.* Milton Keynes: Open University Press.

Talan, Jamie. 2002. "Seeking Damage Control: Brain Scans of Schizophrenics Indicate that Loss Is Ongoing." *Newsday,* February 19. http://www.loni.ucla.edu/~thompson/MEDIA/PNAS/newsday_sz.htm (accessed October 15, 2005).

Tate, Sarah, and David Goldstein. 2004. "Will Tomorrow's Medicines Work for Everyone?" *Nature Genetics* 36:S34–S42.

Taubes, Gary. 1995. "Epidemiology Faces Its Limits." *Science* 269 (5221): 164–69.

————. 2007. "Unhealthy Science: Why Can't We Trust Much of What We Hear about Diet, Health, and Behavior-Related Diseases?" *New York Times Magazine,* September 16.

Taussig, Karen-Sue, Rayna Rapp, and Deborah Heath. 2003. "Flexible Eugenics: Technologies of the Self in the Age of Genetics." In *Genetic Nature/Culture,* ed. Alan H. Goodman, Deborah Heath, and M. Susan Lindee, 58–76. Berkeley: University of California Press.

Taussig, Mark, Michael J. Selgelid, Sree Subedi, and Janardan Subedi. 2006. "Taking

Sociology Seriously: A New Approach to the Bioethical Problems of Infectious Disease." *Sociology of Health and Illness* 28 (6): 838–49.

Taylor, Anne, Susan Ziesch, Clyde Yancy, et al. 2004. "Combination of Isosorbide Dinitrate and Hydralazine in Blacks with Heart Failure." *New England Journal of Medicine* 351:2049–57.

Taylor, Janelle. 1992. "The Public Fetus and the Family Car: From Abortion Politics to a Volvo Advertisement." *Public Culture* 4 (2): 67–80.

———. 2003. "Confronting 'Culture' in Medicine's 'Culture of No Culture.'" *Academic Medicine* 78 (6): 555–59.

Temkin, Owsei. 1968. "Comparative Study of the History of Medicine." *Bulletin of the History of Medicine* 42:362–71.

Tesh, Sylvia Noble. 1990. *Hidden Arguments: Political Ideology and Disease Prevention Policy.* New Brunswick, N.J.: Rutgers University Press.

Theberge, Nancy. 2008. "The Integration of Chiropractic into Healthcare Teams: A Case Study from Sport Medicine." *Sociology of Health and Illness* 30 (1): 19–34.

Thompson, Charis. 2005. *Making Parents: The Ontological Choreography of Reproductive Technologies.* Cambridge: MIT Press.

———. 2006. "Race Science." *Theory, Culture, and Society* 23 (2–3): 547–49.

———. 2007. "Why We Should, in Fact, Pay for Egg Donation." *Regenerative Medicine* 2 (2): 203–9.

———. 2008. "Stem Cells, Women, and the New Gender and Science." In *Gendered Innovations in Science and Engineering,* ed. Londa Schiebinger, 109–30. Stanford: Stanford University Press.

———. Forthcoming. *Good Science: Lessons for the Twenty-first Century from the End of the Beginning of Human Pluripotent Stem Cell Research.* Cambridge: MIT Press.

———. In preparation. *Charismatic Megafauna and Miracle Babies: Essays in Selective Pronatalism.*

Thomson, L. Katherine. 2009. "Transdisciplinary Knowledge Production of Endocrine Disruptors: 'Windows of Vulnerability' in Breast Cancer Risk." Ph.D. diss., University of California, San Francisco.

Tiefer, Leonore. 1986. "In Pursuit of the Perfect Penis: The Medicalization of Male Sexuality." *American Behavioral Scientist* 29:579–99.

———. 1994. "The Medicalization of Impotence — Normalizing Phallocentrism." *Gender and Society* 8 (3): 363–77.

———. 1995. *Sex Is Not a Natural Act and Other Essays.* Boulder, Colo.: Westview.

———. 2006. "Female Sexual Dysfunction: A Case Study of Disease Mongering and Activist Resistance." *PLoS Medicine* 3:436–39.

Timmermans, Stefan. 1999. *Sudden Death and the Myth of CPR.* Philadelphia: Temple University Press.

———. 2000. "Technology and Medical Practice." In *Handbook of Medical Sociology,* ed. C. E. Bird, P. Conrad, and A. M. Fremont, 309–21. Upper Saddle River, N.J.: Prentice Hall.

————. 2006. *Postmortem: How Medical Examiners Explain Suspicious Deaths*. Chicago: University of Chicago Press.

Timmermans, Stefan, and Marc Berg. 1997. "Standardization in Action: Achieving Local Universality through Medical Protocols." *Social Studies of Science* 27 (2): 273–305.

————. 2003. *The Gold Standard: The Challenge of Evidence-Based Medicine and Standardization in Health Care*. Philadelphia: Temple University Press.

Timmermans, Stefan, and Emily Kolker. 2004. "Clinical Practice Guidelines and the Implications of Shifts in Knowledge for Sociological Accounts of Professional Power." *Journal of Health and Social Behavior* 45 (Extra Issue): 177–93.

Tjornhoj-Thomsen, Tine, Helene Goldberg, and Maruska La Cour Mosegaard, eds. 2005. *The Second Sex in Reproduction: Men, Sexuality, and Masculinity*. Berkeley: University of California Press.

Tomatis, Lorenzo. 1995. "Ethical Aspects of Prevention." *Scandinavian Journal of Work and Environmental Health* 21:245–51.

Tomes, Nancy. 1998. *The Gospel of Germs: Men, Women, and the Microbe in American Life*. Cambridge: Harvard University Press.

————. 2001. "Merchants of Health: Medicine and Consumer Culture in the United States, 1900–1940." *Journal of American History* 88 (2): 519–47.

————. 2005. "The Great American Medicine Show Revisited." *Bulletin of the History of Medicine* 79:627–63.

Tone, Andrea, and Elizabeth Siegel Watkins, eds. 2007. *Medicating Modern America: Prescription Drugs in History*. New York: New York University Press.

Towghi, Fowziah. 2007. "Scales of Marginalities: Transformations of Women's Bodies, Medicines, and Land in Postcolonial Balochistan, Pakistan." Ph.D. diss., Medical Anthropology (Joint) Program, University of California, San Francisco and Berkeley.

Trafford, Abigail. 2002. "'Final Word' on Mammograms? Not Yet." *Washington Post*, March 5.

Traweek, Sharon. 1988. *Beamtimes and Lifetimes: The World of High Energy Physicists*. Cambridge: Harvard University Press.

Traynor, Michael. 2000. "Purity, Conversion, and the Evidence Based Movements." *Health* 4 (2): 139–58.

Treichler, Paula. 1999. *How to Have Theory in an Epidemic: Cultural Chronicles of AIDS*. Durham, N.C.: Duke University Press.

Treichler, Paula, Lisa Cartwright, and Constance Penley, eds. 1998. *The Visible Woman: Imaging Technologies, Gender, and Science*. New York: New York University Press.

Tremain, Shelley Lynn, ed. 2005. *Foucault and the Government of Disability*. Ann Arbor: University of Michigan Press.

TSC Staff. 2004. "NitroMed Soars on Trial." TheStreet.Com. July 19. http://www.thestreet.com/_tscs/stocks/biotech/10171994.html (accessed July 7, 2005).

Tsing, Anna Lowenhaupt. 2005. *Friction: An Ethnography of Global Connection*. Princeton, N.J.: Princeton University Press.

Turner, Bryan S. 1984. *The Body and Society*. Oxford: Basil Blackwell.

————. 1990. "Outline of a Theory of Citizenship." *Sociology* 24:189–214.

————. 1992. *Regulating Bodies: Essays in Medical Sociology*. London: Routledge.

————. 1997. "From Governmentality to Risk: Some Reflections on Foucault's Contribution to Medical Sociology." In *Foucault, Health, and Medicine*, ed. Alan Petersen and Robin Bunton, ix–xxii. New York: Routledge.

Turner, Leslie. 2007. "'First World Health Care at Third World Prices': Globalization, Bioethics, and Medical Tourism." *BioSocieties* 2 (3): 303–25.

Turow, J. 1989: *Playing Doctor: Television, Storytelling, and Medical Power*. New York: Oxford University Press.

Tye, Larry. 1998. *The Father of Spin: Edward L. Bernays and the Birth of Public Relations*. New York: Crown.

U.S. Office of Management and Budget (OMB). 1997. *Revisions to the Standards for the Classification of Federal Data on Race and Ethnicity* http://www.whitehouse.gov/omb/fedreg/ombdir15.html (accessed April 16, 2002).

Vaillant, George E. 1984. "The Disadvantages of *DSM-III* Outweigh Its Advantages." *American Journal of Psychiatry* 141 (4): 542–45.

van Balen, F., and Martha C. Inhorn. 2003. "Son Preference, Sex Selection, and the 'New' New Reproductive Technologies." *International Journal of Health Services* 33:235–52.

Vandenbroucke, Jan P. 1988. "Is 'The Causes of Cancer' a Miasma Theory for the End of the Twentieth Century?" *International Journal of Epidemiology* 17:708–9.

van Dijck, José. 2000. "The Language and Literature of Life: Popular Metaphors in Genome Research." In *Wild Science: Reading Feminism, Medicine, and the Media*, ed. Janine Marchessault and Kim Sawchuk, 66–79. London: Routledge.

————. 2005. *The Transparent Body: A Cultural Analysis of Medical Imaging*. Seattle: University of Washington Press.

van Heteren, G. M., M. Gijswijt-Horstra, and E. M. Tansey, eds. 2002. *Biographies of Remedies: Drugs, Medicines, and Contraceptives in Dutch and Anglo-American Healing Cultures*. Atlanta: Rodopi.

van Kammen, Jessika, and Nelly Oudshoorn. 2002. "Gender and Risk Assessment in Contraceptive Technologies." *Sociology of Health and Illness* 24 (4): 436–61.

van Ommeren, Mark, Shekhar Saxena, and Benedetto Saraceno. 2005. "Mental and Social Health during and after Acute Emergencies: Emerging Consensus?" *Bulletin of the World Health Organization* 83 (1): 71–75.

Vaughan, Diane. 1996. *The Challenger Launch Decision*. Chicago: University of Chicago Press.

————. 1999. "The Role of the Organization in the Production of Techno-scientific Knowledge." *Social Studies of Science* 29 (6): 913–43.

Vaughan, Megan. 1991. *Curing Their Ills: Colonial Power and African Illness*. Stanford: Stanford University Press.

Vedantam, Shankar. 2004. "Racial Disparities Played Down." *Washington Post*, January 14, A17.

Veronesi, U., P. Maisonneuve, A. Costa, V. Sacchini, C. Maltoni, C. Robertson, N. Rotmensz, and P. Boyle. 1998. "Prevention of Breast Cancer with Tamoxifen: Preliminary Findings from the Italian Randomised Trial among Hysterectomised Women." *Lancet* 352:93–97.

Vogel, Kathleen M. 2008. "Framing Biosecurity: An Alternative to the Biotech Revolution Model?" *Science and Public Policy* 35 (1): 45–54.

Wahl, Otto F. 2004. "Stop the Presses: Journalistic Treatment of Mental Illness." In *Cultural Sutures: Medicine and Media*, ed. Lester D. Friedman, 55–72. Durham, N.C.: Duke University Press.

Wahlberg, Ayo. 2006. "Bio-politics and the Promotion of Traditional Herbal Medicine in Vietnam." *Health* 10 (2): 123–47.

———. 2008. "Pathways to Plausibility: When Herbs Become Pills." *BioSocieties* 3 (1): 37–56.

Wailoo, Keith. 2001. *Dying in the City of the Blues: Sickle Cell Anemia and the Politics of Race and Health*. Chapel Hill: University of North Carolina Press.

———. 2003. "Inventing the Heterozygote: Molecular Biology, Racial Identity, and the Narratives of Sickle Cell Disease, Tay-Sachs, and Cystic Fibrosis." In *Race, Nature, and the Politics of Difference*, ed. Donald Moore, Anand Pandia, and Jake Kosek, 235–53. Durham, N.C.: Duke University Press.

———. 2004. "Sovereignty and Science: Revisiting the Role of Science in the Construction and Erosion of Medical Dominance." *Journal of Health Politics, Policy, and Law* 29 (4–5): 643–59.

Wailoo, Keith, and Stephen Pemberton. 2006. *The Troubled Dream of Genetic Medicine: Ethnicity and Innovation in Tay-Sachs, Cystic Fibrosis, and Sickle Cell Disease*. Baltimore: Johns Hopkins University Press.

Waitzkin, Howard. 1989. "Social Structures of Medical Oppression: A Marxist View." In *Perspectives in Medical Sociology*, ed. P. Brown, 166–78. Belmont, Calif.: Wadsworth.

———. 1991. *The Politics of Medical Encounters: How Patients and Doctors Deal with Social Problems*. New Haven: Yale University Press.

———. 2000. *The Second Sickness: Contradictions of Capitalist Health Care*. Rev. ed. Lanham, Md.: Rowman and Littlefield.

———. 2001. *At the Front Lines of Medicine*. Blue Ridge Summit, Pa.: Rowman and Littlefield.

Waitzkin, Howard, and Jennifer Fishman. 1997. "Inside the System: The Patient-Physician Relationship in the Era of Managed Care." In *Competitive Managed Care: The Emerging Health Care System*, ed. J. D. Wilkerson, K. J. Devers, and R. S. Given, 136–62. San Francisco: Jossey-Bass.

Waldby, Catherine. 2000. *The Visible Human Project: Informatic Bodies and Posthuman Medicine*. New York: Routledge.

Waldby, Catherine, and Robert Mitchell. 2006. *Tissue Economies: Blood, Organs, and Cell Line in Late Capitalism*. Durham, N.C.: Duke University Press.

Waldram, James B. 2004. *Revenge of the Windigo: The Construction of the Mind and*

Mental Health of North American Aboriginal Peoples. Toronto: University of Toronto Press.

Wann, Marilyn. 1998. *Fat! So? Because You Don't Have to Apologize for Your Size.* Berkeley: Ten Speed.

Warhurst, A. Michael. 2000. *Crisis in Chemicals.* London: Friends of the Earth.

Warner, John Harley. 2004. "Grand Narrative and Its Discontents: Medical History and the Social Transformation of American Medicine." *Journal of Health Politics, Policy, and Law* 29 (4–5): 757–80.

Washburn, Rachel. 2009. "Measuring the 'Pollution in People': Biomonitoring and Constructions of Health and Environment." Ph.D. diss., University of California, San Francisco.

Washington, Harriet A. 2007. *Medical Apartheid: The Dark History of Medical Experimentation on Black Americans from Colonial Times to the Present.* New York: Doubleday.

Watkins, Elizabeth Siegel. 2007. *The Estrogen Elixir: A History of Hormone Replacement Therapy in America.* Baltimore: Johns Hopkins University Press.

Watson-Verran, Helen, and David Turnbull. 1995. "Science and Other Indigenous Knowledge Systems." In *Handbook of Science and Technology Studies,* ed. Sheila Jasanoff, Gerald E. Markle, James C. Petersen, and Trevor Pinch, 115–39. Thousand Oaks, Calif.: Sage.

Wayne, Leslie, and Melody Petersen. 2001. "A Muscular Lobby Rolls Up Its Sleeves." *New York Times,* November 4, section 3, p. 1.

Weber, Lynn. 2006. "Reconstructing the Landscape of Health Disparities Research: Promoting Dialogue and Collaboration Between Feminist Intersectional and Biomedical Paradigms." In *Gender, Race, Class, and Health: Intersectional Approaches,* ed. Amy J. Schulz and Leith Mullings, 21–59. San Francisco: Jossey-Bass.

Weinstein, Deena, and Michael A. Weinstein. 1999. "McDonaldization Enframed." In *Resisting McDonaldization,* ed. B. Smart, 57–69. London: Sage.

Weisman, Carol S. 1998. *Women's Health Care: Activist Traditions and Institutional Change.* Baltimore: Johns Hopkins University Press.

Weiss, R. B., R. M. Rifkin, F. M. Stewart, R. L. Theriault, L. A. Williams, A. A. Herman, and R. A. Beveridge. 2000. "High-Dose Chemotherapy for High Risk Primary Breast Cancer: An On Site Review of the Bezwoda Study." *Lancet* 355:999–1003.

Weissman, Myrna. 1996. Interview by Jackie Orr. New York, March 26.

Weissman, Myrna, Roger Bland, Glorisa Canino, Carlo Faravelli, Steven Greenwald, Hai-Gwo Hwu, Peter Joyce, Elie Karam, Chung-Kyoon Lee, Joseph Lellouch, Jean-Pierre Lepine, Stephen Newman, Mark Oakley-Browne, Maritza Rubio-Stipec, Elisabeth Wells, Priya Wickramaratne, Hans-Ulrich Wittchen, and Eng-Kung Yeh. 1997. "The Cross-National Epidemiology of Panic Disorder." *Archives of General Psychiatry* 54:305–9.

Weisz, George. 2006. *Divide and Conquer: A Comparative History of Medical Specialization.* Oxford: Oxford University Press.

Weiten, Wayne. 2003. *Psychology: Themes and Variations.* 6th ed. Belmont, Calif.: Wadsworth.

Wells, Stacey. 2001. "Industry Outlook: Biotechnology — Why Bioinformatics Is a Hot Career." *San Francisco Chronicle*, March 4, J1.

Wheatley, Elizabeth E. 2005. "Disciplining Bodies at Risk: Cardiac Rehabilitation and the Medicalization of Fitness." *Journal of Sport and Social Issues* 29 (2): 198–221.

White, Kevin. 2002. *Introduction to the Sociology of Health and Illness.* London: Sage.

Whiteis, David G., and J. Warren Salmon. 1990. "The Proprietarization of Health Care and the Underdevelopment of the Public Sector." In *The Corporate Transformation of Health Care: Issues and Directions,* ed. J. W. Salmon, 117–31. Amityville, N.Y.: Baywood.

Whitmarsh, Ian. 2008. *Biomedical Ambiguity: Race, Asthma, and the Contested Meaning of Genetic Research in the Caribbean.* Ithaca, N.Y.: Cornell University Press.

Whyte, Susan Reynolds. 2009. "Health Identities and Subjectivities: The Ethnographic Challenge." *Medical Anthropology Quarterly* 23 (1): 6–15.

Whyte, Susan Reynolds, Sjaak van der Geest, and Anita Hardon, eds. 2002. *Social Lives of Medicines.* Cambridge: Cambridge University Press.

Wiener, Carolyn. 2000. *The Elusive Quest: Accountability in Hospitals.* Hawthorne, N.Y.: Aldine de Gruyter.

Wiener, Norbert. 1948. *Cybernetics, or Control and Communication in the Animal and the Machine.* New York: Technology Press.

———. 1950. *The Human Use of Human Beings: Cybernetics and Society.* Boston: Houghton Mifflin.

Wilder-Smith, Annelies, Eli Schwartz, and Marc Shaw, eds. 2007. *Travel Medicine: Tales Behind the Science.* Amsterdam: Elsevier.

Williams, Gary M. 1995. "Tamoxifen Experimental Carcinogenicity Studies: Implications for Human Effects." *Proceedings of the Society for Experimental Biology and Medicine* 208:141–43.

Williams, Raymond. 1976/1985. *Keywords: A Vocabulary of Culture and Society.* New York: Oxford University Press.

———. 1980. "Advertising: The Magic System." In *Problems in Materialism and Culture,* by Raymond Williams, 170–195. London: Verso.

Williams, Simon J. 1998. "Health as Moral Performance: Ritual, Transgression, and Taboo." *Health* 2:435–57.

———. 1999. "Transgression for What? A Response to Robert Crawford." *Health* 3:367–78.

———. 2004. "Beyond Medicalization-Healthicization? A Rejoinder to Hislop and Arber." *Sociology of Health and Illness* 26 (4): 453–59.

Williams, Simon J., and Michael Calnan. 1994. "Perspectives on Prevention: The Views of General Practitioners." *Sociology of Health and Illness* 16 (3):372–93.

Williams, Simon J., Clive Seal, Sharon Boden, Pam Lowe, and Deborah Lynn Steinberg. 2008. "Medicalization and Beyond: The Social Construction of Insomnia and Snoring in the News." *Health* 12 (2): 251–68.

Willis, Deborah. 2005. "Cancer Diaries." In *Family, History, Memory: Recording African American Life*. New York: Hylas.

Wilson, Mitchell. 1993. "DSM-III and the Transformation of American Psychiatry: A History." *American Journal of Psychiatry* 150 (3): 399–410.

Wilson, Robert A. 1966. *Feminine Forever*. New York: J.B. Lippincott.

Witz, Anne. 2000. "Whose Body Matters? Feminist Sociology and the Corporeal Turns in Sociology and Feminism." *Body and Society* 6 (2): 1–24.

Wobarst, Anthony. 1999. *Looking Within: How X-ray, CT, MRI, Ultrasound, and Other Medical Images Are Created and How They Help Save Lives*. Berkeley: University of California Press.

Wohl, Stanley. 1984. *The Medical Industrial Complex*. New York: Harmony.

Wolfe, Sidney. 2004. "Petition to the FDA to Remove the Cholesterol-Lowering Drug Rosuvastatin (CRESTOR) from the Market." March 4. http://www.citizen.org/publications/print_release.cfm?ID=7305 (accessed December 20, 2004).

Wöllmann, Torsten. 2004. "Andrology." In *Men and Masculinities: A Social, Cultural, and Historical Encyclopedia*, vol. 1, ed. M. Kimmel and A. Aronson, 33–34. Santa Barbara, Calif.: ABC-Clio.

Woloshin, Steve, Lisa M. Schwartz, Jennifer Tremmel, and H. Gilbert Welch. 2001. "Direct-to-Consumer Advertisements for Prescription Drugs: What Are Americans Being Sold?" *Lancet* 358:1141–46.

Wolski, C. A. 2006. "In Contrast." *Medical Imaging Magazine*. http://www.medicalimagingmag.com/issues/articles/2006-05_02.asp (accessed October 15, 2006).

Woodward, Jean M. B., Steven L. Hass, and Paul J. Woodward. 2002. "Reliability and Validity of the Sexual Life Quality Questionnaire." *Quality of Life Research* 11:365–77.

Worcester, Nancy, and Marianne Whatley. 1988. "The Response of the Health Care System to the Women's Health Movement: The Selling of Women's Health Centers." In *Feminism within the Science and Health Care Profession: Overcoming Resistance*, ed. S. V. Rosser, 117–51. New York: Pergamon.

World Health Organization. 2001. *World Health Report: Mental Health; New Understanding, New Hope*. Geneva: World Health Organization.

Wright, Peter, and Andrew Treacher, eds. 1982. *The Problem of Medical Knowledge: Examining the Social Construction of Medicine*. Edinburgh: Edinburgh University Press.

Writing Group for the Women's Health Initiative Investigators. 2002. "Risks and Benefits of Estrogen plus Progestin in Healthy Postmenopausal Women: Principal Results from the Women's Health Initiative Randomized Controlled Trial." *Journal of the American Medical Association* 288 (3): 321–33.

Wu, Chia-Ling, Yu-Ling Huang, Young-Gyung Park, Azumi Tsuge, and Adele E. Clarke. 2009. "Gender and Reproductive Technologies in East Asia: A Partial Bibliography of Works in English." *EASTS: East Asian Science, Technology, and Society: An International Journal* 3 (1): 327–34.

Yancy, Clyde. 2002. "The Role of Race in Heart Failure Therapy." *Current Cardiology Reports* 4:218–25.

Yates, Joanne, and John van Maanen, eds. 2001. *Information Technology and Organizational Transformation: History, Rhetoric, and Practice.* Thousand Oaks, Calif.: Sage.

Young, James Harvey. 1999. "The Development of the Office of Alternative Medicine in the National Institutes of Health, 1991–1996." *Bulletin of the History of Medicine* 72 (2): 279–98.

Yoxen, Edward. 1981. "Life as a Productive Force: Capitalizing the Science and Technology of Molecular Biology." In *Science, Technology, and the Labour Process*, ed. Les Levidow and Robert Young, 66–122. London: Free Association.

———. 1982. "Giving Life a New Meaning: The Rise of the Molecular Biology Establishment." In *Scientific Establishment and Hierarchies*, ed. N. Elias, H. Martins, and R. Whitley. Boston: D. Reidel.

———. 1987. "Seeing with Sound: A Study of the Development of Medical Images." In *The Social Construction of Technological Systems: New Directions in the Sociology and History of Technology*, ed. W. Bijker, T. Hughes, and T. Pinch, 131–54. Cambridge: MIT Press.

Zetka, James R., Jr. 2003. *Surgeons and the Scope.* Ithaca, N.Y.: Cornell University Press.

Zimberg, Robyn. 1993. "Food, Needs, and Entitlement: Women's Experience of Emotional Eating." In *Consuming Passions: Feminist Approaches to Weight Preoccupation and Eating Disorders*, ed. C. Brown and K. Jasper. Toronto: Second Story.

Zola, Irving Kenneth. 1972. "Medicine as an Institution of Social Control." *Sociological Review* 20:487–504.

———. 1991. "Bringing Our Bodies and Ourselves Back In: Reflections on a Past, Present, and Future 'Medical Sociology.'" *Journal of Health and Social Behavior* 32 (1):1–16.

Zones, Jane S. 2000. "Profits from Pain: The Political Economy of Breast Cancer." In *Breast Cancer: Society Constructs an Epidemic*, ed. S. J. Ferguson and A. S. Kasper, 119–51. New York: St. Martin's Press.

Zwillich, Todd. 2004. "Scientist: FDA Incapable of Protecting Safety." WebMD, November 18. http://my.webmd.com/content/article/97/104115.htm (accessed February 8, 2008).

ABOUT THE CONTRIBUTORS

NATALIE BOERO is an assistant professor in the Department of Sociology at San José State University. She received her Ph.D. from the University of California, Berkeley, in 2006. She has published articles in the journal *Qualitative Sociology* and the *Fat Studies Reader* (2009). Her first book, *Fat Panic: Media, Medicine, and Morals in the American Obesity Epidemic*, is forthcoming.

ADELE E. CLARKE is a professor of sociology and history of health sciences at the University of California, San Francisco. Her books include *Situational Analysis: Grounded Theory after the Postmodern Turn* (2005), *Disciplining Reproduction: Modernity, the American Life Sciences, and "Problems of Sex"* (1998), and (co-edited) *Revisioning Women, Health, and Healing* (1999), *Women's Health: Differences and Complexities* (1997), and *The Right Tools for the Job: At Work in Twentieth Century Life Sciences* (1992).

JENNIFER R. FISHMAN is an assistant professor in the So-cial Studies of Medicine Department at McGill University. Her research explores issues of commodification and commercial-ization in science and biomedicine, with her most recent work exploring the biology of aging and the emergence of anti-aging medicine. Her next project will explore the controversies of per-sonalized genomic medicine.

JENNIFER RUTH FOSKET is a principal and founder of Social Green where she does research and writes on the intersections of health, the built environment, and sustainability. She recently published *Living Green: Communities That Sustain* (2009) with Laura Mamo.

KELLY JOYCE is an associate professor of sociology at the College of William and Mary. She is the author of *Magnetic Appeal: MRI and the Myth of Transparency* (2008) and is coeditor of the forthcoming monograph *Technogenarians: Studying Health and Illness through an Aging, Science, and Technology Lens*.

JONATHAN KAHN is a professor of law at Hamline University School of Law. He is the author of *Budgeting Democracy: State Building and Citizenship in the United States, 1890–1928* (1997) and numerous articles on the intersections of law, race, and genetics. His work focuses particular attention on how regulatory mandates interact with scientific, clinical, and commercial practices in producing genetic information in relation to racial categories.

LAURA MAMO is an associate professor at the Health Equity Institute for Research, Practice, and Policy at San Francisco State University. She is the author of *Queering Reproduction: Achieving Pregnancy in the Age of Technoscience* (Duke University Press, 2007) and (with Jennifer Fosket) *Living Green: Communities That Sustain* (2009). Her research and publications address a wide array of topics related to health, science, politics, social inequalities, and gender, aging, and sexuality. She is currently researching the gendered politics of HPV vaccines. She formerly held a position of associate professor of sociology at the University of Maryland.

JACKIE ORR is an associate professor of sociology at Syracuse University. She teaches and writes in the fields of cultural politics, contemporary and feminist theory, and critical studies of technoscience and psychiatry. Her book *Panic Diaries: A Genealogy of Panic Disorder* (Duke University Press, 2006) chronicles the entanglements of bodies, pills, war, computers, power, capital, and technoscientific discourses that have shaped the social management and production of "panic."

ELIANNE RISKA is a professor of sociology at the Swedish School of Social Science, University of Helsinki, Finland. Her research focuses on gender and health and the medical profession. Two of her books explore the gendered aspects of health: (with Elizabeth Ettorre) *Gendered Moods: Psychotropics and Society* (1995) and *Masculinity and Men's Health: Coronary Heart Disease in Medical and Public Discourse* (2004).

JANET K. SHIM is an assistant professor of sociology in the Department of Social and Behavioral Sciences at the University of California, San Francisco. Her research examines the relationships between conceptions of risk, social inequalities, and biomedical science and clinical definitions of difference, particularly within epidemiology. Her publications have appeared in *American Sociological Review, Health, Social Studies of Science*, and *Sociology of Health and Illness*, among others.

SARA SHOSTAK is an assistant professor in the Department of Sociology at Brandeis University. Her research focuses on the relationships between science, subjectivity, and social order. In 2008, she was an associate editor for a special issue of the *American Journal of Sociology* on "Exploring Genetics and Social Structure."

INDEX

Library of Congress Cataloging-in-Publication Data

Biomedicalization : technoscience, health, and illness in the U.S. /
Adele E. Clarke ... [et al.], eds.

p. cm.

Includes bibliographical references and index.

ISBN 978-0-8223-4553-4 (cloth : alk. paper)

ISBN 978-0-8223-4570-1 (pbk. : alk. paper)

1. Medical innovations—Social aspects—United States.

2. Biotechnology—Social aspects—United States.

3. Medical technology—Social aspects—United States.

I. Clarke, Adele.

RA418.5.M4B556 2010

610.28′4—dc22   2010011136